# Artificial Reefs

*Artificial Reefs: Refining Their Role in the Environment and Local Economies* represents the accumulation of over 50 years of research on artificial reefs, as presented at the 12th International Conference on Artificial Reefs and Related Aquatic Habitats (CARAH) held in Colombia in the spring of 2025. The edited, multi-authored chapters offer an up-to-date examination of broad areas of artificial reefs. These include fisheries, environmental management, survey techniques, sophisticated applications of statistical analysis, socio-economics, and the role of citizen scientists in further expanding future research efforts.

A general theme throughout the volume is the consideration given to historical trends in methods, developments, and changes that have occurred during the past 50 years. This book builds on the strong foundation laid by our earlier colleagues and mentors. Each chapter brings together a long and fabled history of research but, more importantly, demonstrates the progress and development that artificial reef research has taken with the inclusion of modern technology, sophisticated analyses, and broadened goals of artificial reef deployments.

Careful consideration is given in every chapter to the relationships artificial reefs have with the local communities with which they are proximate. The subtitle of the book thus emphasizes the modern role that artificial reefs are playing in the economies of coastal communities.

This foundational material will provide guidance for the next 50 years of research on artificial reefs.

**CRC Marine Biology Series**

**The late Peter L. Lutz, Founding Editor**

**Julie Mondon and Stephen Bortone, Series Editors**

**Artificial Reefs in Fisheries Management**
*edited by Stephen A. Bortone, Frederico Pereira Brandini, Gianna Fabi, and Shinya Otake*

**The Biology of Sea Turtles, Volume III**
*edited by Jeanette Wyneken, Kenneth J. Lohmann, and John A. Musick*

**The Physiology of Fishes, Fourth Edition**
*edited by David H. Evans, James B. Claiborne, and Suzanne Currie*

**Interrelationships Between Coral Reefs and Fisheries**
*edited by Stephen A. Bortone*

**Impacts of Oil Spill Disasters on Marine Habitats and Fisheries in North America**
*edited by J. Brian Alford, PhD, Mark S. Peterson, and Christopher C. Green*

**Hagfish Biology**
*edited by Susan L. Edwards and Gregory G. Goss*

**Marine Mammal Physiology: Requisites for Ocean Living**
*edited by Michael A. Castellini and Jo-Ann Mellish*

**Shark Research: Emerging Technologies and Applications for the Field and Laboratory**
*edited by Jeffrey C. Carrier, Michael R. Heithaus, and Colin A. Simpfendorfer*

**Red Snapper Biology in a Changing World**
*edited by Stephen T. Szedlmayer, Stephen A. Bortone*

**The Physiology of Fishes**
*edited by Suzanne Currie, David H. Evans*

**Whale Sharks**
Biology, Ecology, and Conservation
*edited by Alistair D. M. Dove and Simon J. Pierce*

**Biology of Sharks and Their Relatives, Third Edition**
*edited by Jeffrey C. Carrier, Colin A. Simpfendorfer, Michael R. Heithaus, Kara E. Yopak*

**Physiology of Marine Mammals**
Adaptations to the Ocean
*edited by Michael Castellini, Jo-Ann Mellish*

**Artificial Reefs**
Refining Their Role in the Environment and Local Economies
*edited by Stephen A. Bortone, Oscar Delgadillo-G, Ilana Rosental Zalmon*

For more information about this series, please visit: https://www.crcpress.com/CRC-Marine-Biology-Series/book-series/CRCMARINEBIO?page=&order=pubdate&size=12&view=list&status=published,forthcoming

# Artificial Reefs
## Refining Their Role in the Environment and Local Economies

Edited By
Stephen A. Bortone
Oscar Delgadillo-G
Ilana Rosental Zalmon

CRC Press
Taylor & Francis Group
Boca Raton London New York

CRC Press is an imprint of the
Taylor & Francis Group, an **informa** business

Designed cover image: Top image: Artificial reef 6 in Tolú, Gulf of Morrosquillo, Sucre, Colombian Caribbean Credit: Oscar Delgadillo-G; Bottom left: Bullwinkle Oil Platform being towed out to sea near Port Aransas, Texas in 1988. Source: Bullwinkle Oil Platform, Author: Jay Phagan (CC BY 2.0); Bottom right image: Shipwreck in Riohacha, La Guajira, Colombian Caribbean Credit: Oscar Delgadillo-G

First edition published 2027
by CRC Press
2385 NW Executive Center Drive, Suite 320, Boca Raton FL 33431

and by CRC Press
4 Park Square, Milton Park, Abingdon, Oxon, OX14 4RN

CRC Press is an imprint of Taylor & Francis Group, LLC

ISBN: 978-1-041-14682-7 (hbk)
ISBN: 978-1-041-14914-9 (pbk)
ISBN: 978-1-003-67677-5 (ebk)

DOI: 10.1201/9781003676775

Typeset in Times
by Apex CoVantage, LLC

*We dedicate this volume to our recently deceased colleagues who contributed to our understanding of artificial reefs and whose contributions will certainly have long-lasting impacts on the future of artificial reefs as an option in aquatic ecosystem management. In alphabetical order:*

*Gianna Fabi—Institute for Biological Resources and Marine Biotechnology (CNR-IRBIM), Italy, for her career-long efforts to establish standards and criteria for artificial reefs.*

*Patrice Francour—Ecology and Conservation Science for Sustainable Seas, University of Nice-Sophia Antipolis, France, for his studies on the function and analysis of artificial reef properties.*

*Benny J. Galloway—LGL Ecological Research Associates, Texas USA, for his work on the understanding of petroleum platforms as artificial reefs.*

*Joseph G. Halusky—University of Florida, Florida Sea Grant Extension Program, USA, for his work on promoting citizen scientists as monitors of artificial reef function and stability.*

*Joseph J. Kimmel, United States National Marine Fisheries Service, USA, for his long-standing effort to develop underwater survey techniques on reefs.*

*Shimrit Perkol-Finkel—ECOncrete, Israel, for her pioneering work in using artificial reefs to complement natural reefs.*

*Robert L. Shipp—University of South Alabama, USA, for his long career in helping to establish one of the largest artificial reef complexes in the United States.*

# Contents

# About the Editors

**Stephen A. Bortone, PhD,** retired as Executive Director of the Gulf of Mexico Fishery Management Council and now serves as a principal at an environmental consulting firm, Osprey Aquatic Sciences, LLC, which specializes in fisheries and other aquatic sciences. Formerly, he was Director of the Minnesota Sea Grant College Program located in Duluth, where he held an appointment as Professor of Biology at the University of Minnesota Duluth. Additionally, he held positions as Director of the Marine Laboratory at the Sanibel-Captiva Conservation Foundation in Sanibel, Florida; Director of Environmental Science at the Conservancy of Southwest Florida; and Director of the Institute for Coastal and Estuarine Research while serving as Professor of Biology at the University of West Florida. Dr. Bortone received a BS degree (Biology) from Albright College in Reading, Pennsylvania; an MS degree (Biological Sciences) from Florida State University in Tallahassee; and a PhD (Marine Science) from the University of North Carolina at Chapel Hill in 1973.

For the past 57 years, he has conducted research on fisheries and the life history of aquatic organisms, especially fishes, chiefly in the southeastern United States and the Gulf of Mexico. He has published over 180 scientific articles on the broad aspects of the aquatic sciences, but mostly on fisheries. In addition, he has edited nine books on the aquatic sciences: *Sea Grasses*, *Biology of the Spotted Sea Trout*, *Estuarine Indicators*, *Artificial Reefs in Fisheries Management* (with F. Brandini, G. Fabi, and S. Otake), *Interrelationships between Corals and Fisheries*, and *Red Snapper Biology in a Changing World* (with S. Szedlmayer), and *Modern Fisheries Engineering* (with S. Otake), as well as this volume—all with CRC Press—and *Marine Artificial Reef Research and Development: Integrating Fisheries Management Objectives* published by the American Fisheries Society.

Through his research and teaching activities, Dr. Bortone has traveled widely. He has served as Visiting Scientist at Johannes Gutenberg University (Mainz, Germany) and conducted extensive field surveys with colleagues from the University of La Laguna in the Canary Islands. He was a Mary Ball Washington Scholar at University College Dublin in Ireland. He has received numerous teaching and research awards, including the title "Fellow" from the American Institute of Fishery Research Biologists, as well as being recognized as a "Certified Fisheries Professional" by the American Fisheries Society and "Certified Senior Ecologist" by the Ecological Society of America.

**Oscar Delgadillo-G, MS,** holds a degree in marine biology from the Universidad de Bogotá Jorge Tadeo Lozano and a master's degree in marine biology from the Universidad Nacional de Colombia. He is the founder and principal of MoAm S.A.S., a Colombian research firm, where he has formulated and developed 44 projects (leading 37 directly through MoAm) in collaboration with 48 public environmental and management institutions, private companies, NGOs, research institutes, and universities.

His career has largely focused on the design, implementation, and assessment of artificial reefs for artisanal fisheries, ecological restoration, and biodiversity conservation along the Caribbean coast of Colombia, with recent expansion into the Pacific region. His research disciplines also encompass marine and coastal ecology, fish and invertebrate biology, mariculture, invasive and endangered species, protected areas, monitoring program design, and both biophysical and socio-economic assessments.

To date, he has authored 14 research articles, book chapters, and institutional books, as well as 41 academic abstracts, two biodiversity datasets, and 46 technical reports. He has directed and advised seven undergraduate theses in marine biology and served as a peer reviewer for eight international scientific journals. Most recently, he served as the chair and organizer of the 12th International Conference on Artificial Reefs and Related Aquatic Habitats (CARAH), held in Santa Marta, Colombia.

**Ilana Rosental Zalmon, PhD**, has a degree in biological sciences from the University of Rio de Janeiro (1984), a master's degree in biological sciences (zoology) from the Federal University of Rio de Janeiro (1988), a PhD in biological sciences (Zoology/Biosciences) from the University of São Paulo (1995), and completed a postdoctoral fellowship at the Marine Science Institute, University of California, Santa Barbara (2007–2008), where she studied fish associated with biogenic structures. She is Senior Professor at the University of Northern Rio de Janeiro, where she has worked since 1995 as a professor and researcher. Her studies focus on the ecology of natural marine communities (rocky shores and sandy beaches) and experimental communities (biofouling and artificial reefs), as well as the management and conservation of coastal and fishery resources (specifically surf zone fish and artisanal fisheries). Since 1996, artificial reefs have become one of her main research topics, encompassing studies on materials, dimensions, functions, analytical methods, reef design, ecological theories, socioeconomic aspects, and productive potential. She has authored more than 130 scientific papers and 17 book chapters related to artificial reefs, coastal ecosystems, and their anthropogenic impacts. One of her major scientific contributions has been the supervision of around 70 undergraduate, master's, PhD, and postdoctoral students.

# Preface

This volume is chiefly the result of the presentations presented at the 12th International Conference on Artificial Reefs and Related Aquatic Habitats (CARAH), held in Santa Marta, Colombia, from 9 to 14 March 2025 at the Mercure Santa Marta Emile Hotel conference hall by Accor, in Pozos Colorados. Since its inception in 1974, when the first event was held in Houston, Texas, United States, CARAH has served as a crucial platform for the exchange of knowledge and experiences in artificial reef research, development, and management. Over the past five decades, the conference has grown in scope, adapting to advancements in technology, changing environmental policies, and evolving global challenges. Celebrating this 50-year legacy highlights the enduring importance of artificial reefs in sustainable fisheries management, aquatic ecosystem conservation, and research. This event not only acknowledges the progress made but also sets the stage for future innovations in marine habitat applications.

This conference marked an important milestone in the long-standing history of CARAH conferences with this event's motto: "50 Years of Innovation, Oceans of Opportunity." The chapters in this volume offer an updated and forward-looking perspective on artificial reef science and its applications. The CARAH conference included 33 presentations and two forums covering a wide range of thematic areas such as new approaches for ecological assessments, nature-based solutions, fisheries and restoration ecology, oceanography, engineering, design and materials, blue economy and cultural dimensions, artisanal fishers' experiences, and future directions for artificial reef research and management.

# Acknowledgments

The chapters offered here are primarily the result of the 12th CARAH (Conference on Artificial Reefs and Related Aquatic Habitats). The conference was made possible thanks to the commitment and financial support of numerous organizations and individuals. The conference was hosted and organized by Oscar Delgadillo-G and Stefannia Doria, while Stephen A. Bortone was instrumental in encouraging participation by researchers from around the world. We also acknowledge the non-government organization GOAL (Global Organization for Aid and Livelihoods, Dublin, Ireland—www.goalglobal.org) for its crucial support. We thank GOAL executives Luigi Loddo (Country Director/Acting Deputy Regional Director) and Jorge Tejeda (Country Coordinator, Colombia), as well as the technical team: Daniela Torrez (Program Coordinator, Resilience of the Blue Economy), Mario Argeñal (Program Adviser, Resilience of the Blue Economy), and Robert Toro (Livelihoods Specialist). Their support for the 12th CARAH was provided through the project Alliance of the Ireland Civil Society for a Better World, funded by the Government of Ireland in partnership with GOAL. The content presented during the conference is solely the responsibility of the organizers and does not necessarily reflect the views of the Government of Ireland.

We express our appreciation to ARC Marine, LLC, for its support, and especially thank Tom Birbeck (CEO), Samuel Hickling (CSO), Dr. Ferrante Grasselli, and Hannah McFadyen for their contributions. We also acknowledge MoAm S.A.S. for coordinating and executing the conference and especially Oscar Delgadillo-G (CEO), Stefannia Doria (Communities and Communications Coordinator), and researchers Adriana Méndez, Evelyn Pérez, and Dr. Paola Flórez.

Our gratitude extends to the keynote speakers whose outstanding presentations enriched the conference. These included:

- Dr. Ilana Rosental Zalmon—Center of Biosciences and Biotechnology, University of North Rio de Janeiro, Brazil);
- Dr. Tao Tian—Center for Marine Ranching Engineering Science Research of Liaoning and Industry Institute of Marine Ranching, Dalian Ocean University, China;
- Dr. Avery B. Paxton, National Oceanic and Atmospheric Administration, National Marine Fisheries Service (USA);
- Dr. Matthew D. Taylor—Port Stephens Fisheries Institute, NSW Department of Primary Industries and Regional Development; Centre for Marine Science and Innovation, University of New South Wales, Australia;
- Dr. Jeremy Claisse—Department of Biological Sciences, California State Polytechnic University (USA);
- Dr. Jorge Ramos—CinTurs Research Centre for Tourism, Sustainability, and Well-Being, University of Algarve, Portugal;
- Dr. Stephen A. Bortone—Osprey Aquatic Sciences, LLC, Florida (USA).

We also thank Alice Oven from Taylor & Francis/CRC Press and her colleagues (Venkatesh Sundaram, Shikha Garg, and Paul Boyd) for their editing and production skills that facilitated the completion of this volume.

Finally, we extend special thanks to the fishers and local managers who participated in the forum "Fishers' Perceptions on Artificial Reef Applications and Management in the Caribbean (Colombia) and the Eastern Tropical Pacific (Honduras)." Participants included Johan Urieles (Asociación de Pescadores Artesanales y Turismo de Pozos Colorados—Apezcoltur, Santa Marta, Colombia); Ardis Hinojosa (Artisanal Fishing Association Altamar, Wayuu Indigenous Community of Riohacha,

Colombia); and representatives of the Comité para la Defensa y Desarrollo de la Flora y Fauna del Golfo de Fonseca (Coddeffagolf), Honduras, including fisher Franklin Gómez, sub-director Ely Gutiérrez, and technical team member Nohelia Bustillo. Their contributions brought invaluable local perspectives to the conference discussions.

# Contributors

**Umarfarooq A. Abdulwahab**
ExxonMobil Biomedical Sciences, Inc., Spring, Texas (USA)

**Alistair Becker**
Port Stephens Fisheries Institute, New South Wales Department of Primary Industries and Regional Development, Nelson Bay, NSW, Australia

**Victor Blanco**
Florida Sea Grant College, University of Florida, Perry, Florida (USA)

**Stephen A. Bortone**
Osprey Aquatic Sciences, LLC, Palm Coast, Florida (USA)

**Igor David da Costa**
Federal University of Fluminense (UFF), Santo Antônio de Pádua, Rio de Janeiro (Brazil)

**Jean-Claude Dauvin**
Normandie University, Caen Normandy University, UNICAEN, Laboratoire Morphodynamique Continentale et Côtière Caen, France

**Oscar Delgadillo-G**
MoAm S.A.S., Pozos Colorados, Santa Marta, Colombia
and
Universidad Nacional de Colombia, Instituto de Estudios en Ciencias del Mar—CECIMAR, Sede Caribe. Invemar, Santa Marta, Colombia

**Camilo B. García**
Universidad Nacional de Colombia, Departamento de Biología, Facultad de Ciencias Bogotá D.C., Colombia

**Ferrante Grasselli**
ARC Marine Ltd., Torquay (UK)

**Samuel Hickling**
ARC Marine Ltd., Torquay (UK)

**Carolyn Kalinowski**
Florida Fish and Wildlife Conservation Commission, Division of Marine Fisheries Management, Tallahassee, Florida (USA)

**Juliano Silva Lima**
Federal Institute of Sergipe (IFS), São Cristóvão, Sergipe, Brazil

**Neil R. Loneragan**
School of Environmental and Conservation Sciences, College of Environmental and Life Sciences, Murdoch University, Murdoch, Australia
and
Centre for Sustainable Aquatic Ecosystems, Harry Butler Institute, Murdoch University, Murdoch, Australia

**Milton S. Love**
Marine Science Institute, University of California, Santa Barbara, California (USA)

**Keith Mille**
Florida Fish and Wildlife Conservation Commission, Division of Marine Fisheries Management, Tallahassee, Florida (USA)

**James Murphy**
ARC Marine Ltd., Torquay (UK)

**Mary M. Nishimoto**
Marine Science Institute, University of California, Santa Barbara, California (USA)

**Zheyang Ou**
Center for Marine Ranching Engineering Science Research of Liaoning Province, Dalian Ocean University, Dalian, China
and
College of Fisheries and Life Science, Dalian Ocean University, Dalian, China

**Avery B. Paxton**
Southeast Fisheries Science Center, National Marine Fisheries Service, National Oceanic and Atmospheric Administration, Beaufort, NC (USA)

**Sylvain Pioch**
University Montpellier Paul Valéry, Laboratory of Geography and Planning Management of Montpellier, LAGAM, Montpellier, France

**Christopher M. Prosser**
ExxonMobil Biomedical Sciences, Inc., Annandale, New Jersey (USA)

**Jorge Ramos**
CinTurs—Research Centre for Tourism, Sustainability and Well-Being, University of Algarve | Campus de Gambelas, Algarve, Portugal

**Jessica Salaün**
National Center for Scientific Research, EPHE Université Paris, CRIOBE, Maison de l'Océan, Paris, France

**Ann Scarborough-Bull**
Marine Science Institute, University of California, Santa Barbara, California (USA)

**Liwei Si**
Center for Marine Ranching Engineering Science Research of Liaoning Province, Dalian Ocean University, Dalian, China
and
College of Fisheries and Life Science, Dalian Ocean University, Dalian, China

**Matthew D. Taylor**
Port Stephens Fisheries Institute, New South Wales Department of Primary Industries and Regional Development, Nelson Bay, NSW, Australia
and
School of Biological, Earth and Environmental Science, University of New South Wales, New South Wales, Australia

**Tao Tian**
Center for Marine Ranching Engineering Science Research of Liaoning, Dalian Ocean University, Dalian, China
and
Marine Ranching Industry Research Institute, Dalian Ocean University, Dalian, China
and
College of Fisheries and Life Science, Dalian Ocean University, Dalian, China

**James R. Tweedley**
School of Environmental and Conservation Sciences, College of Environmental and Life Sciences, Murdoch University, Murdoch, Australia
and
Centre for Sustainable Aquatic Ecosystems, Harry Butler Institute, Murdoch University, Murdoch, Australia

**Zhongxin Wu**
Center for Marine Ranching Engineering Science Research of Liaoning Province, Dalian Ocean University, Dalian, China
and
College of Fisheries and Life Science, Dalian Ocean University, Dalian, China
and
Key Laboratory of Environment Controlled Aquaculture, Ministry of Education, Dalian Ocean University, Dalian, China

**Ilana Rosental Zalmon**
University of North Rio de Janeiro (UENF), Campos dos Goytacazes, Rio de Janeiro, Brazil

**Sven Zea**
Universidad Nacional de Colombia, Instituto de Estudios en Ciencias del Mar—CECIMAR, Sede Caribe. Invemar, Santa Marta, Colombia

**Yanchao Zhang**
Center for Marine Ranching Engineering Science Research of Liaoning Province, Dalian Ocean University, Dalian, China
and
College of Fisheries and Life Science, Dalian Ocean University, Dalian, China

# 1 Introduction: a Brief History of CARAH, Some Artificial Reef Terms, and Chapter Overviews

*Stephen A. Bortone*, Oscar Delgadillo-G, and Ilana Rosental Zalmon*
* Corresponding author: steve.bortone@gmail.com

## 1.1 CARAH AS A FORUM FOR ARTIFICIAL REEF INVESTIGATIONS

The 12th International Conference on Artificial Reefs and Aquatic Habitats (CARAH) marked a significant milestone—the 50th anniversary of the conference series. Celebrating this 50-year legacy highlights the enduring importance of artificial reefs in sustainable fisheries management, aquatic ecosystem conservation, and research. This event not only acknowledged the progress made but also set the stage for future innovations in marine habitat applications.

Since its inception in 1974, when the first event was held in Houston, Texas (USA), CARAH has served as a crucial platform for the exchange of knowledge and experiences in artificial reef research, development, and management. Table 1.1 provides a summary of CARAHs by year, location, the number of presentations (both oral and written) when available, and references where the published results of many of the presentations can be found. We note that, thanks to the efforts of one of our colleagues, Daniel Sheehy from AquaBio Environmental Technologies, the "lost" CARAH agenda (Bortone, 2011) has been found. A typed replica of this agenda is provided in Figure 1.1. Due to the poor quality of the original, it is not duplicated here, but the copied agenda includes all the remarks, spacing, errors, and content of the original.

The initial conference (Clark et al., 1974) had as its main objective the purpose of including artificial reefs as an option to improve and sustain fisheries. Over the past five decades, the conferences and the field of artificial reef research have grown in scope, adapting to advancements in technology, changing environmental policies, and evolving global challenges. Moreover, the objectives for artificial reef deployment have broadened over the past several decades as well. Some of these new objectives beyond fisheries enhancement include habitat improvement, recreational activities (snorkeling, SCUBA diving, and surfing), beach shoreline management, wave energy attenuation, nature-based solutions to coastal issues, and a focus on local community economic development, among others.

## 1.2 A DEFINITION OF AN ARTIFICIAL REEF

Before proceeding, it is important to have a clear operational definition that will allow everyone to know exactly what the term "Artificial Reef" means. Examining the plethora of generally applied definitions of artificial reefs (see Seaman 2023, for an expanded summary of definitions for an artificial reef), Bortone (2018) began the search for the "ultimate" definition. However, here we begin with Seaman's (2019) definition: "Artificial reefs are benthic structures made of natural or man-made materials, deployed in thousands of locations worldwide for the protection, enhancement, or restoration of components of ocean ecosystems". Given some anticipated misunderstandings, we modify that definition slightly as: "Artificial Reefs—one or more objects of natural or human origin

DOI: 10.1201/9781003676775-1

**TABLE 1.1**

**Summary of past international artificial reef conferences (i.e., CARAH). Number of presentations includes both oral and posters.**

| Conf. No. | Year | Locations | Presentations | References |
|---|---|---|---|---|
| 1st | 1974 | Houston, Texas (USA) | 35 | Colunga and Stone, 1974 |
| 2nd | 1977 | Brisbane (Australia) | 16 | Figure 1.1 |
| 3rd | 1983 | Newport Beach, California (USA) | 57 | Buckley et al., 1985 |
| 4th | 1987 | Miami, Florida (USA) | 77 | Seaman et al., 1989 |
| 5th | 1991 | Long Beach, California (USA) | 212 | Grove and Wilson, 1994 |
| 6th | 1995 | Tokyo, Japan | 120 | Sako and Nakamura, 1995 |
| 7th | 1999 | San Remo (Italy) | 116 | Jensen, 2002 |
| 8th | 2005 | Biloxi, Mississippi (USA) | 102 | Maher, 2006 |
| 9th | 2009 | Curitiba (Brazil) | 72 | Bortone et al., 2011 |
| 10th | 2015 | Izmir (Turkey) | 66 | Lok, 2015 |
| 11th | 2017 | Terengganu (Malaysia) | 17 | Bortone, 2018 |
| 12th | 2025 | Santa Marta (Colombia) | 33 | This volume |

DOWNUNDER 77

SECOND INTERNATIONAL CONFERENCE ON ARTIFICIAL REEFS

Crest Hotel

Sunday 1lth September, 1977.

Morning Session

| | | |
|---|---|---|
| 9-00 - 10-00 | Registration/Formalities | |
| 10-00 - 10-15 | Welcome Speech | Mr C. Wharton (Australia), Queensland State Minister for Fisheries |
| 10-15 - 11-00 | KEYNOTE | "Artificial Reefs and Fisheries Management" |
| | Mr Richard Stone (U.S.A.), National Marine Fisheries Service | |
| 11-00 - 11-45 | KEYNOTE | "An Introduction to the Concept of Artificial Reefs and a Summary of Development to Date in Australia" |
| | Dr David Pollard (Australia), N.S.W. State Fisheries | |
| 11-45 - 12 noon | Discussion | |
| 12 noon - 1-30 pm. | LUNCH | |
| Afternoon Session | Topic | "Regional Studies" |
| | Chairman | Dr David Pollard (Australia), N.S.W. State Fisheries |
| 1-30 - 2-00 | "Artificial Reefs in Taiwan<br>Dr K-H Chang (Taiwan), Academia Sinica, Institute of Zoology | |
| 2-00 - 2-30 | "Artificial Reefs in South Australia"<br>Mr Kevin Branden (Australia), Department of Agriculture and Fisheries, South Australia | |
| 2-30 - 3-00 | "Biological Observations on a Temperate Marine Artificial Reef"<br>Dr R.O. Parker (U.S.A.) and Mr R. Stone (U.S.A.) and Mr C.C. Buchanan (U.S.A.), National Marine Fisheries Service | |
| 3-00 - 3-15 | BREAK | |
| 3-15 - 3-45 | "Macroalgae on California's Artificial Reefs"<br>Dr John Duffy ((U.S.A.), California Department of Fish and Game | |

**FIGURE 1.1** Transposed copy of the agenda for the second international conference on artificial reefs (DOWNUNDER 1977) held in Brisbane, Australia. The version typed above is a direct copy and includes all spacing, errors, and notations. It is offered here courtesy of Daniel Sheehy.

| | |
|---|---|
| 3-45 - 4-15 | |
| 4-15 - 4-45 | "Puget Sound Artificial Reef Study"<br>Mr Jim Walton (U.S.A.), University of ~~Rhode, island~~ Washington |
| 4-45- 5-00 | Discussion |
| 6-00 p.m. | Reception |
| 8-00 p.m. | Downunder 77 International Underwater Film Festival<br>Master of Ceremonies Stanton Waterman (U.S.A.)<br>Special Guests Ron & Valerie Taylor (Australia)<br>David Pilosof (Israel)<br>Victor de Sanctis (Italy) |

DOWNUNDER 77
SECOND INTERNATIONAL CONFERENCE ON ARTIFICIAL REEFS

Crest Hotel

Monday 12th September, 1977

| | |
|---|---|
| Morning Session | Topic "Artificial Reefs and Fisheries Management"<br>Chairman Mr R. Stone (U.S.A.), National Marine Fisheries Service |
| 9-00 - 9-30 am. | "The Development of Artificial Reefs for the American Lobster"<br>Dr Daniel Sheehy (U.S.A.), University of Rhode Island |
| 9-30 - 10-00 | |
| 10-00 - 10-30 | "Benefit-Cost Study of Pinellas County, Florida, Artificial Reefs"<br>Dr Eila Hanni, Dr Heyward Mathews (U.S.A.), University of South Florida |
| 10-30 - 10-45 | BREAK |
| 10-45 - 11-15 | "Artificial Reef Enhancement of Puget Sound Recreational Fishing"<br>Dr Raymond Buckley (U.S.A.), Department of Fisheries |
| 11-15 - 11-45 | "Floating Artificial Reefs"<br>Dr Richard Candle (U.S.A.), Research Division, Goodyear Tyre Co. |
| 11-45 - 12 noon | Discussion |
| 12 noon - 1-30 pm | LUNCH |
| Afternoon Session | Topic "Problems and Techniques"<br>Chairman Mr K. Branden (Australia), Department of Agriculture and Fisheries, South Australia. |
| 1-30 - 2-00 | "A Natural Example of Artificial Reefs"<br>Professor A. Stefanon (Italy), Scientific Committee, CMAS |
| 2-00 - 2-30 | "Problems with Queensland's Artificial Reefs"<br>Mr Eddie Hegerl (Australia), Queensland Conservation Council<br>Mr Jim Davie (Australia), Australian Littoral Society |
| 2-30 - 3-00 | "Some Problems Associated with the Establishment of N.S.W. First Off-shore Artificial Reef"<br>Mr John Matthews (Australia), N.S.W. State Fisheries |
| 3-00 - 3-15 | BREAK |
| 3-15 - 3-45 | "Observations on the Fish Community of the Artificial Reef off Swansea Channel, L. Macquarie N.S.W."<br>Mr Johann Bell (Australia), N.S.W. Scate Fisheries |
| 3-45 - 4-15 | "Evaluation of Artificial Reefs by Remote Sensing and Photogrammetric Surveys"<br>Mr Bill Raymond (U.S.A. and Mr Greg McIntosh (U.S.A.), D.E. Britt and Associates, Inc. |
| 4-15 - 5-00 | Discussion |

**FIGURE 1.1** (Continued)

deployed to influence physical, biological, and/or socioeconomic processes related to living and non-living aquatic resources". This definition can be further and more succinctly abbreviated to: "Artificial reefs—objects deployed in an aquatic environment to influence physical, biological, and/or socioeconomic processes related to aquatic resources".

Why these modifications? Since the objects can be natural or man-made, there are no other alternatives, so the objects do not require a qualifier of natural or man-made. Additionally, it is important to note that benthic structures are objects on the bottom, while many artificial reefs and habitats are positioned midwater or at the surface. This deployment on the bottom (i.e., benthos) is too specific, as some artificial reefs are deployed above the bottom, midwater, or even at the surface (i.e., FADs—Fish Attracting Devices or Fish Aggregating Devices).

Additionally, the plethora of objects (rock jetties, trees, shell materials, automobile bodies, pipelines, docks, piers, seawalls, etc.) deployed in a variety of habitats (ponds, lakes, streams, seas, oceans, etc.) are other than oceanic. Lastly, the modified definition includes both known and anticipated purposes (fisheries, restoration, tourism, surfing, upwelling, etc.).

Rock jetties are not usually thought of as artificial reefs, but then again, there are many other structures that have been and continue to serve as artificial reefs. More inclusively, one might consider any structure, whether placed inadvertently or purposely in the water, can serve as an artificial reef. Thus, pipelines, platforms, underwater cables, seawalls, ports, aquaculture facilities, etc. could also be considered as artificial reefs. This clearly makes the calculation of how much artificial reef material that has been deployed extremely problematic. Bugnot et al. (2022) provided an approximate minimum seafloor area of marine built structures (all types included) up until 2018 of about 32,000 $km^2$, of which designated artificial reefs occupied 3,600 $km^2$. This issue is not addressed here, but future efforts should consider all allochthonous structures as artificial reefs regardless of their composition, location, and origin.

## 1.3 MORE VOCABULARY

Throughout the literature, there are terms related to the configuration of deployments that have become problematic, leading to confusion. To clarify, in this volume, we have sought to provide consistency in the usage of reef deployments, especially when more than one structure is deployed. This volume follows the example of Grove and Sonu (1983) as referenced in Grove et al. (1991). A modification of their figure is included here as Figure 1.2 and also as Figure 1.9 in Seaman (2023).

Comments regarding the hierarchy of artificial reef deployments are warranted. It should be noted that the terms "unit" and "module" are synonymous. Also, readers should note that not all deployment configurations fit neatly into the categories of the hierarchy offered here. The purpose of this explanation is to provide some reference to the terms (i.e., module, unit, set, group, and complex) to establish commonality in their usage.

## 1.4 SOME ADDITIONAL TERMS THAT REQUIRE EXPLANATION

Bohnsack et al. (1991) presented a cogent explanation for a preference to use the term "assemblage" when referring to a multispecies group of organisms on an artificial reef. This distinction is important when applying these terms to artificial reefs as the term "community" implies a coevolved, deterministic system that has emergent properties. Since artificial reefs are relatively "new" habitats in terms of evolutionary time, it is important to make the distinction between groups of organisms associated with an artificial reef as an assemblage and groups of organisms on natural structures as being a community.

This follows previous suggestions to denote that a group of organisms associated with an artificial reef is not a well-tuned and refined result of evolutionary processes (i.e., community) but a short-term, in some ways serendipitous, conglomeration of organisms that are the result of ephemeral

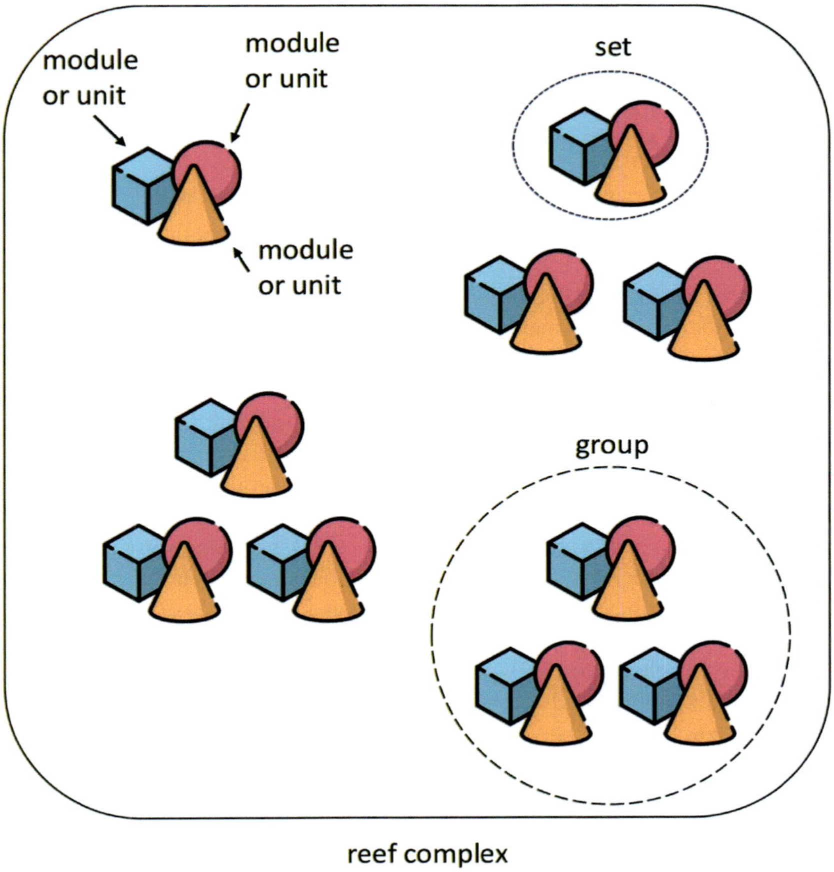


**FIGURE 1.2** Terms used to describe the hierarchy of artificial reef deployments (after Grove and Sonu, 1983, and Grove et al., 1991).

conditions and lack the influence of long-term natural selection at least in short-term. Incorporating this descriptive difference helps give focus to the analyses and interpretations of the results of an artificial reef deployment.

## 1.5 THE CHAPTERS

Below is a very brief overview of the main points and attributes of each of the chapters included in this book. This will, hopefully, give the reader a quick and overall appreciation of the diversity of the book's contents but also serve as a guide on how best to proceed with understanding the nature of artificial reef research.

Chapter 2—Trends in Fish Diversity at an Artificial Reef Complex in the Caribbean Sea Off Colombia by Oscar Delgadillo-G, Sven Zea, and Camilo B. García provides insight into artificial reefs as part of an institutional and community-based effort to offer alternatives to standard fishery management and ecological restoration in a coastal area off Colombia. The results of this study offer deep insight into the specific features (both dependent and independent variables) that influence artificial reef assemblages in a part of the world that is growing in its use of artificial reefs as a fishery management option.

Chapter 3—The Historical Context, Current Status, Management Practices, and Future Development of Artificial Reefs in China by Tao Tian offers a large-scale, long-term perspective on perhaps the largest application of artificial reefs in the world. This chapter also includes a perspective on the strategic development that has taken place in China over the past few decades, with an examination of the policies and standards necessary to fully implement artificial reefs as a fishery management alternative.

Chapter 4—Two Decades of Artificial Reef Research in New South Wales, Australia, by Matthew D. Taylor and Alistair Becker presents an opportunity to closely examine the efforts of Australian researchers. The application of artificial reefs off New South Wales offers insight into the multifaceted approach that modern artificial reef projects must include in order to become a useful and acceptable part of fisheries management.

Chapter 5—Understanding the Ecological Role of Artificial Structures amidst Global Change: A Perspective by Avery B. Paxton expands our modern understanding of how artificial reefs function ecologically. The series of questions offered also provides specific directions for future artificial reef researchers. These directions will help guide artificial reef research far into the future.

Chapter 6—Enhancing Nektonic Communities and Juvenile Fish Biodiversity with Reef Cubes® Deployment in Torbay, United Kingdom, by Samuel Hickling, James Murphy, and Ferrante Grasselli shows how using a standardized reef module (unit) can help us understand the functional aspects of artificial reef designs and deployments. The focus here is on the entire assemblage with special insight into how artificial reefs can be used to benefit the life history features of fishes, with a special emphasis on juveniles.

Chapter 7—New Insights into the Use of Baited Video and Baited Traps for Assessing the Structure and Function of Fish and Mobile Invertebrate Communities Associated with Artificial Reefs by Zhongxin Wu, Liwei Si, James R. Tweedley, Neil R. Loneragan, Yanchao Zhang, Zheyang Ou, and Tao Tian is offered here as an extraordinary opportunity to take advantage of a previously published work (i.e., Zhang, Y., Ou, Z., Tweedley, J. R., Loneragan, N. R., Zhang, X., Tian, T., and Wu, Z. 2024. Evaluating the effectiveness of baited video and traps for quantifying the mobile fauna on artificial reefs in northern China. *Journal of Experimental Marine Biology and Ecology*, 573, 152001. https://doi.org/10.1016/j.jembe.2024.152001). We took the unprecedented approach of offering the authors of the original paper an opportunity to further examine their data in light of the discussions that occurred as a result of the 12th CARAH. Further, this chapter, along with the original paper (from which much of the text has been derived), will give special insight into the application of new methods of surveying fish assemblages on artificial reefs.

Chapter 8—The Ecological Importance of Offshore Pipelines as Artificial Reefs and Their Potential in Fishery Management: A Case Study in the Northwest Gulf of Mexico by Ann Scarborough-Bull, Milton S. Love, Mary M. Nishimoto, Umarfarooq A. Abdulwahab, and Christopher M. Prosser is the only chapter offered in this book that was not part of the 12th CARAH but we believe in the importance of its inclusion. Pipelines have not necessarily been a formal part of the discussion of the materials normally included as artificial reefs. This underappreciation must be acknowledged in future research on artificial reefs, as pipelines (as noted in this chapter) comprise a significant amount and vast array of artificial reef material in many parts of the world.

Chapter 9—Changes in Florida's (USA) Artificial Reef Program: Managing Expectations and Challenges to Benefit Diverse Stakeholders and Ecosystems by Carolyn Kalinowski and Keith Mille provides an opportunity to highlight one of the most extensive artificial reef programs in the United States. Florida has a plethora of artificial reef deployments consisting of a variety of materials in widely diverse habitats. The interesting feature in Florida is that, while under the general auspices of a central governmental coordinator (i.e., the State of Florida), each county (i.e., governmental subdivisions within the State) is the management authority for reefs in its jurisdiction. While this management structure has naturally had its problems, the overall, long-term projection is that these organizational authorities have met the needs of their respective communities within their jurisdictions. The program in Florida can serve as a model for other governmental authorities with less centralized control of environmental management. Additionally, this chapter also provides a glimpse into the discussions that take place at CARAHs and will give readers an appreciation for the advantage of having a dialogue among researchers and managers in the application of artificial reefs to manage aquatic natural resources.

Chapter 10-Citizen Science-Community Engagement in Research, and Coastal Stewardship in Florida's (USA) Northern Gulf of America: The Taylor County Artificial Reef Monitoring Program by Victor Blanco is a perspective from environmental managers of a governmental subdivision in Florida (see Chapter 9 for a non-subordinate view of this management perspective). Here, we also note the novel approach of using "citizen scientists" to assist artificial reef managers in assessing reef deployments. While still in its infancy, monitoring artificial reefs using the efforts of citizen scientists may be an acceptable alternative to standard "professional" scientists in environmental management. The advantage of including "citizen scientists" is not only cost-effective but also offers local residents a stake in the outcome of artificial reef projects.

Chapter 11—Social Sciences in Artificial Reefs: Past "Gaps" and Future Directions by Jorge Ramos expands on the modern emphasis of artificial reefs by examining the important inclusion of the social sciences in artificial reef research. Modern times require modern solutions to management issues, and artificial reef managers need to consider the impacts that reefs can have on human communities as well as the individuals living in those coastal communities. This chapter provides insight to students on the importance of including the social sciences in the decision-making aspects of environmental management, which includes artificial reefs as part of its objectives.

Chapter 12—The "Gap" in Artificial Reef Assessments: The Need for a Combination of Social and Ecological Approaches by Jessica Salaün, Sylvain Pioch, and Jean-Claude Dauvin continues the theme presented in Chapter 11 by providing specific methods to more objectively include aspects of the social sciences into artificial reef assessments. The case study provided offers an example of how to objectively incorporate both ecological and social science aspects of artificial reefs in an evaluation, especially when monitoring and evaluating whether the objectives of a reef's deployment have been met.

Chapter 13—Artificial Reefs: A Perspective of What We Learned and What We Should Do Next by Juliano Silva Lima, Igor David da Costa, and Ilana Rosental Zalmon encapsulates our current understanding of artificial reef science from a variety of perspectives. Their historical approach to considering past research on artificial reefs gives greater insight into the future directions that artificial reef research should take. The examination for the future includes perspectives of missing elements with consideration of the development and adoption of the latest technology into environmental management.

## 1.6 THE FUTURE ARTIFICIAL REEF CONFERENCES

While CARAHs are not the only venue for scientists interested in exploring outcomes, aspirations, and possibilities of artificial reef research, it does stand out as the longest continuous avenue for these endeavors. It is hoped that a more formally established future version of CARAH will provide a permanent opportunity for scientists to regularly and predictably explore and refine their interests in the use of artificial reefs in aquatic resource management.

## LITERATURE CITED

Bohnsack, J.A., D.L. Johnson, and R.F. Ambrose. 1991. Ecology of artificial reef habitats and fishes. Pages 61–105. In W. Seaman and L.M. Sprague, eds. *Artificial Habitats for Marine and Freshwater Fisheries*. Academic Press, San Diego, CA.

Bortone, S.A. 2011. Introduction to the role artificial reefs play in fisheries management. Pages 1–6. In S.A. Bortone, F.P. Brandini, G. Fabi, and S. Otake, eds. *Artificial Reefs in Fisheries Management*. CRC Press, Boca Raton, FL.

Bortone, S.A., R. Brandini, G. Fabi, and S. Otake. 2011. *Artificial Reefs in Fisheries Management*. CRC Press, Boca Raton, FL.

Bortone, S.A. 2018. Preface. Pages vii–viii. In S.A. Bortone, ed. *Marine Artificial Reef Research and Development, Integrating Fisheries Management Objectives*. American Fisheries Society Symposium 86. The American Fisheries Society, Bethesda, MD.

Buckley, R., J. Grant, and J. Stephens, Jr. 1985. Third international artificial reef conference 3–5 November 1983, Newport Beach, California, foreward [sic]. *Bull. Mar. Sci.* 37(1):1–2.

Bugnot, A.B., M. Mayer-Pinto, L. Airoldi, E.C. Heery, E.L. Johnston, L.P. Critchley, E.M.A. Strain, R.L. Morris, L.H.L. Loke, M.J. Bishop, E.V. Sheehan, R.A. Coleman, and K.A. Daftorn. 2022. Current and projected global extent of marine built structures. *Nat. Sustain.* https://doi.org/10.1038/s41893-020-00595-1

Clark, W., J. Mosely, and R. Stone. 1974. Foreward [sic]. Page 3 (abst.). In L. Coulunga and R. Stone, eds. *Proceedings of the International Conference on Artificial Reefs. Houston, TAMU-SG-74–103*. Texas A & M Sea Grant College, College Station, TX.

Colunga, L., and R. Stone. 1974. *Proceedings of the International Conference on Artificial Reefs, TAMU-SG-74–103*. Texas A & M University Sea Grant College, College Station, TX.

Grove, R.S. and C.J. Sonu. 1983. Fishing reef planning in Japan. Pages 189–251. In *Report 83-RD–137*. Southern California Edison Company, Rosemead, CA.

Grove, R.S., C.J. Sonu, and M. Nakamura. 1991. Designing and engineering of manufactured habitats for fisheries enhancement. Pages 109–152. In W. Seaman and L.M. Sprague, eds. *Artificial Habitats for Marine and Freshwater Fisheries*. Academic Press, San Diego, CA.

Grove, R.S., and C.A. Wilson. 1994. Introduction. *Bull. Mar. Sci.* 55(2–3):265–267.

Jensen, A.C. 2002. The seventh international conference on artificial reefs and related aquatic habitats. Introduction. *ICES Journal of Marine Science/Journal du Conseil* 59:S1.

Lok, A. 2015. Forward. *J. Appl. Ichthyol.* 31(Supplement3):2.

Maher, T., ed. 2006. Proceedings of the 8th international conference on artificial reefs and artificial habitats (CARAH), Palace Casino and Resort, Biloxi, Mississippi. *U.S.A. Bull. Mar. Sci.* 78(1):1–226.

Sako, H., and M. Nakamura, eds. 1995. ECOSET E'95. Proceedings of the international conference on ecological system enhancement technology for aquatic environments. *Japan International Marine Science and Tecenology [sic] Federation Tokyo* I–III.

Seaman, W. 2019. Artificial reefs. In J.K. Cochran, J.H. Bokuniewicz, and L.P. Yager, eds. *Encyclopedia of Ocean Sciences*, 3rd Edition. Elsevier, pp. 662–670.

Seaman, W. 2023. *Structure in the Sea*. Elsevier, Amsterdam.

Seaman, Jr., W., R.M. Buckley, and J.J. Polovina. 1989. Advances in knowledge and priorities for research technology and management related to artificial aquatic habitats. *Bull. Mar. Sci.* 44(2): 527–532.

# 2 Trends in Fish Diversity at an Artificial Reef Complex in the Caribbean Sea Off Colombia

*Oscar Delgadillo-G*, Sven Zea, and Camilo B. García*
* Corresponding author: odelgadillo@moam.com.co

## 2.1 INTRODUCTION

The exploitation of natural resources by humans, probably long before recorded history, has significantly altered climatic conditions and shaped the biodiversity trajectory of marine, aquatic, and terrestrial ecosystems (Rick and Erlandson, 2009; Jackson et al., 2011). We now live in the Anthropocene epoch, an era denoted by the biodiversity crisis, ecosystem degradation, accelerated species extinction, global warming, and ocean acidification (Luypaert et al., 2020). Driven by the desire to manipulate and exploit natural resources, *Homo sapiens* has developed techniques for enhancing marine and coastal ecosystems for at least 150,000 years (Rick and Erlandson, 2009). It is suggested that during the Neolithic period in Africa (10,000 to 4,000 BCE), people deliberately submerged objects to create artificial habitats, likely based on observations of increased fish abundance around floating and submerged structures (Seaman and Sprague, 1991). These rudimentary interventions gradually developed into what is now termed "ocean sprawl". This refers to one of the most significant modifications of the seascape, which by 2018 covered an estimated bottom surface area of 32,000 $km^2$. Of this area, artificial reefs account for approximately 3,600 $km^2$ (Bugnot et al., 2022). The proliferation of these human modifications to the seascape has had significant effects on biological productivity, marine diversity, and species distributions (Bishop et al., 2017).

Although artificial reefs function under the same ecological principles as natural reefs and other aquatic ecosystems (Bohnsack et al., 1991), the species associated with them display differential affinities for these habitats depending on their ecological requirements. Marine organisms depend on reef attributes such as intrinsic factors (e.g., reef design) and extrinsic factors (e.g., location, environmental conditions) across different spatial and geographic scales. Consequently, artificial reefs have the capacity to enhance biomass through attraction and/or production and to modify biodiversity, depending on the relationships and feedback processes between the species and their habitat (Bohnsack, 1989; Bortone, 2011; Claisse et al., 2014; Strain et al., 2020). To date, most assessments of the factors influencing biological assemblages have been conducted in subtropical and temperate waters, whereas studies in tropical regions, particularly in the Caribbean Sea, remain limited and largely descriptive (Lima et al., 2019; Strain et al., 2020; Hylkema et al., 2021; Ramm et al., 2021).

Presently, the world is experiencing biodiversity loss, environmental deterioration, and climate change. Thus, the increasing establishment of built structures across diverse applications, geographic regions, and socio-ecological settings necessitates the exploration of different approaches to assess their prospective benefits on marine community dynamics and biodiversity. As a result, recent studies have applied a range of diversity metrics for ecological analysis to evaluate the potential of artificial reefs as effective options in marine management (Cardoso et al., 2020; Paxton et al., 2020; Medeiros et al., 2021; Ibanez-Erquiaga et al., 2025), as well as to better understand the patterns of diversity, ecological interactions, and productivity trends in coral reef ecosystems (Yeager et al., 2017; Yan et al., 2025).

DOI: 10.1201/9781003676775-2

In these respects, research on artificial reefs in Colombia remains incipient, with most efforts since the 1970s limited to descriptive and fragmented ecological and fishery studies. Nevertheless, in certain areas of the Caribbean, long-term artificial reef initiatives have been implemented through inter-institutional agreements led by Ecopetrol S.A. (formerly the Colombian Petroleum Company), in collaboration with public entities, private companies, non-governmental organizations (NGOs), and local community stakeholders. These projects were designed to enhance fisheries, facilitate conservation of biodiversity, and promote ecotourism for community benefit. Implementing these projects has involved the deployment of steel-pipe artificial reefs and the monitoring of abiotic and biotic components across different project stages (Delgadillo et al., 2004; Delgadillo-Garzón and García, 2009; Delgadillo-Garzón et al., 2010; Delgadillo-G and Toro, 2018). Such initiatives provide an ideal scenario to test ecological hypotheses regarding the incidence and interactions of diverse ecological variables in shaping the biodiversity trajectories of these artificial habitats.

To better understand the functionality and continuity of artificial reef applications in tropical environments, this research seeks to elucidate the importance of several reef attributes. For instance, which aspects of structural complexity and reef age (intrinsic factors) and which environmental and location conditions (extrinsic factors) determine or correlate with trends in fish diversity? What is the relative, particular, or synergistic influence of these factors in shaping diversity? Accordingly, the following predictions are proposed: 1) fish faunal diversity on artificial reefs is not primarily determined by physical structural complexity or the biogenic components that develop during reef succession, 2) location and environmental variables may exert differential effects on diversity, and 3) diversity is fundamentally modulated by the interaction of intrinsic and extrinsic variables.

## 2.2 MATERIALS AND METHODS

### 2.2.1 Study area—The Gulf of Morrosquillo

The Gulf of Morrosquillo is located in the southern Caribbean Sea of Colombia, within the Departments of Córdoba and Sucre (9° 22′ 00″—9° 45′ 00″ N and 75° 33′ 00″—75° 55′ 00″ W), under the jurisdiction of the municipalities of San Onofre, Santiago de Tolú, Coveñas, and San Antero (Figure 2.1). The Gulf contains the widest continental shelf in the Colombian Caribbean, extending up to 70 km and down to depths of 55 m (Patiño and Flórez, 1993; Molina et al., 1994). The Gulf of Morrosquillo is bordered by coastal lagoons and mangrove forests, while its northern and southern extremes host coral reefs, seagrass meadows, and rocky shores (Díaz et al., 2003; Quirós-Rodríguez et al., 2010; MoAm, 2013; Delgadillo-G and Ruíz-Álvarez, 2016; Ruidíaz-Ruidíaz et al., 2021). The central portion is characterized by a flat topography with a mid-slope of less than 2%, and its seabed is primarily composed of mud and sand of varying composition. In contrast, the northern and southern portions have structural coralline terraces. Approximately 80% of the seafloor is covered by calcareous material derived from nearby coral reefs, while organic input is limited to coastal areas and is primarily due to fluvial and sediment inflow from the Sinú River, local streams, and coastal lagoons (Patiño and Flórez, 1993; Molina et al., 1994). Surface water temperature ranges from 27.5°C to 32°C, and salinity varies from 25 to 35 (Delgadillo et al., 2004; Ricaurte-Villota and Bastidas-Salamanca, 2017; Invemar, 2022).

### 2.2.2 Artificial Reef Features

The artificial reefs in the Gulf were constructed as modules or units (following Grove et al., 1991) using materials of opportunity, such as discarded carbon steel pipes (previously used for oil transport) and flexible underwater rubber hoses. Three basic cubic frame designs were used for the reefs, with volumes of 350 and 216 m$^3$, and a seafloor coverage area (i.e., footprint) of 70 and 36 m$^2$, respectively. These designs varied in the number, length, configuration, and diameter of

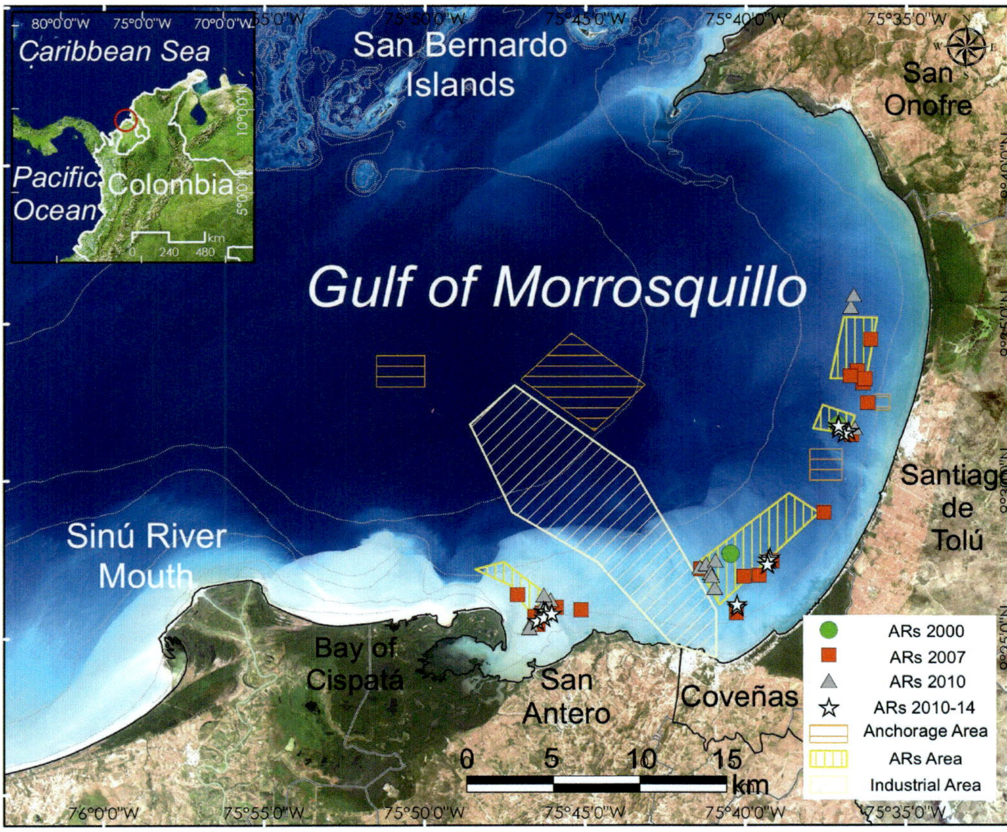


**FIGURE 2.1** General physiographic features and location of the artificial reefs by year in the Gulf of Morrosquillo, Colombian Caribbean.

pipes (Table 2.1, Figure 2.2). Larger reef modules were assembled using T-joints, while the remaining reefs were connected by inserting the pipes from opposite sides through pre-drilled perforations at specific points and securing them with metal plates that acted as reinforcement bars.

The first two modules were deployed in 2000, followed by 20 modules in 2007, and an additional 30 modules between 2010 and 2014. The initially deployed modules included extra pipes arranged at the base of the cubic frame (Delgadillo-Garzón and García, 2009; Delgadillo-Garzón et al., 2010). Reefs were deployed at depths ranging from 9.3 to 16.6 m, within three designated permit areas covering a total of 3,200 ha (Figure 2.1).

### 2.2.3 Field Surveys

Fish assemblages were assessed between 2008 and 2016 at 45 artificial reefs using underwater visual censuses conducted by two in situ researchers using SCUBA. The method involved swimming around and inside the modules, checking holes and crevices, identifying individual fish species in the assemblage, and estimating fish abundance and size during five regular, 6-minute intervals, totaling 30 minutes per diver at each artificial reef, between 08:00 and 13:00 hours local Colombian time (Bortone et al., 2000; Delgadillo et al., 2004). For abundance estimates, small groups of fewer than 50 individual fish were directly counted, while more abundant fish in larger schools were assigned to abundance categories (50–99, 100–199, 200–499, 500–999, 1000–1499, etc.) (Sale, 1997; Delgadillo et al., 2004). At each survey, environmental conditions (i.e., sea state, cloud cover)

**TABLE 2.1**
**Summary of structural complexity features of the steel pipe artificial reefs deployed in the Gulf of Morrosquillo.**

| | | Complexity Scale | | |
|---|---|---|---|---|
| **Artificial Reef Features** | | **1** | **2** | **3** |
| Spatial scale | Measurements (m) | 6 × 6 × 6 | 6 × 6 × 6 | 10 × 5 × 7 |
| | Volume ($m^3$) | 216 | 216 | 350 |
| | Area ($m^2$) | 36 | 36 | 50 |
| Diversity of the elements | Materials (number) | 1 | 1 | 2 |
| | Pipes (number) | 16 | 16 | 47 |
| | Holes (number) | 52 | 92 | 34 |
| Location of elements | Vertical pipes (number) | 4 | 4 | 8 |
| | Horizontal pipes (number) | 8 | 8 | 9 |
| | Diagonal pipes (number) | 4 | 4 | 30 |
| Abundance of the elements | Total (number) | 68 | 108 | 81 |
| Size of the elements | Pipes length (m) | 6 | 6 | 10, 7, 5, 3 |
| | Holes diameters (cm) | 46 | 46, 40, 30 | 60, 46, 40, 30 |

**FIGURE 2.2** Designs of the steel pipe artificial reefs deployed in different years in the Gulf of Morrosquillo (A: 2000; B: 2007; C: 2010–2014).

and physicochemical water variables (i.e., transparency, salinity, temperature, dissolved total solids, pH) were recorded using a combination of electrometric instruments and manual techniques (Sheng, 2000). Survey frequency varied by year and logistical circumstances, ranging from weekly and monthly intervals to quarterly or semiannual monitoring events.

### 2.2.4 Data Treatment for Modeling

Using the information on fish assemblage composition and species abundance for each date and artificial reef, the relative abundance was calculated as the average of the ten time intervals for each census day and site (Delgadillo et al., 2004). Using the statistical program R (R Core Team, 2024) and the Vegan v 2.6–8 package (Oksanen et al., 2024), three diversity metrics were estimated as response variables (Magurran, 1988; Wheater et al., 2011):

1) Species richness (S): total number of species.
2) Shannon-Wiener diversity (H'): calculated as $H' = -\Sigma_i \, p_i \log_b p_i$ (where $p_i$ is the proportional abundance of species $i$ and $b$ is the natural logarithm base).
3) Simpson diversity (D): calculated as $D = \Sigma_i \, p_i^2$ (where $p_i$ is the proportional abundance of species $i$), with the final index expressed as 1–D.

A set of predictor variables associated with the factors under assessment was selected:

1) Structural complexity: Physical features of the artificial reef, such as spatial metrics (volume, area), diversity of physical features (material, holes, cracks), spatial arrangement of features (vertical, horizontal, diagonal), measurements of elements (diameter, length), and abundance or density of features (Tokeshi and Arakaki, 2012; Loke and Chisholm, 2022). An ordinal scale was created based on the principle that a higher quantity, density, and variety of these characteristics corresponded to a higher heterogeneity of the artificial reef (Bohnsack, 1991). Through binary comparison (high vs. low), the features were scored for the three designs and included in an ascending complexity scale (Table 2.1).
2) Location variables (Sheng, 2000): Distances (in km) of the artificial reef to major ecosystems (nearest coastal point, main river, coastal lagoon or stream entrances, coral reefs, and seagrass meadows) and depth were estimated using GIS software in combination with marine and coastal cartography.
3) Environmental variables (Sheng, 2000): Water transparency (vertical and horizontal, m), surface and bottom temperature (°C), salinity, concentration of total dissolved solids ($mg*L^{-1}$), pH, sea state, cloud cover, survey time.
4) Reef age: Time since the artificial reef was initially deployed, used as a measure of development.

### 2.2.5 Ecological Modeling

Ecological models were developed following the procedure described by Geary et al. (2020) and Tredennick et al. (2021):

1) Covariates selection: Collinearity was assessed among independent variables using Spearman rank correlation (Quinn and Keough, 2002). Covariates with high collinearity ($r_s > 0.6$), statistical significance ($p < 0.05$), and lower ecological relevance were excluded. The relationships and functional forms (linear, nonlinear, exponential, etc.) between the selected variables and the response variables were explored by fitting univariate generalized additive models (Wood, 2017). Covariates without significant relationships with the response variable were also excluded. The final models were fitted with the remaining selected covariates.
2) Model establishment: Diversity metrics were modeled on each artificial reef *i(yi)* by fitting a hierarchical generalized additive model with the mgcv package v. 1.9–1 in R (Wood 2023), according to the following scheme (Wood, 2017; Pedersen et al., 2019)

$$E(\mu_i) = f_1\ (\text{time } i) + f_2\ (\text{location}_i) + f_3\ (\text{environmental}_i) + f_4\ (\text{interactions}_i) + f_5\ (\text{interactions}_i) + AA_{k(i)}$$

where $\mu i = E(yi)$ and $yi$ followed a Poisson distribution for species richness and a Gaussian distribution for Shannon-Wiener and Simpson diversity indices. $AA_{k(i)}$ represented the random effect of artificial reef complexity, while *f1* to *f5* were the smooth functions of the covariates.

For each response variable, six model structures were evaluated, ranging from the simplest to the most complex: a) global smoother for all observations, b) main effects, c) group-level smoothers with either shared or individual wiggliness, d) smoothers with or without interactions. Smoothers were fitted using thin plate regression splines (*f1* to *f3*) for main effects in all models; factor smoother interaction with low-order penalized second derivatives ($m = 2$) for groups with shared penalties, and low-order penalized first derivatives ($m = 1$) for groups with individual penalties, thereby reducing collinearity between global and group smoothers. Tensor product smoothers were used for multiple covariate interactions with differing measurement units (*f4* to *f5*). Model fitting for coefficient and smooth-term estimation was performed with the restricted maximum likelihood method, and term significance was assessed with chi-square tests (species richness) and analysis of variance (diversity indices).

The best model was validated on the basis of the convergence of smoother optimization, basis function dimensions, and concurvity. To compare the weight of evidence in the model, the Akaike Information Criterion (AIC) was applied to select the model that best fulfilled these conditions (Burnham et al., 2011; Wood, 2017; Tredennick et al., 2021).

## 2.3 RESULTS

Between 2008 and 2016, 45 artificial reefs were assessed resulting in 93 species recorded from 35 families of 22 orders. In general, the more heterogeneous artificial reefs (GM3) exhibited the highest diversity metrics, followed by the least complex reefs (GM1), while the intermediate reefs (GM2) supported assemblages with lower diversity.

Species richness was best explained by a model of global smoother with main effects and group-level smoothers, assuming the same level of wiggliness, plus interactions. A global trend was detected with a positive linear main effect of distance from natural reefs and seagrass beds (EDF = 1; $\chi^2 = 17.207$; $p < 0.001$; Figure 2.3a). A nonlinear trend also emerged as a function of days since deployment; the functional response was similar across complexity levels but with different shapes (EDF = 4.23; $\chi^2 = 22.839$; $p < 0.001$; Figure 2.3b). No significant relationships were found with depth, transparency, and pH, and no significant interactions were detected. This model explained 61.8% of the deviance.

The Shannon-Wiener diversity index was also best explained by the same type of model as species richness. A global trend was detected with a positive nonlinear main effect as a function of days since deployment (EDF = 3.72; $F = 8.163$; $p < 0.001$; Figure 2.4a), while a global trend with a positive linear main effect was observed as a function of pH (EDF = 1; $F = 12.536$; $p < 0.001$; Figure 2.4b). A variable semi-linear pattern emerged as a function of distance from streams and coastal lagoon entrances; the functional response was similar across complexity levels but with different shapes (EDF = 2.56; $F = 0.758$; $p < 0.05$; Figure 2.4c). Additionally, significant interactions were found between days since deployment and reef complexity with distance from entrances and temperature (EDF = 4.21; $F = 0.091$; $p < 0.01$), as well as with time-of-day and pH (EDF = 6.26; $F = 5.26$; $p < 0.05$). This model explained 49.1% of the deviance.

The Simpson diversity index was best explained by a global smoother model with main effects and group-level smoothers assuming different levels of wiggliness, plus interactions. A global trend was detected with a positive nonlinear main effect as a function of days since deployment (EDF = 1; $F = 4.147$; $p < 0.05$; Figure 2.5a) and a positive linear main effect of pH (EDF = 1; $F = 5.671$; $p < 0.05$; Figure 2.5b). In contrast, a negative linear main effect was identified for surface temperature (EDF = 1; $F = 8.799$; $p < 0.01$; Figure 2.5c). Moreover, a distinct negative trend with reef age was found in the intermediate artificial reefs (EDF = 1.95; $F = 6.918$; $p < 0.001$), whereas a positive effect of pH was detected at the more recently deployed and least complex reefs (EDF = 0.779; $F = 2.158$; $p < 0.05$). Significant interactions were found between higher-complexity artificial reefs and days since deployment, distance from coral reefs and seagrass beds, and temperature (EDF = 1.5; $F = 0.33$; $p < 0.01$). This model explained 43.9% of the deviance.

## 2.4 DISCUSSION

The artificial reef modules assessed between 2008 and 2016 supported a high diversity of fishes, with 93 recorded species. This represents about 7.7% of the approximately 1,200 marine fish species known for the Colombian Caribbean (A. Acero and J. Tavera, *pers. comm.*). It is notable that in total, the structures occupy a seafloor coverage of only 1,688 m$^2$ (less than 0.20 ha). Presumably, these built reefs provide important habitats that sustain both taxonomic and functional diversity, as

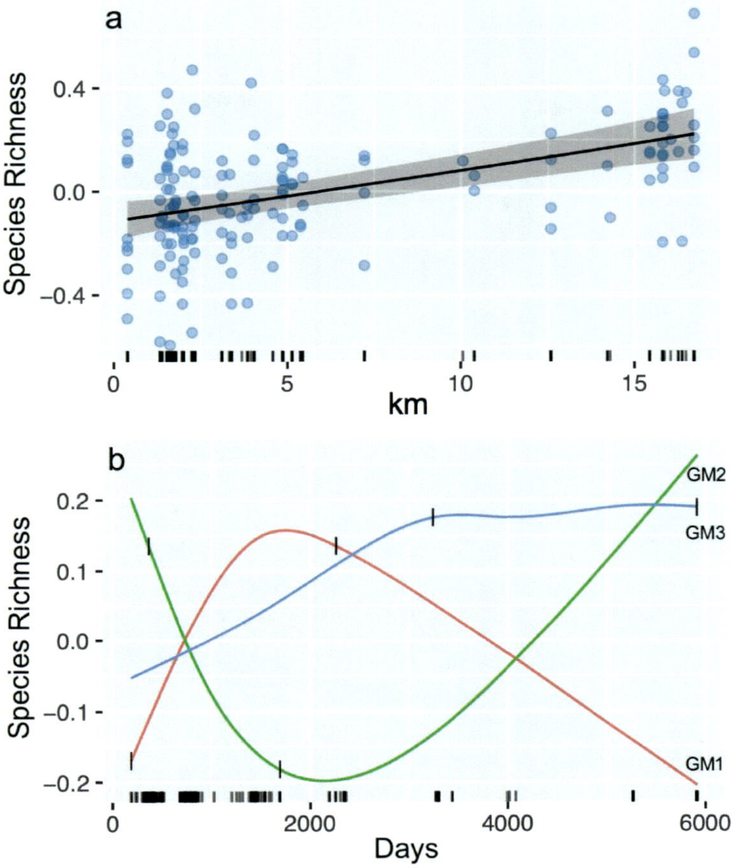


**FIGURE 2.3** Species richness variation as a function of (a) distance (km) from coral reefs and seagrass beds, and (b) days since deployment, grouped by structural complexity (GM1, GM2, GM3) among artificial reefs in the Gulf of Morrosquillo (black bars along the abscissa indicate the data extent, while black bars over the curves indicate the data range for each factor).

reflected in the relatively high number of fish families (35) and orders (22), compared to different natural reefs located on the continental coast of Colombia (Reyes-Nivia et al., 2004; Delgadillo-Garzón and Zapata-Ramírez, 2009; López-Victoria et al., 2023).

### 2.4.1 Species Richness

The positive trend of species richness with increasing distance from natural reefs and seagrass meadows (0.39 to 17.2 km), regardless of reef complexity or age, is consistent with studies documenting the positive effects of isolation on higher diversity, abundance, density, and fish production (Walsh, 1985; Bohnsack, 1989; Belmaker et al., 2005; Claisse et al., 2014). This trend likely arises because artificial reefs in more isolated settings function as uniquely available habitats, characterized by a high perimeter-to-surface-area ratio relative to their footprint on otherwise bare substrates lacking physical complexity and associated structuring organisms. Such conditions reduce resource competition with nearby natural ecosystems and decrease predation, therefore enhancing habitat functionality (e.g., recruitment, shelter, feeding, orientation, growth, reproduction). Generally,

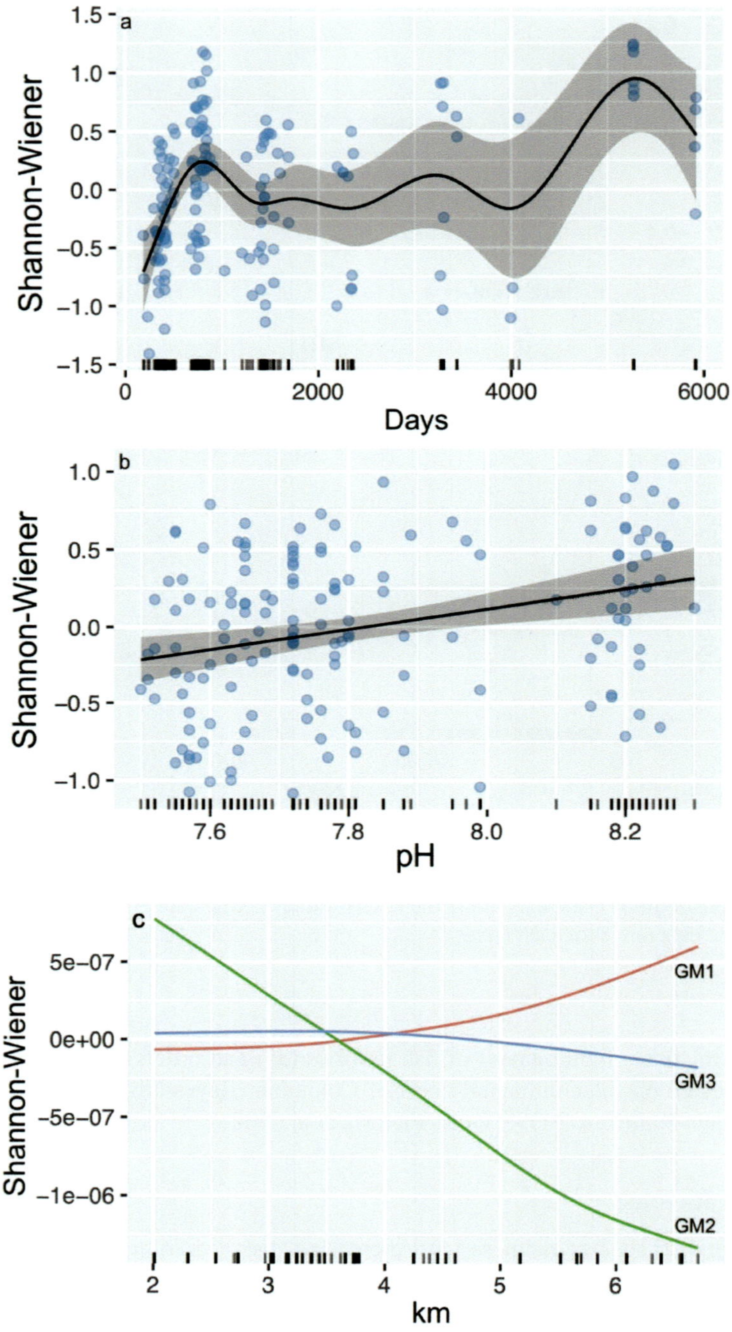


**FIGURE 2.4** Shannon-Wiener diversity variation as a function of (a) days since deployment, (b) pH, and (c) distance (km) from streams and coastal lagoon entrances grouped by structural complexity (GM1, GM2, GM3) among artificial reefs in the Gulf of Morrosquillo.

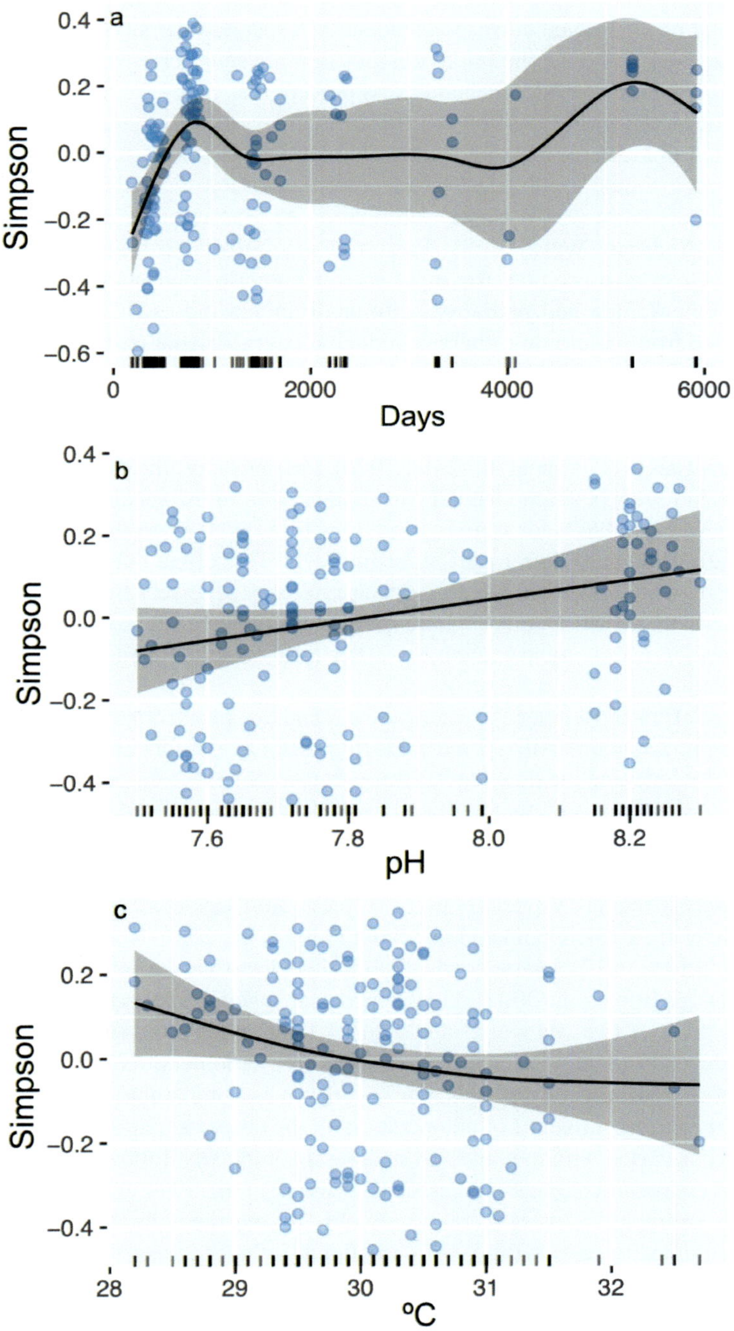


**FIGURE 2.5** Simpson diversity variation as a function of (a) days since deployment, (b) pH, and (c) surface temperature (°C) in the artificial reefs of the Gulf of Morrosquillo.

isolated artificial reefs attract both pelagic immigrants and species from surrounding open sand or muddy areas (Walsh, 1985; Belmaker et al., 2005; Jordan et al., 2005; Caddy, 2011; Folpp et al., 2013; Claisse et al., 2014; Becker et al., 2019). Conversely, Koeck et al. (2014) and Sensurat-Genç et al. (2022) found no significant correspondence in fish species diversity between natural and artificial reefs, suggesting that once the initial colonization phase is achieved, the influence of natural nearby habitats on fish assemblages becomes less perceptible.

The pattern observed in the Gulf of Morrosquillo contrasts with the predictions of island biogeography applied to artificial reefs by Bohnsack et al. (1991), which stated that habitats closer to recruitment sources should exhibit faster rates of colonization and higher species richness than more distant habitats. However, a closer examination of MacArthur and Wilson's (1967) theory, and more recent findings (Walsh, 1985; Paxton et al., 2018, 2020), suggests that proximity to natural resources can also promote the rapid establishment of dominant species. This may lead to intensifying competition and predation, ultimately increasing extinction rates and preventing additional species from colonizing, thereby reducing overall richness. By contrast, more distant artificial reefs may avoid a situation of early fish dominance, allowing a more diverse assemblage to develop over time. Lower richness on nearby artificial reefs could also be attributed to their comparatively limited biogenic and structural complexity, which provides less shelter and food than natural ecosystems (Paxton et al., 2018, 2020). Furthermore, the "stepping-stone" effect of a network of artificial reefs positioned at varying distances from the species source, as suggested by MacArthur and Wilson (1967) for islands, may constrain diversity to the closest structures while delaying the establishment of dominance at farther reefs. This effect may be accentuated when the species that are regionally available to colonize are small or relatively limited (Walsh, 1985). This may be the case within Morrosquillo, where the natural reefs harbor comparatively low diversity levels (10–78 species, averaging 40) relative to other coral ecosystems in the Colombian Caribbean (MoAm, 2013).

The positive relationship between species richness and distance from the source may be mediated by colonization processes on artificial reefs. During this stage of faunal development, species and individuals rapidly colonize the new habitat through recruitment from pelagic waters or surrounding ecosystems. This influx is likely facilitated by the high vertical profile of the structures (5–6 m), particularly when they are isolated (Rilov and Benayahu, 2000, 2002; Sheng, 2000; Claisse et al., 2014; Becker et al., 2019). In tropical waters, faunal colonization occurs relatively slowly (~5 years) and is variable across regions, especially in the tropics (Bohnsack et al., 1991; Golani and Diamant, 1999; Delgadillo et al., 2004; Thanner et al., 2006; Folpp et al., 2011). Notably, a large proportion of the visual surveys here (60.7%) were conducted on reefs less than 3 years old. Moreover, for some of these reefs, Delgadillo et al. (2004) documented that species richness and abundance were still accumulating even after two years after deployment, though at a reduced rate. Although colonization dynamics depend on multiple factors (i.e., dispersal, settlement, recruitment, seasonality), the high richness observed on younger reefs, while sometimes comparable or even higher than older ones, indicates the strong influence of colonization on the observed patterns.

Species richness varied with reef age and structural complexity, following patterns broadly consistent with successional ecological processes (Bohnsack et al., 1994; Thanner et al., 2006; Folpp et al., 2011). For example, richness increases steadily during the early colonizing phase on newly deployed, low complexity reefs (GM1), with additional influence from functional nearby artificial reefs (22). This highlights the role of spatial configuration, as areas with a greater density of artificial reefs and moderate separation between them often show enhanced richness and abundance (Frazer and Lindberg, 1994; Jordan et al., 2005; Paxton et al., 2018; Becker et al., 2019). Thus, species richness can sometimes exceed levels observed on more structurally complex reefs, as documented here. In contrast, high richness and its subsequent stabilization on older, mature, and more complex reefs (GM3) underscore the positive influence of structural and accumulated biogenic complexity over time (Bohnsack et al., 1994; Golani and Diamant, 1999; Cardoso et al., 2020; Strain et al., 2020). Conversely, intermediate reefs (GM2) displayed a markedly different pattern, with a decline in richness after structural degradation began in the second year. This was presumably the result of structural instability and hydrodynamic stress, which progressively eroded vertical relief and habitat integrity until eventual collapse and burial in sediments.

### 2.4.2 Shannon-Wiener Diversity

Shannon-Wiener diversity exhibited a nonlinear trend as a function of reef age, which is consistent with the gradual increase in biogenic structural complexity over time on most artificial reefs. As artificial reefs develop, they create greater heterogeneity of ecological niches, providing shelter and food resources that support a broader range of fishes (Thanner et al., 2006; Koeck et al., 2014; Cardoso et al., 2020). Because this index is particularly sensitive to rare species (Magurran, 1988; Wheater et al., 2011), colonization by fishes with specific ecological requirements is more likely. These less abundant species tend to fluctuate in response to environmental changes, habitat structure, interactions, and disturbances, thus producing irregularities in the index's trajectory (Leitão et al., 2016; Säterberg et al., 2019).

In addition to increased reef age and biotic structural development, the Shannon-Wiener diversity index also showed a positive linear relationship with pH, pointing to the ecological relevance of environmental stressors under climate change scenarios. Globally, low coastal pH values (<8) are commonly due to increases in freshwater run-off, nutrient enrichment, organic matter inputs, and metabolic activity from biological communities owing to the influence of $CO_2$ (Hendriks et al., 2015; Carstensen and Duarte, 2019). In the Gulf of Morrosquillo this increase in pH is influenced by freshwater inputs from the Sinú River, coastal lagoons (i.e., Cispatá, La Caimanera, La Leche, La Boquilla), and at least 23 creeks that flow directly or through the lagoons (Patiño and Flórez, 1993). This has likely led to the pH values observed on the artificial reefs here, averaging 7.82, ranging between 7.51 and 8.3 (<1 pH unit). This span falls within the expected range typical of pH values along low-latitude coastal waters (Carstensen and Duarte, 2019). Although aquatic organisms in these environments are often adapted to variable pH (Hendriks et al., 2015), the trend observed here suggest that lower values negatively influence fish diversity. This is likely linked to the physiological and behavioral sensitivity of larvae and juveniles which, under low pH conditions, suffer reduced growth, feeding capacity, and survival, and thus become more vulnerable to predation (Cattano et al., 2018). Juveniles represent a substantial fraction of the total fish assemblage on artificial reefs in Morrosquillo (Delgadillo et al., 2004), with abundances during these surveys ranging from 8.2% to 28.7% of the total number of individuals in the assemblage. Consequently, their sensitivity to pH fluctuations may contribute to the overall decline in diversity, reducing recruitment and persistence on these structures.

The Gulf of Morrosquillo has been reported to experience episodes of elevated pollution and organic matter inputs from untreated wastewater. These conditions are closely linked to the hydrological regime, runoff pulses, and sedimentation associated with seasonal climatic variability (Patiño and Flórez, 1993; Ruiz et al., 2010; Invemar, 2022). Such conditions negatively affect the eggs and larvae of fish in estuaries and coastal wetlands. This can modify the substrates required for their settlement and influence retention, dispersion, and physiological performance (Miller and Kendall, 2009). These effects, in turn, constrain the movement of early life stages and limit their effective recruitment into the artificial reefs. In this context, the Shannon-Wiener diversity index revealed different responses depending on reef type and distance from riverine and lagoonal entrances. On the most structurally complex reefs (GM3), diversity remained relatively stable, whereas on the least heterogeneous reefs (GM1), it increased with greater distance from freshwater sources. This pattern was expected to occur with a reduction in exposure to runoff-driven stressors at more isolated sites. By contrast, intermediate reefs (GM2) showed a decline in diversity with increasing distance from freshwater inputs, likely reflecting the compounded vulnerability of their fish assemblages. In these reefs, fewer species were available to buffer against even minor environmental effects on the diversity index. Thus, it is suspected that these assemblages were more strongly affected by poor water quality conditions. This trend was perhaps amplified by the physical degradation of GM2 reefs previously described, judging by the consistent difference with the other artificial reefs.

The significant interactions of the Shannon-Wiener index with some of the covariates are not straightforward to interpret. This is because interpretation should involve multiple ecological dimensions and overlapping influences, each carrying a different weight in shaping community trends (Geary et al., 2020; Tredennick et al., 2021). Even so, the results indicate that fish diversity on these artificial reefs is modulated simultaneously by intrinsic factors (reef complexity and development) and extrinsic factors (environmental variability and spatial location). Moreover, since the Shannon-Wiener index is particularly sensitive to species richness and rare taxa (Magurran, 1988; Wheater et al., 2011), it can reveal subtle ecological trends and fluctuations that are not always evident when considering richness alone.

### 2.4.3 Simpson Diversity

A comparable pattern was observed in the positive yet irregular trend of Simpson diversity as a function of reef age, which reflects both the ecological role of artificial reefs and the dynamics of the associated species. During biotic succession on artificial reefs (and on other colonizable habitats), some organisms can acquire dominance by exploiting ecological niches more efficiently than others (MacArthur and Wilson, 1967; Bohnsack et al., 1991; Chase and Leibold, 2003). On artificial reefs, some fish species recruit from their earliest life stages, while others appear later to establish residency as the reef matures (Bohnsack et al., 1991, 1994; Kellinson and Sedberry, 1998; Delgadillo et al., 2004; Jordan et al., 2005; Cardoso et al., 2020). In the Gulf of Morrosquillo, resident fishes (as opposed to visitors or transients) represent between 8.9% and 36% of recorded richness (mean 20.8%) and between 17% and 96.6% of total abundance (mean 59.3%). Here, highly variable abundances were observed among reefs and across years, with a subtle tendency to increase under local conditions. This variability explains the oscillatory nature of the Simpson diversity index during reef development, as this index is sensitive to changes in dominance within the community.

The underlying mechanisms behind these patterns relate to the vertical profile and spatial configuration of the artificial reefs, which enhance recruitment, aggregation, and persistence of fishes. Taller structures are more easily detected (i.e., enhancing colonization) by larvae and juveniles and provide shelter for demersal immigrants. High-profile reefs also offer orientation cues and prey resources for adults, including midwater and pelagic fishes (Rilov and Benayahu, 2000, 2002; Becker et al., 2019; Schram et al., 2024). Additionally, these artificial structures may be influenced by prevailing currents, generating localized upwellings, vortices, eddies, and wakes that resuspend nutrients and sediments, thereby producing turbulence that alters plankton availability. This dynamic situation attracts foraging pelagic and midwater fishes while some demersal species exploit vertical movements in the water column (Sheng, 2000; Ito, 2011; Okano et al., 2011; Williams-Grove and Szedlmayer, 2017; Becker et al., 2019). These processes were evident in the Gulf of Morrosquillo, where pelagic and midwater species were consistently abundant and frequent on the reefs. Furthermore, the presence of multiple reef modules enhances connectivity, thus facilitating movement, aggregation, and colonization across a relatively continuous reef network (Jordan et al., 2005; dos Santos et al., 2010; Becker et al., 2019). Frequent schooling behavioral activity reinforces these dynamics, modifying the patterns of dominance within the assemblages and giving rise to the nonlinear trend of Simpson diversity.

The effect of pH on Simpson diversity followed a trend similar to that observed for the Shannon-Wiener index, but was expressed at the level of common and dominant species. The most pronounced response occurred in the least complex and youngest artificial reefs (GM1). This likely reflects the higher proportion of juvenile stages during the early phases of colonization (Golani and

Diamant, 1999; Delgadillo et al., 2004; Thanner et al., 2006; Folpp et al., 2011). These early stages are particularly vulnerable to fluctuations in pH (Cattano et al., 2018), which increases their sensitivity to environmental stress and, in turn, influences community dominance patterns. In contrast, the negative impact of rising temperatures on Simpson diversity had not been previously recorded for any fish assemblage ecological attribute in Morrosquillo (Delgadillo et al., 2004). This was likely due to the smaller spatial scale of earlier studies. However, multiple studies have shown that fish abundance and diversity vary considerably with temperature changes. This is particularly notable in temperate and subtropical systems, with trends opposite to those observed here (Fabi et al., 2002; Folpp et al., 2011). In tropical artificial reefs, information regarding fish assemblages responding to temperature remains scarce. Interesting, tropical fishes are thought to be more susceptible to ocean warming because they inhabit more thermally stable environments, possess narrower tolerance ranges, and often live close to their upper thermal limits, reducing their adaptive capacity (Rummer and Munday, 2017).

Sea surface average temperatures in the Gulf of Morrosquillo typically oscillate between 28 °C and 29.5 °C (Delgadillo et al., 2004; Ricaurte-Villota and Bastidas-Salamanca, 2017; INVEMAR, 2022). During the period of this research, however, average temperatures reached 30 °C and peaked at 32.7 °C, more than 3 °C above the long-term mean. Notably, 71.5% of recorded temperature values exceeded the previously reported maximum mean of 29.5 °C. According to Rummer and Munday (2017), increases in temperature of 1.5–3 °C above the summer average can compromise aerobic performance in fishes, particularly in early life stages, through imbalances between oxygen supply and demand. This affects swimming capability, feeding capacity, predator avoidance, growth, and reproduction, ultimately increasing mortality risks. Such physiological constraints drive variability in recruitment, trigger movements to cooler areas, and increase local extinction rates, especially in species with restricted distributions (Munday et al., 2008). These mechanisms are consistent with the negative trend of Simpson diversity observed in the Gulf, highlighting the compound effects of warming and acidification on artificial reef assemblages.

### 2.4.4 Final considerations

The ecological models here indicate that structural complexity, whether physical or attained through biogenic development, is not the primary determinant of fish diversity. Instead, location and environmental variability play a predominant role in shaping diversity trends. This underscores the importance of local, patch, and fine-scale processes. Each process exerts influence independently at different temporal and spatial scales, depending on the diversity metric analyzed, since each reflects specific ecological dimensions. In the Gulf of Morrosquillo, the artificial reefs are dispersed across ~3,200 ha and ~25 km of coastline. Subsequently, ecological dynamics in the Gulf operate at fine spatial scales due to the influence of heterogeneous ecosystem mosaics, allowing for environmental gradients. This is further modulated by artisanal fishing pressure that directly affects at least 18% of the reefs. Consequently, the observed trends in the diversity of the fish assemblages reveals the interplay of intrinsic and extrinsic factors, with effects varying in both magnitude and direction according to local environmental conditions and the scale of analysis. These findings challenge the assumption that habitat heterogeneity alone controls diversity on artificial reefs. Here, we demonstrate that environmental and location factors can limit or even override the role of structural complexity at localized scales.

In short, artificial reefs in this region function as a diverse component of the ecosystems with a preeminent role in local ecological dynamics, offering critical habitats that fulfill niche requirements of a wide range of species. The optimization of this functionality will depend on continued research and experimentation under the intricate conditions of the marine environment and socio-ecological systems.

## 2.5 ACKNOWLEDGMENTS

This research was carried out through a series of agreements among public, private, and NGOs, led by Ecopetrol S.A., in collaboration with the Corporación Autónoma Regional de los Valles del Sinú y del San Jorge (CVS), the Corporación Autónoma Regional de Sucre (CARSUCRE), Fundación Oleoductos de Colombia, and MoAm S.A.S. This work is part of the first author's Master of Science thesis and is contribution number 593 of the Instituto de Estudios en Ciencias del Mar (CECIMAR), Universidad Nacional de Colombia, Sede Caribe.

## REFERENCES

Becker A, Smith JA, Taylor MD, McLeod J, Lowry MB. 2019. Distribution of pelagic and epi-benthic fish around a multi-module artificial reef-field: Close module spacing supports a connected assemblage. *Fish. Res.* 209:75–85.

Belmaker J, Shashar N, Ziv Y. 2005. Effects of small-scale isolation and predation on fish diversity on experimental reefs. *Mar. Ecol. Prog. Ser.* 289:273–283.

Bishop MJ, Mayer-Pinto M, Airoldi L, Firth LB, Morris RL, Loke LHL, Hawkins SJ, Naylor LA, Coleman RA, Chee SY, Dafforn KA. 2017. Effects of ocean sprawl on ecological connectivity: Impacts and solutions. *J. Exp. Mar. Biol. Ecol.* 492:7–30.

Bohnsack JA. 1989. Are high densities of fishes at artificial reefs the result of habitat limitation or behavioral preference? *Bull. Mar. Sci.* 44(2):631–645.

Bohnsack JA. 1991. Habitat structure and the design of artificial reefs. *In*: Bell SS, McCoy ED, Mushinsky HR, editors. *Habitat structure: The physical arrangement of objects in space*. Springer Science, Bury St Edmunds, UK. p. 412–426.

Bohnsack JA, Harper DE, McClellan DB, Hullsbeck M. 1994. Effects of size on colonization and assemblage structure of fishes at artificial reefs off southeastern Florida, USA. *Bull. Mar. Sci.* 55(2–3):796–823.

Bohnsack JA, Johnson D, Ambrose R. 1991. Ecology of artificial reef habitats and fishes. *In*: Seaman W, Sprague LM, editors. *Artificial habitats for marine and freshwater fisheries*. Academic Press, San Diego, CA, USA. p. 61–107.

Bortone SA. 2011. A pathway to resolving an old dilemma: Lack of artificial reefs in fisheries management. *In*: Bortone SA, Brandini FP, Fabi G, Otake S, editors. *Artificial reefs in fisheries management*. CRC Press, Taylor & Francis Group, Boca Raton, FL, USA. p. 311–321.

Bortone SA, Samoilys M, Francour P. 2000. Fish and macroinvertebrate evaluation methods. *In*: Seaman W Jr, editor. *Artificial reef evaluation with application to natural marine habitats*. CRC Press, Boca Raton, FL, USA. p. 127–164.

Bugnot AB, Mayer-Pinto M, Airoldi L, Heery EC, Johnston EL, Critchley LP, Strain EMA, Morris RL, Loke LHL, Bishop MJ, Sheehan EV, Coleman RA, Daftorn KA. 2022. Current and projected global extent of marine built structures. *Nat. Sustain.* https://doi.org/10.1038/s41893-020-00595-1

Burnham KP, Anderson DR, Huyvaert KP. 2011. AIC model selection and multimodel inference in behavioral ecology: Some background, observations, and comparisons. *Behav. Ecol. Sociobiol.* 65:23–35.

Caddy JF. 2011. How artificial reefs could reduce the impacts of bottlenecks in reef fish productivity within natural fractal habitats. *In*: Bortone SA, Brandini FP, Fabi G, Otake S, editors. *Artificial reefs in fisheries management*. CRC Press, Taylor & Francis Group, Boca Raton, FL, USA. p. 45–64.

Cardoso AP, Matos MR, Rosa RS, Alvarado F, Medeiros AP, Santos BA. 2020. Increased fish diversity over day and night in structurally complex habitats on artificial reefs. *J. Exp. Mar. Biol. Ecol.* 522:151244.

Carstensen J, Duarte CM. 2019. Drivers of pH variability in coastal ecosystems. *Environ. Sci. Technol.* 53:4020–4029.

Cattano C, Claudet J, Domenici P, Milazzo M. 2018. Living in a high CO2 world: A global meta-analysis shows multiple trait-mediated fish responses to ocean acidification. *Ecol. Monogr.* 88(3):320–335.

Chase JM, Leibold MA. 2003. *Ecological niches: Linking classical and contemporary approaches*. University of Chicago Press, Chicago, IL, USA.

Claisse JT, Pondella DJ, Love M, Zahn LA, Williams CM, Williams JP, Bull AS. 2014. Oil platforms off California are among the most productive marine fish habitats globally. *Proc. Natl. Acad. Sci. U.S.A.* 111(43):15462–15467.

Delgadillo O, García CB, Correa J. 2004. Dinámica temporal de la asociación de peces en dos arrecifes artificiales del Golfo de Morrosquillo, Caribe colombiano. *Actual. Biol.* 26(81):219–230.

Delgadillo-G O, Ruíz-Álvarez T. 2016. A bathymetric study to support management plans for La Caimanera Coastal Lagoon, Gulf of Morrosquillo, Colombia. *J. Coast. Conserv.* 20(3):237–244.

Delgadillo-G O, Toro NN. 2018. An integrative artificial reef project for conservation and coastal management in the Colombian Caribbean. *In*: Bortone SA, editor. *Marine artificial reef research and development: Integrating fisheries research objectives. Am. Fish. Soc. Symp. 86*. American Fisheries Society, Bethesda, MD, USA. p. 215–234.

Delgadillo-Garzón O, Delgado-Cuadros F, Henríquez RM, García LM, Ruíz-Álvarez T, Zamora-Guzmán A, Martínez K, Blanco F, Coronado-Luna G, Mogollón MJ, Puerta-Padilla F, Jiménez JC, Navarro L, Vishnoff IMJ, Perdomo-Ramos SN. 2010. Avances y desafíos del programa Diáspora: arrecifes artificiales en el golfo de Morrosquillo, Caribe colombiano. *In*: INVEMAR-ACIMAR, editor. *XIV Seminario Nacional de Ciencias y Tecnologías del Mar (SENALMAR)*. Serie Publ. Esp. INVEMAR No. 21, Santa Marta, Colombia. p. 561–565.

Delgadillo-Garzón O, García CB. 2009. Impacto de dos arrecifes artificiales en la pesca artesanal diurna del golfo de Morrosquillo. *Rev. Biol. Trop.* 57(4):993–1007.

Delgadillo-Garzón O, Zapata-Ramírez P. 2009. Evaluación rápida de peces arrecifales y su relación con la estructura del sustrato en la Islas del Rosario, área marina protegida del Caribe colombiano. *Rev. Acad. Colomb. Cienc.* 33(127):273–283.

Díaz JM, Barrios LM, Gómez-López DI, editors. 2003. Las praderas de pastos marinos en Colombia: estructura y distribución de un ecosistema estratégico. *In*: INVEMAR, editor. *Serie Publ. Esp. No. 10*. Santa Marta, Colombia. 160 p.

dos Santos LN, Shimada D, Rosental I. 2010. Fish responses to increasing distance from artificial reefs on the southeastern Brazilian coast. *J. Exp. Mar. Biol. Ecol.* 386:54–60.

Fabi G, Grati F, Lucchetti A, Trovarelli L. 2002. Evolution of the fish assemblage around a gas platform in the northern Adriatic Sea. *ICES J. Mar. Sci.* 59:309–315.

Folpp H, Lowry M, Gregson M, Suthers IM. 2011. Colonization and community development of fish assemblages associated with estuarine artificial reefs. *Braz. J. Oceanogr.* 59:55–67.

Folpp H, Lowry M, Gregson M, Suthers IM. 2013. Fish assemblages on estuarine artificial reefs: Natural rocky-reef mimics or discrete assemblages? *PLoS ONE* 8(6):e63505.

Frazer TK, Lindberg WJ. 1994. Refuge spacing similarly affects reef-associated species from three phyla. *Bull. Mar. Sci.* 55(2–3):388–400.

Geary WL, Bode M, Doherty TS, Fulton EA, Nimmo DG, Tulloch AIT, Tulloch VJD, Ritchie EG. 2020. A guide to ecosystem models and their environmental applications. *Nat. Ecol. Evol.* https://doi.org/10.1038/s41559-020-01298-8

Golani D, Diamant A. 1999. Fish colonization of an artificial reef in the Gulf of Eilat, northern Red Sea. *Environ. Biol. Fish.* 54:275–282.

Grove RS, Sonu CJ, Nakamura M. 1991. Design and engineering of manufactured habitats for fisheries enhancement. *In*: Seaman W Jr, Sprague LM editors. *Artificial habitats for marine and freshwater fisheries*. Academic Press Inc., San Diego, CA, USA. p. 109–152.

Hendriks IE, Duarte CM, Olsen YS, Steckbauer A, Ramajo L, Moore TS, Trotter JA, McCulloch M. 2015. Biological mechanisms supporting adaptation to ocean acidification in coastal ecosystems. *Estuar. Coast. Shelf Sci.* 152:A1–A8.

Hylkema A, Hakkaart QCA, Reid CB, Osinga R, Murk AJ, Debrot AO. 2021. Artificial reefs in the Caribbean: A need for comprehensive monitoring and integration into marine management plans. *Ocean Coast. Manag.* 209:105672.

Ibanez-Erquiaga B, Baktoft H, Mildenberg TB, Teilmann J, Kleivane L, Kornau LM, Agersted MD, Hüllert SM, Svendsen JC. 2025. Increased fish, biodiversity, and body size near a North Sea oil and gas platform. *Mar. Environ. Res.* 204:106959.

INVEMAR. 2022. Análisis histórico 2001–2021 de la calidad de aguas marinas y costeras en el Caribe y Pacífico colombianos. *In*: Obando P, Espinosa L, editors. *Red Cam: INVEMAR, MinAmbiente, CORALINA, CORPOGUAJIRA, CORPAMAG, CRA, CARDIQUE, CARSUCRE, CVS, CORPOURABÁ, CODECHOCÓ, CVS, CRC, CORPONARIÑO. Informe técnico 2021*. Serie Publ. Periód. No. 4, INVEMAR, Santa Marta. 218 p.

Ito Y. 2011. Artificial reef function in fishing grounds off Japan. *In*: Bortone SA, Brandini FP, Fabi G, Otake S, editors. *Artificial reefs in fisheries management*. CRC Press, Taylor & Francis Group, Boca Raton, FL, USA. p. 239–264.

Jackson JBC, Alexander KE, Sala E. 2011. *Shifting baselines: The past and the future of ocean fisheries*. Island Press, Washington, DC, USA.

Jordan LKB, Gilliam DS, Spieler RE. 2005. Reef fish assemblage structure affected by small scale spacing and size variations of artificial patch reefs. *J. Exp. Mar. Biol. Ecol.* 326:170–186.

Kellinson G, Sedberry G. 1998. The effects of artificial reef vertical profile and hole diameter on fishes off South Carolina. *Bull. Mar. Sci.* 62(3):763–780.

Koeck B, Tessier A, Amour AB, Pastor J, Bijaoui B, Dalias N, Astruch P, Saragoni G, Lenfant P. 2014. Functional differences between fish communities on artificial and natural reefs: A case study along the French Catalan coast. *Aquat. Biol.* 20:219–234.

Leitão RP, Zuanon J, Villéger S, Williams SE, Baraloto C, Fortunel C, Mendonça FP, Mouillot D. 2016. Rare species contribute disproportionately to the functional structure of species assemblages. *Proc. R. Soc. B.* 283:20160084.

Lima JS, Zalmon IR, Love M. 2019. Overview of trends of ecological and socioeconomic research on artificial reefs. *Mar. Environ. Res.* 145:81–96.

Loke HLL, Chisholm RA. 2022. Measuring habitat complexity and spatial heterogeneity in ecology. *Ecol. Lett.* https://doi.org/10.1111/ele.14084

López-Victoria M, Herrera MA, Muñoz-López V, Rodríguez-Moreno M, Puentes-Sayo A, Torres-Rodríguez J, Torres O, Chasqui LH, Tavera J, Acero A. 2023. Fish species richness of a coral reef under suboptimum conditions: The case of Varadero (Cartagena Bay, Colombia). *Acta Biol. Colomb.* 28(1):108–117.

Luypaert T, Hagan JG, McCarthy ML, Poti M. 2020. Status of marine biodiversity in the Anthropocene. *In*: Jungblut S, Liebich V, Bode-Dalby M, editors. *Yourmares 9—The oceans: Our research, our future.* Springer, Switzerland. p. 57–82.

MacArthur RH, Wilson EO. 1967. *The theory of island biogeography.* Princeton University Press, Princeton, NJ, USA.

Magurran AE. 1988. *Ecological diversity and its measurement.* Chapman & Hall, London, UK.

Medeiros APM, Ferreira BP, Betancur-R R, Cardoso AP, Matos MR, Santos BA. 2021. Centenary shipwrecks reveal the limits of artificial habitats in protecting regional reef fish diversity. *J. Appl. Ecol.* https://doi.org/10.1111/1365-2664.14053

Miller BC, Kendall AW. 2009. *Early life history of marine fishes.* University of California Press, Berkeley, CA, USA.

MoAm. 2013. Descubriendo los arrecifes coralinos del golfo de Morrosquillo. *Noticias. MoAm S.A.S.* http://moam.com.co/es/descubriendo-los-arrecifes-coralinos-del-golfo-de-morrosquillo/. Accessed 28 May 2025.

Molina A, Molina C, Giraldo L, Parra C, Chevillot GP. 1994. Dinámica marina y sus efectos sobre la geomorfología del golfo de Morrosquillo. *Bol. Cient. CIOH* 14:93–112.

Munday PL, Jones GP, Pratchett MS, Williams AJ. 2008. Climate change and the future of coral reef fishes. *Fish Fish.* 9:261–285.

Okano T, Takeda M, Nakagawa Y, Hirata K, Mitsuhashi K, Kawaguchi S, Ito J. 2011. Artificial reefs to induce upwelling to increase fishery resources. *In*: Bortone SA, Brandini FP, Fabi G, Otake S, editors. *Artificial reefs in fisheries management.* CRC Press, Taylor & Francis Group, Boca Raton, FL, USA. p. 265–278.

Oksanen J, Simpson GL, Blanchet FG, Kindt R, Legendre P, Minchin P, O'Hara R, Solymos P, Stevens M, Szoecs E, Wagner H, Barbour M, Bedward M, Bolker B, Borcard D, Borman T, Carvalho G, Chirico M, De Caceres M, Durand S, Evangelista H, FitzJohn R, Friendly M, Furneaux B, Hannigan G, Hill M, Lahti L, McGlinn D, Ouellette M, Ribeiro Cunha E, Smith T, Stier A, Ter Braak C, Weedon J. 2024. *Vegan: Community ecology package. R package version 2.7-0.* https://github.com/vegandevs/vegan; https://vegandevs.github.io/vegan/

Patiño F, Flórez F. 1993. *Estudio ecológico del golfo de Morrosquillo.* Universidad Nacional de Colombia, Fondo FEN, Centro Editorial U.N., Bogotá, Colombia. 110 p.

Paxton AB, Revels LW, Rosemond RC, Van Hoeck RV, Lemoine HR, Taylor JC, Peterson CH. 2018. Convergence of fish community structure between a newly deployed and an established artificial reef along a five-month trajectory. *Ecol. Eng.* 123:185–192.

Paxton AB, Shertzer KW, Bacheler NM, Kellison GT, Riley KL, Taylor JC. 2020. Meta-analysis reveals artificial reefs can be effective tools for fish community enhancement but are not one-size-fits-all. *Front. Mar. Sci.* 7:282. https://doi.org/10.3389/fmars.2020.00282

Pedersen EJ, Miller DL, Simpson GL, Ross N. 2019. Hierarchical generalized additive models in ecology: An introduction with mgcv. *PeerJ* 7:e6876. https://doi.org/10.7717/peerj.6876

Quinn GP, Keough MJ. 2002. *Experimental design and data analysis for biologists.* Cambridge Univ. Press, Cambridge, UK.

Quirós-Rodríguez JA, Arias-Ríos JE, Ruiz R. 2010. Estructura de las comunidades macroalgales asociadas al litoral rocoso del departamento de Córdoba, Colombia. *Caldasia* 32(2):339–354.

R Core Team. 2024. *R: A language and environment for statistical computing*. R Foundation for Statistical Computing, Vienna, Austria. https://www.R-project.org/

Ramm LAW, Florisson JH, Watts SL, Becker A, Tweedley JR. 2021. Artificial reefs in the Anthropocene: A review of geographical and historical trends in their design, purpose, and monitoring. *Bull. Mar. Sci.* 97(4):699–728. https://doi.org/10.5343/bms.2020.0046

Reyes-Nivia MC, Rodríguez-Ramírez A, Garzón-Ferreira J. 2004. Peces asociados a formaciones coralinas de cinco áreas del Caribe colombiano: listado de especies y primeros registros para las áreas. *Bol. Invest. Mar.* Cost. 33:101–115.

Ricaurte-Villota C, Bastidas-Salamanca M, editors. 2017. *Regionalización oceanográfica: una visión dinámica del Caribe*. Instituto de Investigaciones Marinas y Costeras—INVEMAR. Serie Publ. Esp. No. 14, Santa Marta, Colombia. 180 p.

Rick TC, Erlandson JM. 2009. Coastal exploitation. *Science* 325:952–953.

Rilov G, Benayahu Y. 2000. Fish assemblage on natural versus vertical artificial reefs: The rehabilitation perspective. *Mar. Biol.* 136:931–942.

Rilov G, Benayahu Y. 2002. Rehabilitation of coral reef-fish communities: The importance of artificial reef relief to recruitment rates. *Bull. Mar. Sci.* 70(1):185–197.

Ruidíaz-Ruidíaz R, Quirós-Rodríguez JA, Campos-Campos NH. 2021. Caracterización estructural de las formaciones coralinas en la región de San Antero, Córdoba, Caribe colombiano. *Rev. Acad. Colomb. Cienc. Ex. Fis. Nat.* 45(176):709–720.

Ruiz LE, Gónima L, Pérez M. 2010. Caracterización de la dinámica de la turbidez del agua. *In*: Sierra-Correa PC, Gónima L, editors. *Teledetección aplicada al ordenamiento ambiental del Distrito de Manejo Integrado Cispatá*. Serie Publ. Esp. No. 10. Instituto de Investigaciones Marinas y Costeras (INVEMAR), Santa Marta, Colombia. p. 21–27.

Rummer JL, Munday PL. 2017. Climate change and the evolution of reef fishes: Past and future. *Fish Fish.* 18:22–39.

Sale PF. 1997. Visual census of fishes: How well do we see what is there? *In*: Lessios HA, Macintyre IG, editors. *Proc. 8th Int. Coral Reef Symp.* Smithsonian Tropical Research Institute, Balboa, Panama. Vol. 2. pp. 1435–1440.

Säterberg T, Jonsson T, Yearsley J, Berg S, Ebenman B. 2019. A potential role for rare species in ecosystems dynamics. *Sci. Rep.* 9:11107.

Schram MJ, Emory ME, Kilborn JP, Peake JA, Wall KR, Williams I, Stallings CD. 2024. Reef fish assemblages differ both compositionally and functionally on artificial and natural reefs in the eastern Gulf of Mexico. *ICES J. Mar. Sci.* https://doi.org/10.1093/icesjms/fsae075

Seaman W Jr, Sprague LM. 1991. Artificial habitat practices in aquatic systems. *In*: Seaman W Jr, Sprague LM, editors. *Artificial habitats for marine and freshwater fisheries*. Academic Press Inc., San Diego, CA, USA. pp. 1–29.

Sensurat-Genç T, Lok A, Özgül A, Oruç AC. 2022. No effect of nearby natural reef existence on fish assemblage at shipwrecks in the Aegean Sea. *J. Mar. Biol. Assoc. U.K.* 102:613–626.

Sheng YP. 2000. Physical characteristics and engineering at reef sites. *In*: Seaman W Jr, editor. *Artificial reef evaluation with application to natural marine habitats*. CRC Press, Boca Raton, FL, USA. p. 51–94.

Strain EMA, Steinberg PD, Vozzo M, Johnston EL, Abbiati M, Aguilera MA, et al. 2020. A global analysis of complexity-biodiversity relationships on marine artificial structures. *Glob. Ecol. Biogeogr.* 30:140–153.

Thanner SE, McIntosh TL, Blair SM. 2006. Development of benthic and fish assemblages on artificial-reef materials compared to adjacent natural reef assemblages in Miami-Dade County, Florida. *Bull. Mar. Sci.* 78(1):57–70.

Tokeshi M, Arakaki S. 2012. Habitat complexity in aquatic systems: Fractals and beyond. *Hydrobiologia* 685:27–47.

Tredennick AT, Hooker G, Ellner SP, Adler PB. 2021. A practical guide to selecting models for exploration, inference, and prediction in ecology. *Ecology* 102(6):e03336. https://doi.org/10.1002/ecy.3336

Walsh WJ. 1985. Reef fish community dynamics on small artificial reefs: The influence of isolation, habitat structure, and biogeography. *Bull. Mar. Sci.* 36(2):357–376.

Wheater CP, Bell JR, Cook PA. 2011. *Practical field ecology: A project guide*. Wiley-Blackwell, John Wiley & Sons Ltd., West Sussex, UK.

Williams-Grove LJ, Szedlmayer ST. 2017. Depth preferences and three-dimensional movement patterns of red snapper, *Lutjanus campechanus*, on an artificial reef in the northern Gulf of Mexico. *Fish. Res.* 190:61–70.

Wood S. 2017. *Generalized additive models: An introduction with R*. 2nd ed. CRC Press, Boca Raton, FL, USA.

Wood S. 2023. *mgcv: Mixed GAM computational vehicle with automatic smoothness estimation. R package version 1.9-1.* https://cran.r-project.org/web/packages/mgcv/index.html

Yan HF, Morais RA, Bellwood DR. 2025. Species abundances surpass richness effects in the biodiversity–ecosystem function relationship across marine fishes. *Nat. Commun.* 16:7789.

Yeager LA, Deith MCM, McPherson JM, Williams ID, Baum JK. 2017. Scale dependence of environmental controls on the functional diversity of coral reef fish communities. *Glob. Ecol. Biogeogr.* 26:1177–1189.

# 3 The Historical Context, Current Status, Management Practices, and Future Development of Artificial Reefs in China

*Tao Tian**
* Corresponding author: tian2007@dlou.edu.cn

## 3.1 INTRODUCTION

Artificial reefs refer to man-made structures, made of either natural or artificial materials, that are scientifically designed and systematically deployed on the seabed in specific marine areas to improve the marine ecological environment and enhance biological resources. Ideally, they provide optimal habitats for marine organisms to forage, reproduce, grow, and develop (Zhou et al., 2019). The value of artificial reefs, however, extends far beyond ecological conservation and fisheries development. They also demonstrate tremendous application potential and significant value in fields such as diving and recreation, tourism and sightseeing, recreational fisheries, and coastal zone protection (Chen et al. 2023; Zhu et al. 2020). Artificial reefs not only contribute to the health and sustainable development of marine ecosystems but also bring substantial ecological and economic benefits to humans, promoting the diversified development and utilization of marine resources.

The materials used for artificial reef deployment are diverse and cover multiple fields, including natural materials and solid waste materials (Kuang et al. 2025; Zhang et al. 2022). With the continuous advancement of science and technology, new materials are constantly emerging, offering more options for the deployment of artificial reefs. China is characterized by an expansive and sinuous coastline, extensive maritime territories, rich marine biodiversity, and diverse habitats. However, amidst a history of declining inshore fishery resources and increasing marine ecological pressure, constructing artificial reefs and enhancing fishery resources have become important pathways for adjusting the fisheries structure, conserving resources, and advancing China's "Maritime Power" strategy. In recent years, driven by policies and supported by science and technology, China's artificial reef deployment has transitioned from the experimental stage to a phase of large-scale, systematic, and ecological management. This has led to an accumulation of rich experience in reef design, site selection, layout, and effectiveness assessment.

This study systematically reviews the development history and practical achievements of artificial reefs in China from four aspects: historical context, current status, management practices, and future development (Figure 3.1). It reviews the evolution of China's artificial reefs from early experiments to the integrated deployment of modern marine ranching. Additionally, it analyzes the current distribution of reef areas and their ecological and economic effects and summarizes management experiences in policy systems, engineering technology, monitoring, and assessment. Finally, combined with new directions such as deep-sea aquaculture and ecological restoration, it forecasts the future development of artificial reefs in China. This review provides a basis for promoting the deployment of modern marine ranching in China.

DOI: 10.1201/9781003676775-3

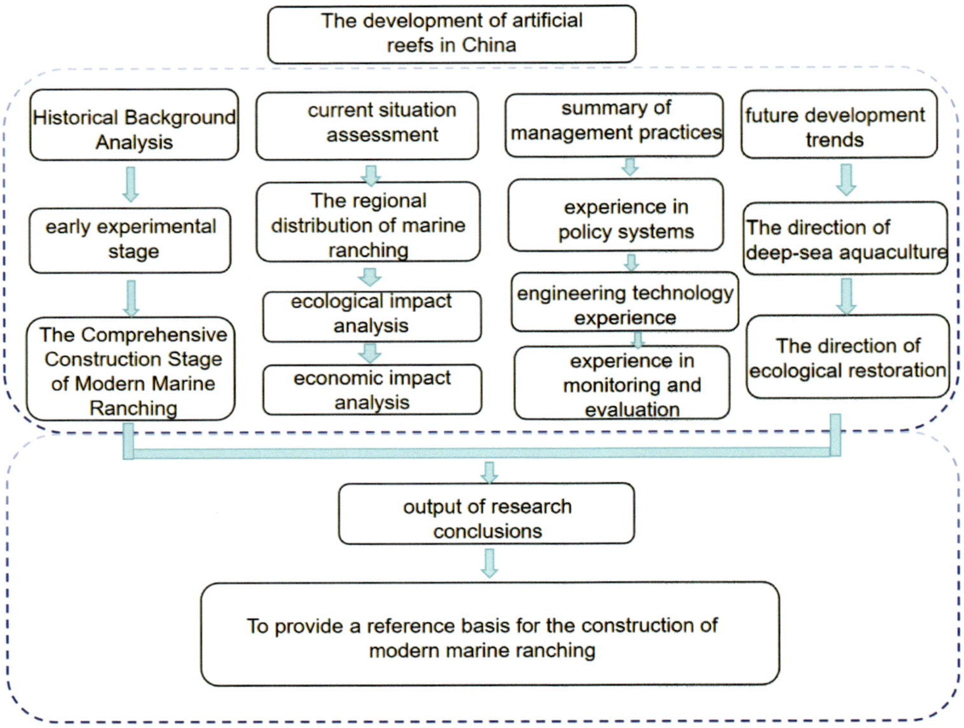


**FIGURE 3.1** Flow chart of the methodology followed for the review of China's artificial reef efforts.

## 3.2 HISTORY OF ARTIFICIAL REEF DEPLOYMENT IN CHINA

China has a long history of fisheries, predating agriculture and serving as a vital means of subsistence for early humans. The earliest written records can be traced back to books unearthed from the Yin Dynasty over 3,000 years ago (Ding et al. 2021). In primitive society, people in coastal areas of China used simple tools to collect shellfish in the intertidal zone and fished with primitive implements such as wooden sticks, stones, bones, and fishing gear (Wang et al. 2025). Written references to artificial reefs first occurred with the application of "Shen Ye" (a relatively ancient form of fishery operation in ancient China). This was an ancient fishing method that used branches and stones to attract fish. Shen Ye served as the prototype of artificial reefs around 2,000 years ago during the Spring and Autumn and Warring States Periods or the Han Dynasty. In the ancient book *Er Ya* from the Jin Dynasty, there are records of Chinese fishermen throwing branches and stones into the sea to attract fish for concentrated fishing (Ding et al. 2020). China's more scientifically based artificial reef deployment started relatively late, beginning in the late 1970s. In 1979, the Guangxi Zhuang Autonomous Region deployed China's first batch of artificial reefs in the Beibu Gulf, marking the dawn of modern artificial reef deployment in China (Qin et al. 2021). Subsequently, coastal provinces such as Guangdong, Shandong, and Liaoning successively launched experimental projects. Building on experimental reef deployments in the 1980s, in the 1990s there was a relatively stable period of reef development. During this period, research into the ecological effects of artificial reefs expanded. Beyond focusing on yield-increasing effects, scientific research began to advance. By studying resource differences between areas modified with artificial reefs and control areas lacking artificial reefs, researchers explored how artificial reefs could provide habitat, refuge, foraging, and breeding grounds for marine organisms. This gave rise to the prototype concept of "marine ranching", gradually uncovering the ecological functions of artificial reefs (Tang et al. 2025). However, this early stage of research faced challenges. The high investment in artificial reef deployment, coupled with unclear short-term economic benefits, led to an unwillingness among individuals in government and society to invest funds. Additionally, the lack of national unified planning, laws, and regulations resulted in issues such as illegal trawling and poaching that damaged reef areas after deployment. This greatly undermined the ecological effectiveness of reef deployments. Additionally, environmental risks associated with waste materials gradually emerged.

At the dawn of the 21st century, the Chinese government began attaching greater significance to the multifaceted roles of artificial reefs (Zeng et al. 2018; Liu et al. 2023). The focus of artificial reefs shifted from a single objective of yield enhancement to being included as a core component of broader marine ranching deployment. Their functions became more diversified, including restoring and optimizing the marine ecological environment, enhancing and conserving fishery resources, protecting biodiversity and aquatic wildlife, and driving the development of recreational fisheries and tourism. At the advent of the 21st century, large-scale use of reinforced concrete in artificial reef deployment began with designs focusing more on simulating natural ecological environments with consideration of flow-field effects, biological attachment, and structural stability/durability.

Since 2015, China has entered a stage of systematic deployment and high-quality artificial reef development. Driven by the national strategies of "Maritime Power" and "Ecological Civilization Deployment", the deployment of artificial reefs and marine ranching has ushered in a period of rapid increase in these deployments. The central and local governments have established a special financial fund with an unprecedented level of investment (Ma et al. 2024). Coastal provinces from north to south have conducted large-scale planning efforts as well as the deployment of national marine ranching demonstration zones. Reef design has become more scientific, modular, and ecological. Presently, modern technologies such as seaweed seed release, satellite navigation, underwater monitoring, surveillance drones, and large-scale real-time monitoring are integrated. The application of these technologies has led to more intelligent management of the ecological environment and biological resources in reef areas. An integrated development model combining reef deployment, resource enhancement, and recreational tourism has taken shape, emphasizing the simultaneous development of economic, ecological, and social benefits. Currently, the core goals of artificial reef deployment in China encompass restoring marine ecosystem, rehabilitating fishery resources, and fostering sustainable marine fisheries (Liu et al. 2025).

## 3.3 CURRENT STATUS OF ARTIFICIAL REEF DEPLOYMENT IN CHINA

### 3.3.1 Classification of Artificial Reefs

More than 70 countries worldwide have conducted research and deployed artificial reefs, with 89% of all reefs distributed in the Northern Hemisphere and a relatively balanced distribution between the Eastern and Western Hemispheres. Most countries deploy reefs to enhance fishery resources or promote fisheries development, with an emphasis on recreational fisheries in North America and Australia (Ramm et al. 2021). Underwater video surveys have become the mainstream monitoring method, and coastal developed countries have increasingly adopted more advanced monitoring technologies over the past decade (Zhao et al. 2023). With strong policy and financial support from governments at all levels, China's marine ranching and artificial reef deployment have transitioned into a phase of rapid, large-scale expansion after more than 40 years of development that synthesized international best practices. Common types of artificial reefs in China are shown in Figure 3.2.

Artificial reefs are classified into categories on the basis of individual reef units or modules. The classification system includes five descriptors: ecological function, target organism, material, deployment water layer, and morphological structure.

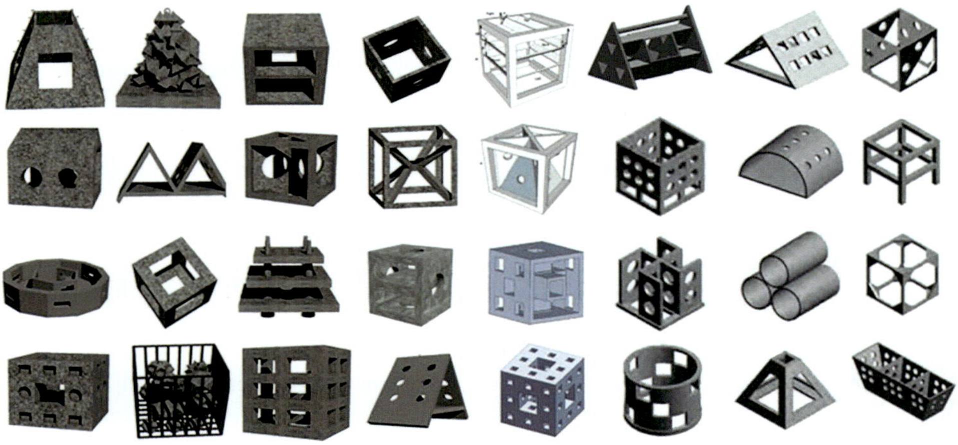

**FIGURE 3.2** Common types of artificial reef modules in China.

### 3.3.2 Classification by Ecological Function

Fish-attracting reefs: Mainly designed to attract fish for exploitation using appropriate fishing gear, and these reefs are generally placed in fish migration channels. Conservation reefs: Primarily intended to cultivate and protect aquatic organisms, improve the aquatic ecological environment, and provide habitats for organisms. Enhancement reefs: Used to enhance specific or multiple fishery resources through the release of specific species of juveniles whose entire life cycle is in the reef area and to harvest catch, typically from shallow waters. Spawning reefs: Provide spawning and hatching sites or substrates for aquatic organisms, facilitating reproduction. Resting reefs: Offer stopover and energy replenishment sites for migratory fish during their migration. Landscape reefs: Focus on constructing benthic landscape ecosystems to develop underwater sightseeing (Lee et al. 2016; Pan et al. 2022).

### 3.3.3 Classification by Target Organism

Fish reefs: Targeting fish as the main organisms. Algal reefs: Focusing on macroalgae as the main organisms. High-value species reefs: Targeting high-value seafood such as sea cucumbers, sea urchins, and abalones. Coral reefs: Centered on corals as the main organisms. Cephalopod reefs: Targeting cephalopods such as octopuses and squids. Crustacean reefs: Focusing on crustaceans as the main organisms. Mollusk reefs: Targeting mollusks as the main organisms (Zeng et al. 2020; Han et al. 2023).

### 3.3.4 Classification by Material

Concrete reefs: Mainly made of concrete. Stone reefs: Primarily constructed from natural stone. Steel reefs: Mainly composed of steel. Wooden reefs: Primarily made of wood. Shell reefs: Mainly constructed from shells. Other materials: Reefs made of materials not listed earlier. If a single material with a specific name is used, it should be named after that material; reefs modified from old ships should be named based on the main material of the hull (Li et al. 2025; Huang et al. 2025).

### 3.3.5 Classification by Deployment Water Layer

Benthic reefs: Deployed on the seabed. Mid-water reefs: Main body deployed in the middle water layer. Floating reefs: Main body deployed in the surface water layer (Zhang et al. 2024).

### 3.3.6 Classification by Morphological Structure

Rectangular reefs: With a rectangular outer contour, subdivided into cubic and rectangular prism shapes. Trapezoidal reefs: With a trapezoidal outer contour, subdivided into triangular trapezoidal, quadrilateral trapezoidal, etc., based on the number of angles. Cylindrical reefs: With a cylindrical outer contour, subdivided into circular, cylindrical, and elliptical cylindrical shapes. Pile-shaped reefs: Formed by stacking multiple small units, with a pile-like outer contour. Spherical reefs: With a spherical or hemispherical outer contour. Conical reefs: With a conical or pyramidal outer contour. Ship-shaped reefs: With a ship-like outer contour. Irregular reefs: With outer contours not falling into any of the above categories (Lima et al. 2019; Watterson et al. 2022).

### 3.3.7 Deployment of Artificial Reefs

By the end of 2015, the deployment area of artificial reefs in China had reached 619.8 square kilometers, with a total deployment volume of 60.94 million cubic meters. Among them, artificial reef complexes (*sensu* Grove and Sonu 1983) in the Yellow Sea and Bohai Sea regions accounted for 157.1 square kilometers of deployment area and 18.054 million cubic meters of deployment volume; the East China Sea region accounted for 206.2 square kilometers and 7.0 million cubic meters; and the South China Sea region accounted for 256.6 square kilometers and 42.191 million cubic meters (Mohamed et al. 2023; Zeng et al. 2021). From 2015 to 2025, 189 national marine ranching demonstration zones had been approved for deployment across China, covering the Bohai Sea, Yellow Sea, East China Sea, and South China Sea, with two-thirds located in the Bohai Sea and Yellow Sea regions. According to incomplete statistics, the scale of artificial reefs deployed along China's coasts has exceeded 70 million cubic meters, driving the efficient and sustainable development of the country's marine ranching (Wu et al. 2019; Yang et al. 2019; Li et al. 2022). Northern China: The artificial reefs constructed in the Yellow Sea and Bohai Sea are mainly aquaculture-enhancement reefs, focusing on the cultivation of high-value seafood such as sea cucumbers, sea urchins, abalones, and conches. They are generally constructed and managed by private enterprises or aquatic companies. Eastern China: The reefs are mainly fish conservation and protection reefs, focusing on the deployment of small individual reefs and the decentralized layout of large reefs. As conservation-oriented reefs, they face management challenges, and maximizing their economic benefits will be a key focus in the future. Southern China: The artificial reefs are mainly composed of large prefabricated concrete components (e.g., 3m × 3m × 3m, 4m × 4m × 4m), with the main functions of attracting fish and enhancing resources. Usually, reef groups (*sensu* Grove and Sonu 1983) are formed in a grid-like pattern on the seabed. Currently, the deployment scale of artificial reefs in the South China Sea has reached 256.6 square kilometers, with a cumulative deployment volume of 42.191 million cubic meters (Liu et al. 2022; Huang et al. 2016, 2019).

### 3.3.8 Research and Technological Development of Artificial Reefs

Research on artificial reefs mainly focuses on core themes or objectives such as fishery resource enhancement, flow-field effect simulation, numerical simulation analysis, the evolution of biological community structure, and the impact on environmental factors. Since 2000, the number of academic papers in this field has increased dramatically, indicating a growing research focus (Figure 3.3). Therefore, this study examined literature published from 2000 to September 2025, using "artificial reefs" as the keyword searched via titles, abstracts, and keywords. A total of 1,122 scientific publications on artificial reefs were retrieved from the CNKI (Chinese National Knowledge Infrastructure) database. After 2015, the number of patent applications related to artificial reefs also showed a gradual increase (Figure 3.4). Patent data were collected from 2015 to September 2025, with "artificial reefs" as the keyword, and 1,004 related patents were identified in the CNKI database.

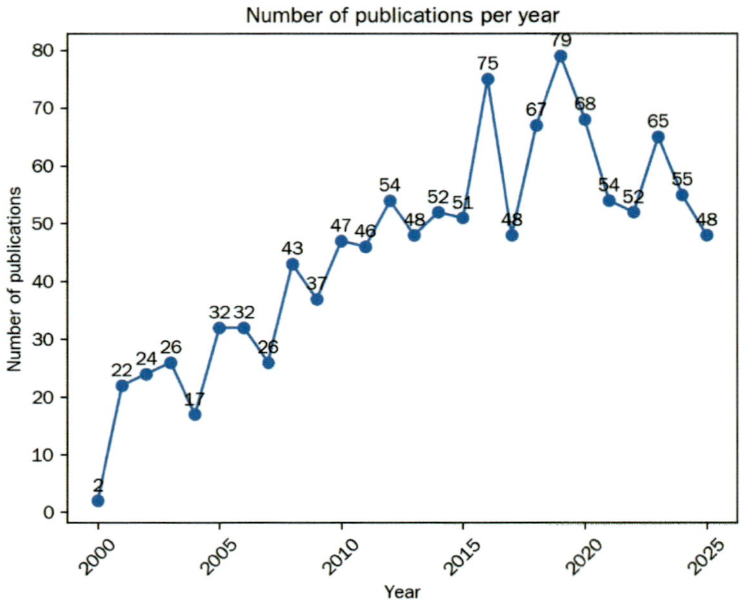


FIGURE 3.3 Number of publications on artificial reefs in China (2000–2025).

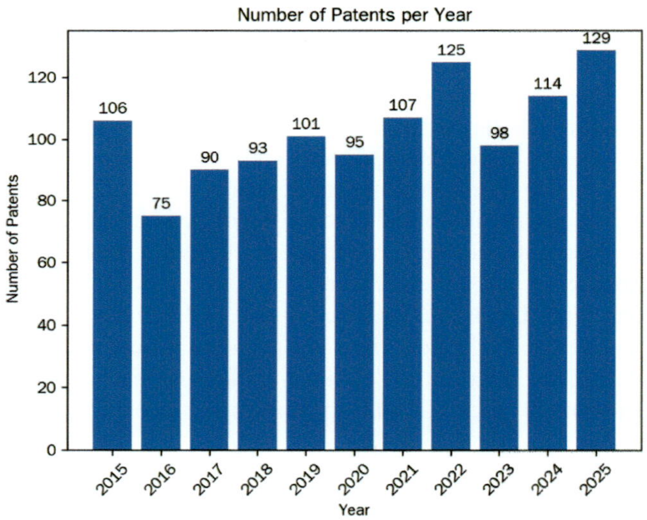


FIGURE 3.4 Number of patents filed on artificial reefs in China (2015–2025).

In deep-sea aquaculture, intelligent decision-making and remote control systems based on intelligent perception and 5G cell phone communication technologies are gradually being applied to deep-sea aquaculture processes, enabling scientific management and intelligent decision-making for various production factors (Li et al. 2023). In addition, the design and development of nursery system reefs to facilitate asexual coral reproduction have been advancing, and in situ underwater coral planting technologies off islands and coastal coral reefs have become more sophisticated.

## 3.4 ARTIFICIAL REEF MANAGEMENT IN CHINA

As a major marine fisheries country, China has promoted the development of marine ranching and the restoration of fishery resources through large-scale artificial reef deployment in recent years. This study systematically summarizes China's experience in artificial reef management, covering key aspects such as policy, planning and layout, technical standards, ecological monitoring, and community participation.

### 3.4.1 Whole-Process, Standardized Management

Before conducting any relevant marine development activities, the right to use sea areas in China must be obtained in accordance with the law, along with a special environmental impact assessment. Additionally, sea-area use demonstration reports must be prepared for different types of sea areas, specifically divided into open aquaculture sea areas and artificial reef sea areas. For artificial reef deployment, special plans and detailed implementation schemes are scientifically formulated to clarify elements such as reef type, deployment area, and deployment scale. Reef construction and deployment are conducted in strict accordance with national and industrial standards, with dynamic supervision implemented throughout the entire process of design, construction, and deployment to ensure deployment quality and ecological compliance.

After the deployment of artificial reefs, a long-term supervision procedure must be followed, which includes annual effectiveness assessments conducted to track the effects of ecological restoration and fishery resource enhancement in reef areas in real time. Additionally, comprehensive reviews are organized every five years to systematically assess key indicators such as the structural stability of reefs and the sustainability of ecological functions. Meanwhile, acoustic surveys and assessments are conducted to accurately grasp dynamic changes in the topography and geomorphology of reef areas and the distribution of biological communities. These activities provide data support for subsequent management optimization and function improvement.

### 3.4.2 Construction and Deployment Policy

At the institutional level for marine ranching, including artificial reef deployment, the central government has issued a number of guiding documents to ensure sound policy. These documents include basic laws such as the Fisheries Law of the People's Republic of China, Environmental Protection Law of the People's Republic of China, and Law of the People's Republic of China on the Administration of Sea Areas, as well as special management documents such as the Regulations on the Administration of Artificial Reef Deployment and Plan for Marine Ranching Deployment.

Local governments have actively responded and advanced the implementation of these policies. To date, multiple coastal cities, including Lianyungang (2017), Haikou (2021), Qingdao (2022), Yantai (2022), Weihai (2022), Weifang (2022), Rizhao (2022), and Binzhou (2024), have launched local deployment work related to artificial reefs and marine ranching. Provinces have formulated targeted implementation rules based on regional marine characteristics and the actual development of fisheries. These rules emphasize the dual goals of "fishery resource enhancement" and "ecological restoration" in policy implementation and incorporate artificial reef projects into the marine ecological compensation mechanism to further strengthen the orientation of ecological protection (Hu et al. 2023; Xu et al. 2023).

### 3.4.3 Establishment of a National Expert Team

To strengthen technical support, the Ministry of Agriculture and Rural Affairs has established an Expert Advisory Committee for Marine Ranching Deployment. Meanwhile, relying on the China Society of Fisheries, a Marine Ranching Branch has been established, which brings together more than 100 experts in the field and sets up 29 technical workstations. As a national academic and technical service organization, the committee provides professional consulting, technical guidance, and scientific research support for the deployment of artificial reefs and marine ranching, contributing to the high-quality development of the industry.

### 3.4.4 Scientific Planning and Regional Layout

China attaches great importance to the carrying capacity of the marine environment in the planning of artificial reefs. The plan for spatial layout is developed based on hydrological characteristics, biological community structures, and ecological needs of the three major marine regions (South China Sea, East China Sea, and Yellow-Bohai Sea). Tailored reef structural designs and material selection schemes are developed for the unique environmental conditions of each region to ensure the compatibility of reefs with regional ecosystems. In terms of ecological function optimization, efforts are made to promote the linkage between artificial reefs and natural ecosystems such as seagrass beds and oyster reefs. By constructing composite ecosystems, the synergistic restoration effect is strengthened, and the overall ecological service functions of marine areas are improved. In terms of resource conservation and management, a linkage mechanism between the fishing moratorium system and reef area management has been established. Through their synergistic efforts, a stable growth and reproduction environment for fishery resources in reef areas is provided, effectively ensuring the effectiveness of resource restoration (Yan et al. 2025; Chong et al. 2024).

### 3.4.5 Technological Innovation and Standardization

China has established a relatively complete set of technical standards for artificial reefs and marine ranching, including 17 standards related to artificial reefs, providing solid technical support for the standardized development of the industry (Table 3.1). Data for these standards are sourced from the China National Standards Information Service Platform, Industry Standard Information Service Platform, and Local Standard Information Service Platform.

On this basis, China continues to advance technological innovation and standard optimization in the field of artificial reefs. In terms of material application, ecological concrete is actively promoted. Through changes in structural materials, the pH value of reefs is reduced and the biological attachment on the reef surface is enhanced, facilitating the rapid colonization and reproduction of marine organisms. In terms of structural design, reef combinations suitable for multiple water layers (benthic, mid-water, floating) have been developed based on the characteristics of different marine water layers. This not only enhances the fish-attracting effect of reefs but also improves their resistance to water flow, ensuring long-term stability. Regarding standard implementation, pivotal industry standards—including the *Technical Specifications for Evaluation of the Effects of Artificial Reefs* (Ministry of Agriculture of the People's Republic of China, 2015)—have been promulgated to delineate technical requirements across the complete project lifecycle. These encompass pre-deployment site selection, in-situ deployment specifications, and post-deployment ecological monitoring and effectiveness evaluation, thereby driving the transition from empirical, experience-based practices toward rigorously standardized approaches in artificial reef deployment (Zhang et al. 2021; Lan et al. 2006).

### 3.4.6 Ecological Monitoring and Effectiveness Assessment

China has gradually established a closed-loop dynamic management mechanism for artificial reefs featuring a "deployment-monitoring-assessment-optimization" process, ensuring the continuous improvement of ecological functions and benefits of reef areas through refined management throughout the entire process. During the deployment phase, high-precision GPS positioning is used to ensure more accurate deployment points. After deployment, acoustic survey equipment is used to investigate and record the actual deployment position deviation and settlement degree of the

**TABLE 3.1**
**Classification of artificial reef standards**

| Standard Type | Standard Number/Name | Issuing Authority/Applicable Scope | Core Content |
|---|---|---|---|
| National Standard | GB/T 40946–2021 *Technical Guidelines for Marine Ranching Construction* | State Administration for Market Regulation | Overall planning of marine ranching, including layout requirements for artificial reefs |
| Industrial Standard | SC/T 9416–2014 *Technical Specifications for Artificial Reef Construction* | Ministry of Agriculture and Rural Affairs | Whole-process technical specifications for artificial reef deployment |
| | SC/T 9442–2022 *Technical Specifications for Quality Evaluation of Artificial Reef Deployment* | Ministry of Agriculture and Rural Affairs | Methods for evaluating the deployment quality of artificial reefs |
| | JC/T 2603–2021 *Ecological Concrete Reefs* | Ministry of Industry and Information Technology | Standards for materials and processes of ecological concrete reefs |
| Group Standard | T/SCSF 0012–2021 *Technical Regulations for Site Selection of Artificial Reef Construction* | China Society of Fisheries | Methods for ecological assessment of artificial reef site selection |
| | T/SCSF 0020–2024 *Technical Specifications for Evaluating Flow Field Effects of Artificial Reefs* | China Society of Fisheries | Methods for evaluating the impact of reefs on surrounding water flow |
| | T/SCSF 0024–2024 *Calculation of Effective Cubic Volume on Land and Drawing of Design Blueprints for Artificial Reefs* | China Society of Fisheries | Specifications for design blueprints and calculation methods of cubic volume for reefs |
| | T/SCSF 0001–2020 *Technical Specifications for Quality Evaluation of Artificial Reef Construction Projects* | China Society of Fisheries | Content, methods, and grade classification for project quality evaluation |
| | T/SCSF 0006–2020 *Technical Specifications for Transportation and Deployment of Artificial Reefs* | China Society of Fisheries | Technical specifications for the transportation and deployment process of reefs |
| | T/SCSF 0005–2020 *Technical Specifications for Manufacturing of Artificial Reefs* | China Society of Fisheries | Standards for materials and processes of concrete reefs |
| | T/SCSF 0021–2024 *Classification and Naming of Artificial Reefs* | China Society of Fisheries | Classification system and naming rules for artificial reefs |
| | T/SCSF 0019–2022 *Technical Guidelines for Layout of Artificial Reef Construction* | China Society of Fisheries | Methods for optimizing the spatial layout of reefs |
| | T/SCSF 0016–2022 *Technical Specifications for Geological Survey of Artificial Reef Construction* | China Society of Fisheries | Technical requirements for geological surveys before reef deployment |
| | T/SCSF 0017–2022 *Technical Specifications for Acoustic Survey and Evaluation of Artificial Reefs* | China Society of Fisheries | Application specifications of acoustic technology in reef survey |

*(Continued)*

**TABLE 3.1** *(Continued)*
**Classification of artificial reef standards**

| Standard Type | Standard Number/Name | Issuing Authority/Applicable Scope | Core Content |
|---|---|---|---|
| Local Standard | DB21/T 2690–2016 *Technical Regulations for Acoustic Domestication of Fish in Artificial Reef Areas* | Local Standardization Authority | Methods for promoting fish aggregation using acoustic technology |
| | DB21/T 2279–2014 *Technical Regulations for Effect Evaluation of Artificial Reefs in Liaoning Province* (2014) | Liaoning Province | Technical specifications for effect evaluation of artificial reefs in Liaoning Province |
| | DB37/T 2090–2012 *Technical Specifications for Artificial Reef Construction in Shandong Province* | Local Standardization Authority | Local technical specifications for artificial reef deployment |
| | DB13/T 1562–2012 *Technical Specifications for Artificial Reef Construction in Hebei Province* | Local Standardization Authority | Local technical specifications for artificial reef deployment |

reefs on the seabed, and evaluate the deployment effect. In the monitoring phase, multiple technical means such as acoustic detection (including multi-beam detection and sonar technology), underwater photography, and diving surveys are comprehensively used to regularly track the structural stability of reefs. In addition, monitoring includes the state of biological attachment on the surface, as well as the composition and dynamic changes of biological communities in surrounding marine areas, ensuring the comprehensiveness and timeliness of data coverage. In the assessment phase, the focus is on core effectiveness indicators of reef areas. Assessments also include key monitoring while analyses are conducted on changes in biomass, species diversity, and resource fluctuations of target species (i.e., economically important fish). This permits a comprehensive evaluation of the ecological restoration effects and fishery resource enhancement benefits of reef areas. In the optimization phase, management strategies for artificial reefs are adjusted in a targeted manner based on monitoring and assessment data (Zhang et al. 2023).

### 3.4.7 Long-Term Operation with Multi-Stakeholder Participation

In the deployment of artificial reefs and as part of its marine ranching initiative, China has actively explored and established a multi-governance model characterized by 'government leadership, enterprise operation, scientific research support, and fisherman participation'. It encourages fishery enterprises or cooperatives to participate in reef deployment and subsequent operation, with revenues from concessionary fishing rights and ecological tourism used to offset maintenance costs. A cooperation mechanism has been established between scientific research institutions and local fishermen to conduct follow-up surveys of biological resources in reef areas and provide training on scientific conservation. Pilot marine ranching demonstration zones have been set up in regions such as Shandong and Liaoning, forming a development model that integrates ecological restoration and industrial integration (Yan & Wei et al. 2025; Lv et al. 2016).

## 3.5 FUTURE DEVELOPMENT TRENDS OF ARTIFICIAL REEFS IN CHINA

The future development of artificial reefs in China will place greater emphasis on the coordination and balance between ecological restoration, marine ranching deployment, fishery resource enhancement, and carbon sequestration. Artificial reefs have evolved from simple fishery enhancement management measures to important components of inshore ecosystems integrating ecological, economic, and social functions (Li et al. 2024).

There are significant differences in hydrological conditions, sediments, and biological communities between China's northern and southern marine areas. Therefore, reef design, material selection, and deployment configuration should be adapted to local conditions. In the coastal areas of southern China, artificial reefs will be designed to prioritize typhoon resistance and water exchange. In northern marine areas, designs will focus on enhancing adaptability to factors such as low temperatures (Komyakova et al. 2021). The deployment of artificial reefs will also be coordinated with the protection and restoration of typical ecosystems such as seagrass beds, algal fields, coral reefs, and oyster reefs. Together, they will form an organically connected habitat network to improve the ecological resilience and biodiversity of the entire inshore region. Future reef deployment will not only focus on short-term resource restoration effects but also emphasize the synergistic improvement of long-term ecological benefits and carbon sequestration capacity to achieve the benign circulation and sustainable utilization of marine ecosystems.

### 3.5.1 Floating Artificial Reefs

Currently, research on artificial reefs in China mainly focuses on benthic reefs, with relatively few practical cases of floating reefs. With the expansion of marine development areas, especially the development of China's fisheries moving toward deeper and more distant marine areas (with depths exceeding 30–50 meters), floating artificial reefs can break through the limitations of water depth, expanding the deployment and application scope of marine ranching. Benthic artificial reefs are ideally deployed at depths of 10-60m (maximum 100 m), on flat seabeds with minimal silt and firm substrates.

Since 2015, Chinese researchers have gradually increased their focus on floating artificial reefs, mainly concentrating on functional research, including overall design, hydrodynamic characteristics, and structural optimization. In the floating artificial reef experiment in Hainan, China, a mid-water landscape floating reef integrating fish resource aggregation, protection, and recreational diving functions was designed, suitable for open waters with depths exceeding 30 meters (Wang et al. 2022).

### 3.5.2 Composite Artificial Reefs

#### 3.5.2.1 New Deployment Method—Prefabricated Assembled Artificial Reefs

Based on the application of RPC (Reactive Powder Concrete) technology, the types of standard reef components should be as few and simple as possible to facilitate deployment, industrial production, and assembly, while ensuring good stability. Artificial reefs will be deployed at different heights in a coordinated manner to increase flow, expand surface area, and improve water flow. Considering that different aquatic organisms have different aggregation preferences and behavioral characteristics, full consideration must be given during the design process. This includes checking foundation bearing capacity and stability to prevent reefs from sliding, overturning, sinking, or being buried due to the action of ocean currents after deployment. This ensures the durability and safety of reef structures. By utilizing BIM (Building Information Modeling) and model clustering analysis, general prefabricated standard component types suitable for ecological artificial reef structures have been extracted, promoting the large-scale and industrial production of standard components (Kong et al. 2022).

#### 3.5.2.2 New Deployment Method—Special-Function Artificial Reefs

Structures such as underwater sculptures, underwater palaces, and underwater cities have expanded the functions of artificial reefs from environmental restoration and optimization to fishery resource enhancement and protection. These reef functions can affect multiple fields, including underwater sightseeing, exploration, and science popularization, enriching the connotation of modern recreational fisheries. Combined with marine ranching, memorial reefs or burial reefs can be developed, which incorporate human cremains into artificial reefs and customizing shapes, specifications, and appearances according to individual needs. This approach not only meets the needs of modern social development but also has a certain special significance, contributing to marine public welfare (Choi et al. 2002).

In response to the unique needs of diving and tourism in marine ranching built in tropical waters, a landscape reef reinforced with corrosion-resistant fiberglass-reinforced plastic has been built in Hainan Province, China.

By making use of new materials and deployment methods for artificial reefs, 3D printing technology for artificial reefs has been developed, with feasibility studies and pilot experiments completed. The uneven surface of 3D-printed reefs is more conducive to biological attachment. In 3D printing technology, labor costs are reduced by 8.45%, while machinery costs increase from 15.57% (traditional method) to 24.72%. The overall deployment cost can be reduced by more than 20%, and the improvement in mechanization level has a significant effect on reducing carbon emissions. Carbon emissions from materials, machinery, transportation, and labor are all lower than those of traditional deployment methods, with total emissions accounting for only 71.08% of the traditional prefabrication method (Kuang et al. 2025).

### 3.5.3 Resource Utilization of Waste Materials

In the field of artificial reefs, an increasing variety of waste materials are being recycled, such as magnesium-based materials, deployment waste, recycled concrete, and wind turbine blades. The utilization of waste materials is integrated with new deployment methods, such as "3D printing + solid waste" artificial reefs. To date, more than 249 different materials have been used for artificial reefs, with six common key materials: natural (wood, stone), deployment (concrete, steel), waste (waste tires, old ships, industrial waste), biological (shell powder, oyster shells), other materials (ceramics, fibers, plastic), and normally installed marine structures with coral reef protection functions. Currently, stone and concrete remain the main materials for artificial reefs. However, in the near future, driven by the need to reduce carbon emissions and recycle waste resources, a greater variety of waste materials will undoubtedly be used in the production of artificial reefs.

## 3.6 CONCLUSION

After years of exploration and practice, modern artificial reef deployments in China have evolved from initial experimental deployments to a systematic and large-scale national ecological project, becoming a key pathway for realizing the sustainable utilization of marine fishery resources and the restoration of inshore ecosystems. Driven by the national strategies of "Maritime Power" and "Ecological Civilization Deployment", the functions of artificial reefs have gradually expanded from a single focus on fishery enhancement to the coordinated development of multiple dimensions such as ecological restoration, biodiversity conservation, recreational fisheries, and carbon sequestration capacity enhancement, forming a modern management model centered on marine ranching.

Thus far, China has accumulated rich experience in policy planning and deployment configuration, technological innovation, and monitoring and assessment of artificial reefs. This has been achieved through whole-process standardized management, multi-level legal and regulatory guarantees, and expert advice in reef design. These designs have emphasized the integration of scientific, ecological, and functional characteristics through the development of diverse reef types adapted to different marine areas and target species. Experience has also led to improved material selection, promoting the resource utilization of waste and low-carbon transformation, embodying the principles of a circular economy and sustainable development.

In the future, the development of artificial reefs in China will expand to deep-sea areas and multiple water column layers, with new structures such as floating reefs and adjustable water-layer reefs breaking through the limitations of traditional water depths. Technological and material innovations will continue, with new technologies such as 3D printing, prefabricated assembly, and ecologically suitable concrete improving reef performance and deployment efficiency. The integration of multiple industries will accelerate, with artificial reefs combining social functions including recreational tourism, science popularization and education, and ecological burials. Greater emphasis will be placed on systematic ecological restoration, with the coordinated restoration of ecosystems such as seagrass beds, coral reefs, and oyster reefs to build a resilient and healthy marine habitat network.

To achieve high-quality development of artificial reefs as a viable management option, it is necessary to further strengthen scientific and technological support, improve the long-term operation mechanism, and promote the learning of international experiences. Ultimately, a modern Chinese marine ranching system integrating ecological benefits, economic value, and social significance will be built. These efforts will undoubtedly provide valuable solutions for addressing global marine ecological challenges.

## REFERENCES

Chen X, Che X, Zhou Y, et al. A numerical simulation study and effectiveness evaluation on the flow field Effect of trapezoidal artificial reefs in different layouts. *Journal of Marine Science and Engineering*, 2023, 12(1): 3.

Choi CG, Takeuchi Y, Terawaki T, et al. Ecology of seaweed beds on two types of artificial reef. *Journal of Applied Phycology*, 2002, 14(5): 343–349.

Chong L, Siders ZA, Lorenzen K, et al. Global synthesis of effects and feedbacks from artificial reefs on socio-ecological systems in recreational fisheries. *Fish and Fisheries*, 2024, 25(2): 303–319.

Ding Q, Shan X, Jin X, et al. Ecological footprint and vulnerability of marine capture fisheries in China. *Acta Oceanologica Sinica*, 2020, 39(4): 100–109.

Ding Q, Shan X, Jin X, et al. A multidimensional analysis of marine capture fisheries in China's coastal provinces. *Fisheries Science*, 2021, 87(3): 297–309.

Grove RS, Sonu CJ. *Review of Japanese Fisheries Reef Technology*. Report 83-RD–137. Southern California Edison Company, Rosemead, CA, 1983.

Han C, Liu K, Kinoshita T, et al. Assessing the attractive effects of floating artificial reefs and combination reefs on six local marine species. *Fishes*, 2023, 8(5): 248.

Hu Y, Wang J, He H, et al. Three-dimensional marine ranching cage inspection path planning integrating the differential evolution and particle swarm optimization algorithms. *IEEE Access*, 2023, 11: 109747–109763.

Huang D, Hoeksema BW, Affendi YA, et al. Conservation of reef corals in the South China Sea based on species and evolutionary diversity. *Biodiversity and Conservation*, 2016, 25(2): 331–344.

Huang J, Lowe RJ, Ghisalberti M, et al. Parametric modelling of wave transformation across porous artificial reefs. *Coastal Engineering*, 2025, 200: 104754.

Huang Y, Yan M, Xu K, et al. Distribution characteristics of microplastics in Zhubi reef from South China Sea. *Environmental Pollution*, 2019, 255: 113133.

Komyakova V, Chamberlain D, Swearer SE. A multi-species assessment of artificial reefs as ecological traps. *Ecological Engineering*, 2021, 171: 106394.

Kong J, Cong G, Ni S, et al. Recycling of waste oyster shell and recycled aggregate in the porous ecological concrete used for artificial reefs. *Construction and Building Materials*, 2022, 323: 126447.

Kuang W, Pavlenko P, Guo H, et al. Utilization of machine-made sand waste in 3D-printed ecological concrete for artificial reefs. *Case Studies in Construction Materials*, 2025, 22: e04494.

Lan CH, Hsui CY. The deployment of artificial reef ecosystem: Modelling, simulation and application. *Simulation Modelling Practice and Theory*, 2006, 14(5): 663–675.

Lee MO, Kim JK, Kim BK. A review-status of development and research of artificial reefs in the east Asian countries. *Journal of Fisheries and Marine Sciences Education*, 2016, 28(3): 630–644.

Li C, Lin H, Guo Y, et al. Fatty acid analysis reveals the trophic interactions among organisms in the Zhelin Bay Marine Ranch. *Frontiers in Marine Science*, 2023, 10: 1132246.

Li M, Chen Z, Li M, et al. Integrated assessment of trace elements in a marine ranching area based on multi-species and multi-level biomarkers: A case study in China's national-level marine ranching demonstration area. *Frontiers in Marine Science*, 2024, 11: 1413847.

Li R, Guo Y, Zang A, et al. Variation in the attached community structure and function of artificial habitats during ecological succession: A case study of the artificial reef area in Bailong Pearl Bay, China. *Frontiers in Marine Science*, 2022, 9: 950151.

Li W, Qu K, Wang C, et al. Hydrodynamic effects of porous trapezoidal artificial reefs on the wave transformation and wave setups at uneven coral reefs. *Physics of Fluids*, 2025, 37(9).

Lima JS, Zalmon IR, Love M. Overview and trends of ecological and socioeconomic research on artificial reefs. *Marine Environmental Research*, 2019, 145: 81–96.

Liu G, Xu Y, Ge W, et al. How can marine fishery enable low carbon development in China? Based on system dynamics simulation analysis. *Ocean & Coastal Management*, 2023, 231: 106382.

Liu J, Cao L, Xu W, et al. Formation and development of coral reefs in the South China Sea. *Palaeogeography, Palaeoclimatology, Palaeoecology*, 2022, 594: 110957.

Liu Z, Lu W, Wang T, et al. The assessment of carrying capacity of marine fishery resources in China. *Frontiers in Marine Science*, 2025, 11: 1518235.

Lv W, Huang Y, Liu Z, et al. Application of macrobenthic diversity to estimate ecological health of artificial oyster reef in Yangtze Estuary, China. *Marine Pollution Bulletin*, 2016, 103(1–2): 137–143.

Ma J, Wu Z, Guo M, et al. Dynamic relationship between marine fisheries economic development, environmental protection and fisheries technological Progress—A case of coastal provinces in China. *Ocean & Coastal Management*, 2024, 247: 106885.

Ministry of Agriculture of the People's Republic of China. *Technical Specifications for Evaluation of the Effects of Artificial Fish Reef* (in Chinese). SC/T 9417–2015, 2015.
Mohamed HF, Abd-Elgawad A, Cai R, et al. Microbial community shift on artificial biological reef structures (ABRs) deployed in the South China Sea. *Scientific Reports*, 2023, 13(1): 3456.
Pan Y, Tong H, Wei D, et al. Review of structure types and new development prospects of artificial reefs in China. *Frontiers in Marine Science*, 2022, 9: 853452.
Qin M, Wang X, Du Y, et al. Influencing factors of spatial variation of national marine ranching in China. *Ocean & Coastal Management*, 2021, 199: 105407.
Ramm LAW, Florisson JH, Watts SL, et al. Artificial reefs in the Anthropocene: a review of geographical and historical trends in their design, purpose, and monitoring. *Bulletin of Marine Science*, 2021, 97(4): 699–728.
Tang S, Xu P, Zhang S, et al. An idea for marine ranching planning based on ocean currents and its practice in Zhongjieshan Archipelago National Marine Ranching, China. *Aquaculture*, 2025, 596: 741780.
Wang L, Liang Z, Chai Z, et al. Construction and evolution of artificial reef ecosystems: Response and regulation of marine microorganisms. *Environmental Pollution*, 2025, 367: 125610.
Wang X, Feng J, Lin C, et al. Structural and functional improvements of coastal ecosystem based on artificial oyster reef construction in the Bohai Sea, China. *Frontiers in Marine Science*, 2022, 9: 829557.
Watterson E, Thompson M, Gainza J, et al. Technology-driven approach to the modelling and design of artificial surf reefs. *Coastal Engineering Proceedings*, 2022 (37): 68–68.
Wu Z, Tweedley JR, Loneragan NR, et al. Artificial reefs can mimic natural habitats for fish and macroinvertebrates in temperate coastal waters of the Yellow Sea. *Ecological Engineering*, 2019, 139: 105579.
Xu P, Jiao M, Li H, et al. Effects of artificial reef and fishing moratorium on trophic structure of biological community in the Pearl River Estuary marine ranch based on stable isotopes. *Marine Environmental Research*, 2023, 190: 106066.
Yan J, Wei X, Si L, et al. Bacterial community composition and its relationship with environmental factors in the artificial reef area for marine ranching in Changhai County. *Animals*, 2025, 15(5): 639.
Yan J, Yue Y, Lin S, et al. Preliminary study on the structure of fungal communities in artificial reef areas in the northern Yellow Sea. *Oceans*, 2025, 6(1).
Yang X, Lin C, Song X, et al. Effects of artificial reefs on the meiofaunal community and benthic environment-A case study in Bohai Sea, China. *Marine Pollution Bulletin*, 2019, 140: 179–187.
Zeng L, Tang Z, Chen P, et al. Bioacoustic application on fisheries management in an artificial reefs' ecological reserve of Bohai Gulf China. *Environmental Earth Sciences*, 2018, 77(21): 728.
Zeng L, Tang Z, Chen P, et al. Optimization of fishery resources assessment methods and ecological effects evaluation of artificial reefs. *Marine Biology Research*, 2021, 17(1): 72–85.
Zeng X, Tanaka KR, Mazur M, et al. Effects of habitat on reef fishes biodiversity and composition in rocky reefs. *Aquatic Biology*, 2020, 29: 137–148.
Zhang NH, Tian T, Shen L, et al. Research and application progress of artificial reef materials (in Chinese). *Journal of Dalian Ocean University*, 2022, 37(2): 352–360.
Zhang R, Zhang H, Liu H, et al. Differences in trophic structure and trophic pathways between artificial reef and natural reef ecosystems along the coast of the North Yellow Sea, China, based on stable isotope analyses. *Ecological Indicators*, 2021, 125: 107476.
Zhang T, Zhao W, Zhou C, et al. Design and modeling of an eco-friendly anchored fish aggregating device with artificial reef subjected to wave and current. *Ocean Engineering*, 2024, 313: 119599.
Zhang Y, Wan H, Zhao Y, et al. Macrobenthic functional group analysis of ecological health of the intertidal artificial oyster reefs in the Yangtze Estuary, China. *Frontiers in Marine Science*, 2023, 9: 1059353.
Zhao Y, Li Y. Blue transition for sustainable marine fisheries: Critical drivers and evidence from China. *Journal of Cleaner Production*, 2023, 421: 138535.
Zhou X, Zhao X, Zhang S, et al. Marine ranching construction and management in East China Sea: Programs for sustainable fishery and aquaculture. *Water*, 2019, 11(6): 1237.
Zhu W, Qin C, Ma H, et al. Response of protist community dynamics and co-occurrence patterns to the construction of artificial reefs: A case study in Daya Bay, China. *Science of the Total Environment*, 2020, 742: 140575.

# 4 Two Decades of Artificial Reef Research in New South Wales, Australia

*Matthew D. Taylor* and Alistair Becker*
* Corresponding author: matt.taylor@dpi.nsw.gov.au

## 4.1 INTRODUCTION

Artificial reefs have long been employed to enhance fisheries (Bohnsack and Sutherland 1985). The deployment of artificial reefs throughout the world has accelerated over the latter part of the 20th century and well into the 2000s (Bao et al. 2025). The science supporting artificial reefs has similarly developed over this period. Historically, artificial reef deployments have involved 'splash and dash' deployments using materials of opportunity (e.g., Branden et al. 1994; Diplock 2011), often with little or no follow-up monitoring. More recently, artificial reef programmes often involve careful planning informed by research and include monitoring and evaluation following deployments (Becker et al. 2018). Contemporary artificial reef structures also include specific design features to benefit desirable species (Ramm et al. 2021) and are constructed from durable and environmentally friendly materials (e.g., Carral et al. 2020). These modern artificial reefs also focus on outcomes for recreational fisheries (Ramm et al. 2021).

Recreational fishing is one of the most popular and widespread pastimes in Australia. According to the most recent national social and economic survey on recreational fishing, participation in recreational fishing can exceed 30% of the population in some locations (Moore et al. 2023). Recreational fishing is thus a highly consequential source of economic activity. This same survey indicates that recreational fishing contributes over AUD 11 billion (~USD 7.25 billion) per annum to the Australian economy, supporting over 100,000 jobs nationwide. For this reason, and given continued population growth, governments across most jurisdictions periodically invest in strategies aimed at improving fishing opportunities. These investments may include funding for planning, deployment and monitoring of artificial reefs. Monitoring programmes often focus on tracking fish colonisation and succession (including confirming colonisation by recreationally important species), habitat usage and behaviour. These programmes can also address specific permit conditions such as debris accumulation, interactions with threatened species and integrity of the structures.

In Australia, artificial reefs have been deployed in most jurisdictions; however, the majority of large-scale deployments have occurred in Victoria, Queensland, Western Australia (WA) and New South Wales (NSW). In Victoria, seven sets (*sensu* Grove et al. 1983) of reefs have been deployed in recent years, primarily focused in Port Phillip Bay, the largest expanse of sheltered waters in the state. Queensland has similarly focused on reef deployments in the sheltered waters of Moreton Bay and Great Sandy Strait, with a mix of purpose-built reefs and materials of opportunity, intended to support both diving and recreational fisheries outcomes. Western Australia has an ongoing programme of artificial reef deployments focused on purpose-built structures in open waters. Deployments to date have numbered seven reefs but this programme is unique in that it is

 DOI: 10.1201/9781003676775-4

driven and project-managed primarily by the state's recreational fishers. This project is supported by a significant investment from the Recreational Fishing Initiatives Fund (which allocates revenue generated from the WA recreational boat fishing licence).

New South Wales operates Australia's largest contemporary artificial reef programme, having established an extensive artificial reef network targeted at enhancing recreational fisheries. Here, 'network' collectively refers to artificial reefs of varying size and complexity deployed across the waters contained within the jurisdictional boundaries of NSW. Planning for this programme commenced in the early 2000s and has steadily developed since then. A key feature of this programme is the inclusion of a robust research and monitoring component, which has supported adaptive decision-making, quantification of benefits and a science-driven evolution of the programme. With over 20 years since its commencement, it is a unique opportunity to consider the context, funding model and scale of the programme and review both the scientific basis and the research achievements over this time. Here, we provide an overview of the NSW Artificial Reefs Program and the scientific advances that have accompanied the development of the programme over the past two decades. We conclude by highlighting opportunities for future research and development to further improve the return on investment in artificial reefs for recreational fisheries in NSW and elsewhere.

## 4.2 FUNDING AND IMPLEMENTATION OF THE NSW ARTIFICIAL REEF PROGRAM

While there have been various ad hoc artificial reef deployments in NSW over the latter half of the 20th century (Branden et al. 1994; Pollard 1989), planning for the first coordinated programme commenced in the early 2000s (Diplock 2011). The development of the programme was tightly linked to the implementation of the NSW saltwater and freshwater Recreational Fishing Licence around 2000, which subsumed the freshwater fishing licence that had been implemented earlier. Unlike other licences, the legislation governing the new Recreational Fishing Licence mandated that all the revenue generated must be placed into a Trust account, administered by the NSW Government (the NSW Recreational Fishing Trusts [RFTs]). The legislation also stipulated that, other than an allowance for administration, investment of RFT funds was to target programmes designed to enhance recreational fishing opportunities and outcomes across both freshwater and saltwater habitats. In addition, all parties, whether they are government agencies, non-government organisations, local councils, universities or other organisations, are required to apply and compete for access to the Trust funds. The expenditure decisions for these funds are the responsibility of an independent panel of representatives from the recreational fishing sector. To date, funding has supported a diverse range of programmes spanning education, access, ecological research (to inform stock assessment and sustainable management), recreational fisheries monitoring, compliance and fisheries enhancements (and their monitoring), including stocking, repair and rehabilitation of natural habitats, and artificial reefs. In recent years, revenue in the RFT has averaged around AUD 20 million (~USD 13 million) per year (Department of Regional New South Wales 2023), with recent statistics indicating up to 166,319 short-term (three days and one month) and 230,000 long-term (one and three years) fishing licences are sold annually (New South Wales Recreational Fishing Trusts 2023).

The most significant investment made from the NSW RFTs was the establishment of Recreational Fishing Havens (RFH) in NSW. The establishment of RFHs commenced soon after the advent of the NSW RFTs and involved using revenue to remove commercial fishing licences from 30 estuaries (25 entire estuaries and five partial estuaries) along the NSW coast. The RFHs included the major estuaries adjacent to some of the most populous regions of the state and greatly reduced intersectoral conflict. Most importantly, these RFHs provided estuaries spanning the entire state where investments in fisheries enhancement could be made with some confidence that recreational (rather than commercial) fishers would be the sole (or major) beneficiaries of these interventions. This largely avoids the complexities that could arise when benefits from investment in fisheries enhancements flow to sectors that have not co-invested in the programme (e.g., Arnason 2008).

### 4.2.1 Learning from Angler Survey Data to Target and Frame Reef Deployments

As noted earlier, the NSW RFTs fund a range of monitoring programmes for recreational fisheries. The most extensive and longest-running monitoring programme is the Recreational Fisheries Monitoring Program (RFMP). Coupled with the 2000–2001 survey of recreational fishing (which pre-dated the RFMP), this programme has provided an ongoing timeseries of statewide and localised indices of recreational fishing activity. These survey data were collected using a standardised approach that now spans 25 years (2000–2025, Henry and Lyle 2003; Murphy et al. 2020, 2022, 2024; West et al. 2015). This valuable information has supported the planning and implementation of the NSW artificial reef program, and the timeseries identifies clear patterns of fishing effort among platforms (i.e., either boat or shore-based fishing), bioregions, habitats and locations. It also identifies important species for different classes of fishers, bait usage and fishing methods. In addition, recreational monitoring efforts have revealed the most important motivational factors for recreational fishers in NSW (e.g., Fowler et al. 2022; Vaishampayan et al. 2025), which are similarly important in planning investments in fisheries enhancements like artificial reefs.

Broadly, the time series of recreational fisheries data outlined earlier clearly indicates that estuaries are the most heavily targeted waterbodies in NSW, followed by inshore oceanic waters. Within these waterbodies, Sparidae, Platycephalidae and Carangidae are the families that are most widely and heavily targeted (Table 4.1). Within estuaries, the most heavily targeted sparids include Yellowfin Bream (*Acanthopagrus australis*, although the reported catch includes a mix of two similar additional species—Black Bream *Acanthopagrus butcheri* and Tarwhine *Rhabdosargus sarba*) and Pink Snapper (*Chrysophrys auratus*). The most heavily targeted carangid is Yellowtail Scad (*Trachurus novaezelandiae*), and the most heavily targeted platycephalid is Dusky Flathead (*Platycephalus fuscus*). Within inshore habitats, the most heavily targeted sparid is Pink Snapper, the most heavily targeted carangids include Yellowtail Kingfish (*Seriola lalandi*), Silver Trevally (*Pseudocaranx georgianus*) and Yellowtail Scad, while the most heavily targeted platycephalid is Bluespotted Flathead (*Platycephalus caeruleopunctatus*). Some attributes of these taxa are summarised in Table 4.1. The motivational characteristics of fishers exploiting saltwater habitats in NSW include both typical catch (i.e., catch and/or harvest) and non-catch-based elements. Important catch- and harvest-based objectives include (1) maximising the number of trips where a fish is caught; (2) maximising the number of fish caught (whether they are retained or released); (3) maximising the number of hits (i.e., bites) by sportfish; and (4) maximising the number of legal-sized fish caught (e.g., Fowler et al. 2022). Collectively, this information has provided an evolving insight into NSW recreational fisheries over the past 25 years, which continues to inform the NSW Artificial Reefs Program in both a general and technical sense.

### 4.2.2 Estuary Reefs in NSW

The estuary artificial reef network in NSW was developed in late 2005, with deployments mostly completed by 2007 (Folpp et al. 2013), and the fish assemblages on these reefs are now around 20 years old. All estuary reef deployments in NSW involved Reef Ball® units (www.reefball.org) units, deployed in sets, groups and complexes (Grove et al. 1983). The deployment of the reefs was focused on five RFHs spread across regional and urban areas of NSW (i.e., Bellinger River, Lake Macquarie, Botany Bay, St. Georges Basin and Lake Conjola, Figure 4.1, Table 4.2). Deployments were also undertaken in one non-RFH estuary—Merimbula Lake. These estuaries represent important recreational fisheries, but all had minimal natural rocky-reef habitat; although Botany Bay

**TABLE 4.1**

**Summary of biological and fishery attributes for the most important recreational species targeted across estuaries and inshore habitats in NSW, including the minimum total length at which the species can be legally harvested (Minimum length), the percentage of recreational catch that is released (% released) and the species appearance in artificial reef Baited Remote Underwater Video footage (% BRUV deployments). See footnotes for further explanation.**

| Species | Family | Key habitats | Minimum length (cm) | % released[1] | % BRUV deployments[2] | |
|---|---|---|---|---|---|---|
| | | | | | Estuary | Ocean |
| Yellowfin Bream (*Acanthopagrus australis*) | Sparidae | Primarily estuarine (occasionally inshore), demersal, reef and seagrass associated | 25 | 80 | 67% | -[4] |
| Pink Snapper (*Chrysophrys auratus*) | Sparidae | Estuarine and oceanic, demersal, reef and seagrass (juv.) associated | 30 | 74 | 55% | 82% |
| Yellowtail Scad (*Trachurus novaezelandiae*) | Carangidae | Estuarine and oceanic, demersal and pelagic | No minimum | 40 | 26% | 82% |
| Dusky Flathead (*Platycephalus fuscus*) | Platycephalidae | Primarily estuarine (occasionally inshore), benthic, soft-sediment associated | 36 | 71 | 2% | -[4] |
| Silver Trevally (*Pseudocaranx georgianus*) | Carangidae | Primarily oceanic (but also bays and inlets) | 30 | 63 | -[3] | 73% |
| Yellowtail Kingfish (*Seriola lalandi*) | Carangidae | Primarily oceanic (but also bays and inlets) | 65 | 79 | -[3] | 55% |
| Bluespotted Flathead (*Platycephalus caeruleopunctatus*) | Platycephalidae | Primarily oceanic (but also bays and inlets), benthic, soft-sediment associated | 33 | 57 | -[3] | 91% |

[1] Refers to the percentage of recreational catch that is released, an indicator of the magnitude of catch-and-release fishing, as reported in Murphy et al. (2024).

[2] Indicates the percentage of baited remote underwater video (BRUV) deployments during the relevant monitoring period in which the species was observed, as a coarse indicator of species presence.

[3] Not calculated for estuary artificial reefs.

[4] Not calculated for ocean artificial reefs.

**FIGURE 4.1** Map of artificial reef deployments in New South Wales, showing the distribution of estuary reefs (grey circles, labelled on the left) and ocean reefs (grey triangles, labelled on the right) along the coast. The inset map shows the location of the main map on the eastern Australian seaboard. Reef names correspond to those listed in Table 4.2.

already contained abundant artificial hard structures (Stelling-Wood et al. 2023), which were augmented by the artificial reefs.

The Reef Ball units were ideal for establishing reefs in NSW estuaries for a number of reasons. Firstly, they provided hard structure with low vertical relief, which is appropriate for use in shallow estuaries. Secondly, they offered design-specific features that were particularly suitable for estuarine sparids (i.e., Yellowfin Bream, Tarwhine and Pink Snapper, Curley et al. 2013). Furthermore, they could easily be deployed as extensive reef 'sets' and 'groups' across soft-sediment habitats in such a fashion to produce a large reef complex area on these habitats (which would benefit Dusky Flathead). Additionally, the Reef Balls could also be easily integrated with other habitats

**TABLE 4.2**

**Summary of estuary and ocean artificial reef deployments under the New South Wales Artificial Reefs Program. Locations of all reefs are shown in Figure 4.1, and schematics of the ocean reef designs are shown in Figure 4.2.**

| Name/Location | Deployment coordinates | Year deployed | Water depth (m) | Material and design |
|---|---|---|---|---|
| Estuary reefs: | | | | |
| Bellinger River | 30°.26.525′ S 153°00.786′ E | 2008 | 3 | 150 Mini-bay concrete Reef Ball[1] units, arranged in groups |
| Lake Macquarie | 33° 05.814′ S 151° 36.891′ E | 2005 | 5 | 520 Mini-bay concrete Reef Ball[1] units, arranged in groups |
| Botany Bay | 33°58.940′ S 151°13.447′ E | 2006 | 6–15 | 400 Mini-bay concrete Reef Ball[1] units, arranged in groups |
| St. Georges Basin | 35° 07.449′ S 150° 37.031′ E | 2007 | 4–6 | 600 Mini-bay concrete Reef Ball[1] units, arranged in groups |
| Lake Conjola | 35° 15.628′ S 150° 28.330′ E | 2007 | 4–8 | 400 Mini-bay concrete Reef Ball[1] units, arranged in groups |
| Merimbula Lake | 36° 53.918′ S 149° 53.175′ E | 2008 | 5 | 400 Mini-bay concrete Reef Ball[1] units, arranged in groups |
| Ocean reefs: | | | | |
| Tweed | 28°13.428′ S 153°35.473′ E | 2020 | 25 | Steel grotto (10 m × 14 m$\phi$) plus reef field of 32 concrete arrowhead units |
| Port Macquarie | 31°25.044′ S 152°58.950′ E | 2016 | 46 | 20 × 5 m cubic concrete reef temple units, 5 with attached steel towers to 9 m height |
| Forster | 32° 10.023′ S 152° 33.310′ E | 2023 | 34 | Two steel reef tower units of 256 m$^2$ area (each), 12 m high |
| Newcastle | 33°04.300′ S 151°42.018′ E | 2019 | 29 | Two steel reef tower units of footprint 85 m$^2$, and 6.4 m height (with tower to 12 m) height |
| Terrigal | 33° 27.607′ S 151° 30.473′ E | 2023 | 48 | Two steel tower units |
| Sydney | 33°50.797′ S 151°17.988′E | 2011 | 38 | Single steel reef tower unit of footprint 180 m$^2$ and 12 m height, with two vertical towers |
| Southern Sydney | 34°05.780′ S 151°10.577′E | 2017 | 30 | 36 × 5 m cubic concrete reef temple units, arranged in groups, 10 with attached steel towers to 9 m height |
| Wollongong | 34°31.150′ S 150°54.840′ E | 2019 | 32 | Steel reef tower units with a vertical profile up to 12 m |
| Shoalhaven | 34°50.955 S 150°47.731 E | 2015 | 33 | 20 × 6.5 m cubic concrete reef temple units, arranged in groups |
| Eurobodalla | 35°53.140′ S 150°12.178′ E | 2021 | 43 | Two steel reef tower units |
| Merimbula | 36°54.850 S 149°56.255′ E | 2018 | 32 | Two steel reef tower units of footprint 85 m$^2$, and 6.4 m height (with tower to 12 m) height |

[1] size: 0.5 m high and 0.7 m $\phi$

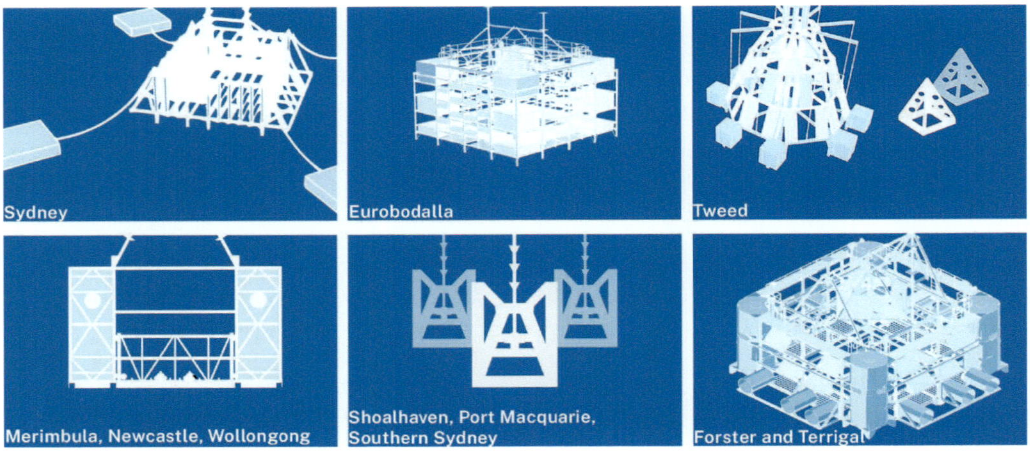


**FIGURE 4.2** Examples of ocean reef units deployed by the NSW Artificial Reefs Program. Graphics were provided by NSW Department of Primary Industries, Recreational Fisheries Management Branch (and are available at http://www.dpi.nsw.gov.au/fishing/recreational/resources/artificial-reef). Labels correspond to Table 4.1 and Figure 4.1.

in the estuarine seascape (e.g., seagrass beds). This overall combination of features aligns with the requirements for the most important recreational species in NSW estuaries, as outlined earlier, and interactions with these species are further described later.

### 4.2.3 Ocean Artificial Reefs in NSW

The deployment of the oceanic artificial reef network in NSW commenced following the completion of the estuary artificial reef network. Notably, the network is still under development with two proposed reefs currently in planning. All reef deployments have occurred within inshore waters (although these are popularly referred to as 'offshore' artificial reefs) at depths of 25–50 m (see ocean reefs in Figure 4.1; Table 4.2). Ocean artificial reefs in NSW have involved much larger structures than estuary reefs and have variably included larger steel units (Figure 4.2) deployed singularly or in groups, as well as smaller concrete units (with or without towers) deployed in groups to form larger 'reef fields' or complexes. These ocean reefs have internal volumes of up to 1,600 $m^3$ and biological footprints as large as 200,000 $m^2$ (C. Weire, pers. comm.).

Similarly to the estuary reefs, all ocean reefs provided design-specific features that were well suited to the aforementioned most important recreational species that are targeted in these inshore habitats. All reef units, regardless of type and material, included complex structures for the benefit of Pink Snapper (e.g., Harasti et al. 2015). The deployment of reef units in groups and complexes also provides a maximal length of reef-soft sediment interface, which is a preferential habitat favoured by benthic and epibenthic species such as Bluespotted Flathead and Silver Trevally (e.g., Moore et al. 2009; Scoulding et al. 2023). The vertical relief of larger steel structures and towers, on steel and concrete structures, provided an ideal habitat for the more midwater pelagic species Yellowtail Scad and Yellowtail Kingfish (e.g., Rees et al. 2018; Scoulding et al. 2023). The response of these recreational species to the reef structures is outlined below.

## 4.3 RESEARCH ON THE IMPACT, ECOLOGICAL ROLE AND FUNCTION OF ARTIFICIAL REEFS

As noted earlier, investment in the NSW Artificial Reef Program has been accompanied by an extended period of investment in research and monitoring of the reefs and the recreational fishers that exploit them (hereafter, collectively referred to as the 'research program'). This has provided data to evaluate the impact and performance of the reefs for important species. However, this investment has also allowed exploration of research on the more fundamental aspects of temperate reef ecology, encompassing both artificial and natural reef habitats. The information generated has supported the adaptive development of the programme over time.

Two primary techniques were employed in the research program. First, underwater video techniques were used to assess the development of the reef assemblages in terms of their succession, diversity and abundance (e.g., Becker et al. 2017b; Folpp et al. 2011; Lowry et al. 2012). Second, acoustic telemetry was implemented to explore how important species (both target species and threatened species) used and interacted with the reefs. These techniques were complemented by a variety of other approaches including hydroacoustics (e.g., Becker et al. 2023) and fisher monitoring (e.g., Becker et al. 2020; Keller et al. 2016). The programme has also led to novel methodological developments of broader relevance (e.g., Becker et al. 2017a, 2021b; Lowry et al. 2011; Smith et al. 2017). Importantly, this included the development of a 'drop cam'-based approach (i.e., a video camera literally dropped on a line to record the reef's fauna) to rapidly conduct surveys over larger areas in and around reef complexes and better examine the broader ecological footprint of artificial reefs. This technique overcame some of the bias inherent in unbaited or baited remote underwater video (BRUV) surveys (Smith et al. 2017). Most recently, remotely operated vehicles (ROVs) have been used which provided a means to better sample cryptic species occupying spaces within the artificial reef structures.

Overall, the research programme has yielded over 40 manuscripts dealing with a range of areas, including evolution and functional characteristics, succession, abundance, residency and connectivity of important recreational species, productivity and attraction/production, ecological footprint, trophic interactions, angler usage patterns and interactions with threatened species. Together, this work represents a significant and unique collection of internationally relevant research, and a selection of highlights from this substantial body of work are outlined below.

## 4.3.1 Outcomes of Artificial Reef Deployments for Key Species

The research programme provided a clear indication of interactions between important recreational fishers and the reef network. Considering species presence in monitoring video recordings as a coarse indicator of the impact of the reefs, the reef network performs well across most important recreational species (Table 4.1). Inshore reefs, however, appeared to outperform estuary reefs in terms of the number of video samples on which key recreational species were observed.

### 4.3.1.1 Estuary Reefs

For estuary reefs, early work on these habitats showed that within a year of deployment, the reef communities had developed to such a point that species diversity greatly exceeded that observed on both sedimentary habitats and the limited natural reef habitats available within the studied estuaries (Lowry et al. 2014). These patterns were reflected in the overall relative fish abundance (using MaxN as a metric [MaxN is an index of abundance equivalent to the maximum number of individuals of a particular species visible in a single frame of video footage]), with abundance values observed on artificial reefs about two to six times the abundance observed on natural reefs, and up to 30 times the abundance observed on sedimentary habitats (Folpp et al. 2013). This latter study identified that three sparid species (i.e., Yellowfin Bream, Pink Snapper and Tarwhine [*Rhabdosargus sarba*]) were key contributors to the reef fish assemblage abundance.

Collectively, recreationally important sparids (Yellowfin Bream and Pink Snapper) have responded particularly strongly to the artificial reef network, being observed on 67% and 55% of BRUV deployments, respectively (Table 4.1). Over the first two years following deployment, the relative abundance of sparids on artificial reefs increased almost linearly over time (Folpp et al. 2020), and these increases were mirrored on adjacent natural reefs in the monitored estuaries. This suggests that the increase in abundance for these species in the developing reef assemblage is not simply due to fish being 'attracted' from other more natural habitats within these systems, but likely reflects new fish production on artificial reefs (Folpp et al. 2020). Long-term studies have also shown that the abundance of important recreational species has been maintained on these artificial reefs over ten years following deployment, indicating that the productivity (as indicated by standing biomass) gains of these structures persist for an extended period (Becker et al. 2022). Despite the extensive edge habitat around artificial reef complexes and the classification of Dusky Flathead as an ecotype that has direct contact with reef habitats (Lowry et al. 2012), BRUV surveys encountered Dusky Flathead far less frequently than sparids (Table 4.1). However, this may largely be an artefact of the sampling method considered, with Dusky Flathead observed far more often through Underwater Visual Census (UVC) on artificial reefs than on BRUV surveys (Lowry et al. 2012).

Acoustic telemetry studies have provided remarkable insight into the usage of estuarine artificial reefs by recreationally important species. The majority of research has focused on the association and interaction of Yellowfin Bream with estuary artificial reefs across a range of temporal scales. Earlier work compared natural and artificial reefs and showed that fish displayed much higher residency on artificial reefs than natural reefs, suggesting that these represent superior habitats for this species (Lowry et al. 2017). Tracking of fine-scale movements and habitat usage patterns on estuary artificial reefs revealed considerable fidelity to particular reef groups by different fish (i.e., fish apparently used only a part of the reef complex to which they returned, Taylor et al. 2018). This work also highlighted the role of the artificial reef in the broader seascape, with fish foraging diurnally over adjacent seagrass habitat and appearing to use the reef primarily for refuge. Other work also showed considerable fidelity to artificial reef habitats for other important recreational species such as Luderick (*Girella tricuspidata*) and Tarwhine (Taylor et al. 2017).

#### 4.3.1.2 Ocean Reefs

The time series monitoring has shown that inshore reefs performed well in terms of presence in video samples for the most important species harvested recreationally from inshore habitats (Table 4.1). Fish responded rapidly to the deployment of the ocean reefs; Yellowtail Kingfish and Yellowtail Scad (pelagic species, which mostly occupy the taller sections of the reef) and Pink Snapper, Silver Trevally and Bluespotted Flathead were observed within weeks of reef deployments (Becker et al. 2017b, 2019). Comparisons between natural and artificial reefs using BRUV surveys revealed distinct differences in fish assemblages, primarily due to high abundances of pelagic, schooling species such as Yellowtail Scad (Becker et al. 2017b) on the artificial reefs. The high relief of the NSW ocean reefs (~6–12 m tall; Table 4.2), designed to extend into the pelagic zone relative to the mostly low-lying natural reef, most likely explains this difference. For reefs that were monitored over a period of up to four years, there appeared to be no clear ecological successional patterns (Woolley et al. 2026). It was expected that assemblages on the structures would initially consist of colonising species followed by a richer, stable assemblage. Interestingly, a far more stochastic pattern was observed and was consistent across six ocean reefs, spanning a broad latitudinal range. This is probably due to the small size of the artificial reefs relative to the surrounding marine ecosystem, the high mobility of many species found on the structures, and oceanographic variability off coastal NSW.

The layout and spacing of structures were important considerations when planning artificial reef deployments from both an ecological and practical perspective. This was because the way in which fish distributed themselves around reef structures informed the planning process as the programme developed. As noted earlier, both drop cameras and hydroacoustic surveys have also been used to map the distribution of benthic and pelagic fish around several artificial reefs. Unsurprisingly, the fish density was consistently higher in close proximity to reef structures and declined with increasing distance (i.e., typically over 50–70 m) from the reef (Becker et al. 2019, 2023; Scott et al. 2015). This decline tended to be more pronounced at the larger steel structures (Scott et al. 2015), whereas at concrete module reef complexes, the larger number of units and their arrangement in groups were believed to reduce competition for space immediately adjacent to the reefs. An important element of ocean artificial reefs in NSW is their exposure to coastal currents along this high-energy coastline, which has important implications for how the reef ecosystem operates, as discussed below.

Acoustic telemetry studies conducted on NSW ocean artificial reefs have revealed the residency, movement patterns, connectivity with nearby habitats and feeding behaviour of fish associated with the reefs. Blue-spotted Flathead, Fiddler Rays (*Trygonorrhina fasciata*) and Port Jackson Sharks (*Heterodontus portusjacksoni*) were found to move between artificial reef structures and natural reefs located 5 km away (Keller et al. 2017a). The observed connectivity patterns highlighted that the planned locations of artificial reefs should consider the proximity of reefs to natural reefs, alongside fish movement patterns, to facilitate productivity on deployed structures for target species. Fine-scale telemetry has also shown interesting patterns in the distribution of Bluespotted Flathead around the reefs, which are described in more detail below (Puckeridge et al. 2021).

Interactions between 'threatened, endangered or protected species' (i.e., TEPS) have been largely overlooked in the artificial reef scientific literature. Given that the most common goal of artificial reef deployments is to increase fishing yield (Becker et al. 2018), it is likely that deployments could concentrate fishing pressure in particular areas, which could create risks for TEPS that associate with the reefs. While there are few marine and estuarine TEPS in NSW, some of these species are mobile sharks that could interact with ocean artificial reefs. In NSW, large numbers of two such species (i.e., White Shark, *Carcharodon carcharias* and Grey Nurse Shark, *Carcharias taurus*) have been tagged with long-life (10 years) acoustic transmitters as part of on-going ecological research and shark mitigation programmes (e.g., Bruce et al. 2019; Butcher et al. 2023; Dwyer et al. 2023). During the period 2017–2021, each new ocean artificial reef deployment included acoustic listening stations, and these in situ receivers were monitored for interactions with tagged individuals for up to 2.5 years. No Grey Nurse Sharks were detected interacting with the reefs, but 590 White Sharks temporarily interacted with various ocean artificial reefs. Most (~90%) of these interactions, however, lasted less than 40 minutes (Becker et al. 2024a) and were likely driven by the inquisitive nature of sharks and their roving migratory behaviour. Importantly, the short duration of interactions indicated that exposure of these TEPS to fishing on the reefs was minimal.

### 4.3.2 Fisher Visitation and Economic Impacts of Artificial Reefs

Substantial monitoring efforts have focused on visitation to ocean artificial reefs by recreational fishers. Both Keller et al. (2016) and Becker et al. (2020) applied remote long-range camera technology to monitor boat-based fishing effort on two ocean artificial reefs in NSW. This allowed continual shore-based daytime monitoring of fishing effort in the area influenced by the artificial reef (Figure 4.3). The majority of fishing effort on the reefs was observed during morning hours when coastal conditions in southeastern Australia are generally calmer. In addition, the intensity of fishing effort tended to be greater on weekends than on weekdays. But most surprising were the levels of fishing effort and harvest that occurred on ocean artificial reefs. For the Sydney artificial reef, Keller et al. (2016) calculated a fishing effort rate of 45,000 hours of fishing $km^{-2}$ $y^{-1}$ (estimates were standardised to a square kilometre of reef area, to allow for comparison with other habitats). This fishing effort was higher than in any other monitored natural ecosystem in NSW. Keller et al. (2017b) combined effort data with catch-rate data from creel surveys and revealed that the reef habitat could similarly support harvest rates that were an order of magnitude greater than in any other monitored natural ecosystem in NSW.

Furthermore, targeted social research on the socio-economic impacts of ocean artificial reefs in NSW (Nous Group 2019) showed that 58% of recreational fishers interviewed had fished at an artificial reef in the previous 12 months. However, it is important to note that this was not a probabilistic survey design, so some bias was likely inherent in these results. Nevertheless, the results showed that this may have led to increased localised expenditures in both fishing-related and non-fishing-related industries. Concomitantly, the presence of artificial reefs encouraged increased recreational

FIGURE 4.3 Image from a long-range camera used to monitor fishing effort on artificial ocean reefs in NSW. The white polygon indicates the reef area (a boat fishing on the reef can be seen).

activity that delivered flow-on economic and social impacts for local communities. Furthermore, the survey indicated that fishing on artificial reefs led to an improved fishing experience, largely through improved catch rates for key species. Quantitative analyses to ascertain the social return on investment indicated that a reef valued at ~AUD 1 million (USD 0.65 million) could return around AUD 56 million (USD 36 million) over the reef's 30-year lifespan (Nous Group 2019). This research provided a novel and unique insight into the economic value associated with the social impacts arising from investment in ocean artificial reefs, compared to artificial reef programmes more generally.

### 4.3.3 Artificial Reef Function: Attraction versus Production

The comparative roles of fish attraction and production on artificial reefs have been of considerable scientific interest throughout the contemporary history of artificial reefs (e.g., Bortone 1998; Layman and Allgeier 2020; Lindberg 1997). The research programme in NSW has provided some novel insight into this issue in the context of ocean artificial reefs. Smith et al. (2016) developed a new approach that estimated legitimate new production arising from artificial reefs via estimation of biomass flux, with the time spent at the artificial reef moderating the level of production that can be considered 'new' and attributed to the presence of the reef. Using the Sydney artificial reef as a case study, the approach modelled a large standing stock biomass at the reef (370 kg, on the reef and within 5 m of the structure). Local production, which includes production at the reef by visiting fish that would still occur at another location if the reef did not exist, was also large at 211 kg $y^{-1}$. However, new production (e.g., additional production) due to the artificial reef itself was only estimated to equate to 4%–5% of local production, much of which was attributed to highly resident fish.

While this study showed that the reef itself was a highly productive habitat, a large proportion of the overall biomass of harvested fish is likely allochthonous. Smith et al. (2016) highlighted that artificial reefs that have a large biomass flux relative to standing biomass may expose regional fish populations to elevated fishing pressure. The consequences of this, however, largely depend on the characteristics of fish species being targeted on the reef system and how these species are distributed across the broader ecosystem (Smith et al. 2015).

### 4.3.4 Planktonic Productivity and Benthopelagic Coupling

The diversity of approaches employed to monitor artificial reefs in NSW has also provided remarkable insight into the functional aspects of reef trophic productivity. The structural complexity of artificial reefs attracts high abundances of zooplanktivorous fish that exploit this structure for refuge. Additionally, these fish help incorporate pelagic productivity from nearby surrounding waters into the reef's food web. However, Champion et al. (2015) showed that the trade-off between foraging gain and predation risk imposes an upper limit on the volume of adjacent water that can be effectively exploited, at least for zooplanktivorous fish. The outcome of this trade-off was an upper limit on the relationship between reef size and the amount of zooplankton productivity that can be effectively exploited, as the biomass of zooplanktivorous fish that the structure can support exceeded the carrying capacity imposed by planktonic productivity in nearby waters.

In fact, further research showed that access to these zooplankton resources actually transcended trophic levels across the reef system and directly influenced the distribution of fish across the reef. Hydroacoustic surveys revealed that pelagic zooplanktivorous fish aggregated on the upcurrent face of the reef structure (Holland et al. 2021), acting as a formidable 'wall of hungry mouths' to intercept zooplankton carried on the prevailing current. This, in turn, affected the distribution of piscivorous Bluespotted Flathead across the structure (which were determined to be predators of zooplanktivorous fishes through analysis of stomach contents) and led to the concentration of foraging activity centres in this location (as revealed through acoustic telemetry, Puckeridge et al. 2021). This integrated suite of research provided detailed insight into how foraging behaviours and trophic relationships can exploit the structural nuance of individual artificial reef structures.

## 4.4 RECENT OPPORTUNITIES AND FUTURE POSSIBILITIES

Twenty years of research on artificial reefs in NSW has provided a solid scientific foundation for the program, and the maturing network of estuary and ocean reefs is delivering tangible outcomes for the recreational fishers who paid for the programme through their fishing licence fees. Present and future work under consideration focuses on specific areas to further increase the enhancement of fisheries productivity by artificial reefs and expand the scope and spread of benefits associated with habitat enhancement in NSW.

### 4.4.1 'Integrated Fisheries Enhancement' and Stocking of Artificial Reefs

In recent years, there has been an increasing focus worldwide on integrated fisheries enhancements that employ both habitat-aided and aquaculture-aided approaches (Lorenzen et al. 2021). As outlined earlier, NSW invests heavily in an adaptively managed marine stocking programme (Blount et al. 2017), and recent efforts have explored the potential to stock artificial reefs with hatchery-reared fish. Initial work investigated the potential for stocking estuary artificial reefs with Yellowtail Kingfish. While nursery habitats are poorly defined for this species, juveniles are frequently observed associating with hard structure in estuaries. Fish were released to evaluate whether the low-relief estuary artificial reefs provided suitable habitat for the species and could potentially serve as a nursery to supply adolescent fish to reefs in adjacent ocean waters. Yellowtail Kingfish were released onto the Botany Bay estuary artificial reef (Table 4.2), with 40 being implanted with acoustic transmitters and a further 470 being externally tagged. Subsurface release was used to ensure that fish were placed directly onto the reef. However, the released fish rapidly dispersed from both the reef and the estuary within a month following release, and only three fish were detected at the adjacent ocean reefs (Becker et al. 2021a). While this outcome was disappointing, it provided a clear result and an indication that stocking efforts with juvenile Yellowtail Kingfish were unlikely to be feasible.

In contrast, later work stocking Yellowfin Bream at the Bellinger River artificial reef (Table 4.2) showed substantial promise. Acoustically tagged, hatchery-reared fish were released and monitored with an estuary-wide detection array (Becker et al. 2024b). After a six-month monitoring period, it was observed that fish stocked on the artificial reef showed substantially higher residency at this location than fish released at a nearby natural reef reference site. However, other work involving stocking the Lake Macquarie estuary artificial reef with adult Yellowfin Bream, which were rescued from a power station canal, was largely unsuccessful. Fish rapidly migrated away from the reefs during the two weeks following stocking (Taylor et al. 2017). However, Luderick (*Girella tricuspidata*) and Tarwhine that were released in the same operation mostly remained resident on the artificial reef for the duration of the six-month monitoring period.

These diverse results highlight a range of outcomes from the preliminary investigation of integrated enhancements in the NSW Artificial Reefs Program. Stocking hatchery-reared juvenile Yellowfin Bream onto these habitats certainly shows promise and is a priority for future development. Pink Snapper have also been identified for future stocking efforts, particularly for estuaries with restricted connectivity to ocean spawning habitats (e.g., Lake Conjola, St. Georges Basin; Table 4.2). While it is unlikely that Yellowtail Kingfish stocking on estuary artificial reefs will be revisited, the model whereby estuary reefs are used as a nursery stock to supply larger, later recruits to ocean reefs (as proposed earlier for Yellowtail Kingfish) presents another opportunity for Pink Snapper, particularly during periods of lower natural recruitment. Residency on estuary reefs and connectivity with ocean reefs will need to be studied in further detail before these efforts are upscaled and potential trophic impacts on local and regional food webs evaluated.

### 4.4.2 Angler Access and Biogenic Reefs

While the NSW artificial reefs network provides substantial fishing opportunities for boat-based fishers, ongoing monitoring of the recreational fishery has shown that (1) shore-based fishing is the most common type of fishing in NSW, and (2) boat-based estuary fishing is far more common than boat-based ocean fishing (Murphy et al. 2022, 2024). Coupled with the potential for integrated habitat- and aquaculture-aided enhancement (note that under current policy, hatchery-reared fish cannot currently be released in ocean waters in NSW; NSW Department of Primary Industries 2014), and integration with NSW's burgeoning biogenic habitat enhancement programme (Cole et al. 2024), there are significant opportunities for further development of estuary reefs within the RFH network. It is notable that a large number of RFHs are recruitment limited (Taylor et al. 2005). Consequently, coupling reef deployments with stocking in recruitment and limited reef habitat systems provides remarkable opportunities for increasing recreational fishing opportunities in these estuaries. Such practices, however, require continued strong investment in research and development to ensure the enhancement system operates as intended.

While designing artificial reefs to provide opportunities for shore-based fishers within tidal estuaries creates some unique challenges, there are opportunities to achieve this. There has been substantial investment by the RFT into accessible estuarine 'fishing platforms' (decks extending out over the water for use by shore-based anglers) in NSW—see https://www.dpi.nsw.gov.au/fishing/recreational/resources/accessible-recreational-fishing-for-everyone). Integrating artificial reefs with these facilities will further improve fishing opportunities for shore-based fishers, especially those with limited physical mobility. However, this will require particular consideration of how reef units can be deployed without leading to excessive fouling of fishing gear for shore-based fishers. Designing specific subtidal 'living seawalls' that provide suitable habitat for key recreational species may offer a suitable option for these situations.

Over the past ten years, the scale and impacts of lost oyster reef habitats have received substantial attention in Australia as well as throughout the world (e.g., Gillies et al. 2020; McAfee et al. 2020).

Consequently, programmes to restore these biogenic habitats are being developed or are already underway in an expanding number of jurisdictions (Howie and Bishop 2021) including NSW (Cole et al. 2024). Recreational fisheries enhancement goals are often proposed as part of the rationale underpinning the need for oyster reef restoration (Coen and Luckenbach 2000; Cook et al. 2021), but the integration of specific design features to benefit important recreational species has not been fully implemented in NSW. Further development of biogenic reefs incorporating design-specific features into NSW RFHs also offers significant opportunities to deliver recreational fisheries outcomes, particularly in conjunction with shore-based access enhancements (as outlined earlier). Further research, however, is required to define novel design-specific elements that will ensure the desired cross section of ecosystem services is delivered from integrated reefs.

## 4.5 CONCLUSION

The unique funding model for investment in recreational fishing opportunities in NSW has provided secure resourcing for the science-based development of the largest artificial reef network in Australia. A comprehensive, long-term research programme has provided a strong basis for the ongoing development of the NSW Artificial Reefs Program. This has led to ongoing improvements in the design-specific features of artificial reefs as new reefs have been deployed and provided insight into how fishers and recreational species interact with these artificial structures. This programme has also contributed fundamental knowledge on how artificial reef communities function and how the reefs operate as productive fisheries. These insights have incrementally improved our appreciation of the ecology and function of artificial reef systems, and the findings are certainly relevant to practitioners elsewhere. Despite two decades of development, additional work remains to be done to ensure that the network can continue to develop and cater to the ever-growing recreational fishing fraternity in the future.

*Acknowledgements:* This chapter arose from the content of an invited keynote presentation by MDT at the 12th Conference on Artificial Reefs and Related Aquatic Habitats (CARAH) in Colombia, 2025. The views offered are those of the authors as scientists and do not necessarily reflect the broader policy, plans or intentions of the organisations for which they work. The authors acknowledge the contributions of M.B. Lowry, H. Folpp, J.A. Smith, I.M. Suthers, C. Champion, M. Holland, M. Scott, S. Woolley, C. Weire, B. Doolan, B. van der Walt, J. Harnwell, M. Gamble, J. McLeod, M. Burns, T. New, B. McCartin, I. Thiebaud, B. Leach and M. Gregson.

## REFERENCES

Arnason R. 2008. On the economics of releasing cultured fish into the aquatic environment. *Rev Fish Sci.* 16(1–3):135–145.

Bao H, Nikolaeva A, Xia J, Ma F. 2025. Evolution trends and future prospects in artificial marine reef research: A 28-year bibliometric analysis. *Sustainability.* 17(1):184.

Becker A, Butcher PA, Morris S, Champion C, Peddemors VM, Lowry MB, Taylor MD. 2024a. Interactions between White Shark (*Carcharodon carcharias*) and artificial reefs along the east-coast of Australia. *Fish Res.* 273:106961.

Becker A, Dworjanyn S, Niella Y, Smith JA, Lowry MB, Taylor MD. 2024b. Residency and movement patterns of juvenile Yellowfin Bream (*Acanthopagrus australis*) stocked on estuarine artificial reefs. *Mar Freshw Res.* https://doi.org/10.1071/MF24041

Becker A, Lowry M, Taylor MD. 2017a. Monitoring of reef associated and pelagic fish communities on Australia's first purpose built offshore artificial reef. *ICES J Mar Sci.* 74(1):277–285.

Becker A, Lowry M, Taylor MD, Folpp H. 2017b. *Assessment of the Sydney Offshore Artificial Reef.* Taylors Beach: NSW DPI Fisheries Final Report Series No. 155.

Becker A, Lowry MB, Fielder DS, Taylor MD. 2021a. Dispersal of yellowtail kingfish (*Seriola lalandi*) from a coastal embayment following a recreational fisheries enhancement stocking program: Attempts to integrate aquaculture and habitat-based initiatives. *Bull Mar Sci.* 97(4):615–630.

Becker A, Lowry MB, Fowler AM, Taylor MD. 2023. Hydroacoustic surveys reveal the distribution of mid-water fish around two artificial reef designs in temperate Australia. *Fish Res*. 257:106509.
Becker A, Lowry MB, Taylor MD. 2021b. Receiver tilt: A scourge for aquatic telemetry or useful predictor variable. *Anim Biotelemetry*. 9:13.
Becker A, Smith JA, Taylor MD, McLeod J, Lowry MB. 2019. Distribution of pelagic and epi-benthic fish around a multi-module artificial reef-field: Close module spacing supports a connected assemblage. *Fish Res*. 209:75–85.
Becker A, Taylor MD, Folpp H, Lowry M. 2018. Managing the development of artificial reef systems: The need for quantitative goals. *Fish*. 19(4):740–752.
Becker A, Taylor MD, Folpp H, Lowry M. 2022. Revisiting an artificial reef after 10 years: What has changed and what remains the same? *Fish Res*. 249:106261.
Becker A, Taylor MD, McLeod J, Lowry MB. 2020. Application of a long-range camera to monitor fishing effort on an offshore artificial reef. *Fish Res*. 228:105589.
Blount C, O'Donnell P, Reeds K, Taylor MD, Boyd S, van der Walt B, McPhee DP, Lincoln Smith M. 2017. Tools and criteria for ensuring estuarine stock enhancement programs maximise benefits and minimise impacts. *Fish Res*. 186:413–425.
Bohnsack JA, Sutherland DL. 1985. Artificial Reef Research: A review with recommendations for future priorities. *Bull Mar Sci*. 37(1):11–39.
Bortone SA. 1998. Resolving the attraction-production dilemma in artificial reef research: Some yeas and nays. *Fisheries*. 23(3):6–10.
Branden KL, Pollard DA, Reimers HA. 1994. A review of recent artificial reef developments in Australia. *Bull Mar Sci*. 55(2–3):982–994.
Bruce BD, Harasti D, Lee K, Gallen C, Bradford R. 2019. Broad-scale movements of juvenile white sharks *Carcharodon carcharias* in eastern Australia from acoustic and satellite telemetry. *Mar Ecol Prog Ser*. 619:1–15.
Butcher PA, Lee KA, Brand CP, Gallen CR, Green M, Smoothey AF, Peddemors VM. 2023. Capture response and long-term fate of White Sharks (*Carcharodon carcharias*) after release from SMART drumlines. *Biology*. 12(10):1329.
Carral L, Camba Fabal C, Lamas Galdo MI, Rodríguez-Guerreiro MJ, Cartelle Barros JJ. 2020. Assessment of the materials employed in green artificial reefs for the Galician Estuaries in terms of circular economy. *International Journal of Environmental Research and Public Health*. 17(23):8850.
Champion C, Suthers IM, Smith JA. 2015. Zooplanktivory is a key process for fish production on a coastal artificial reef. *Mar Ecol Prog Ser*. 541:1–14.
Coen LD, Luckenbach MW. 2000. Developing success criteria and goals for evaluating oyster reef restoration: Ecological function or resource exploitation? *Ecol Eng*. 15(3):323–343.
Cole VJ, Harasti D, Dahle SK, Russell K. 2024. Determining the best practice for Sydney rock oyster, *Saccostrea glomerata*, reef restoration and enhanced ecological benefits. *BMC Ecology and Evolution*. 24(1):114.
Cook PA, Warnock B, Gillies CL, Hams A. 2021. Historical abundance and distribution of the native flat oyster (*Ostrea angasi*) in estuaries of the Great Southern region of Western Australia help to prioritise potential sites for contemporary oyster reef restoration. *Mar Freshw Res*. 73(1):48–56.
Curley BG, Jordan AR, Figueira WF, Valenzuela VC. 2013. A review of the biology and ecology of key fishes targeted by coastal fisheries in south-east Australia: Identifying critical knowledge gaps required to improve spatial management. *Rev Fish Biol Fish*. 23:435–458.
Department of Regional New South Wales. 2023. *Department of Regional New South Wales Annual Report 2022–2023*. Department of Regional New South Wales, Orange.
Diplock J. 2011. *Artificial Reefs—Design and Monitoring Standards Workshops*. Final Report to the Fisheries Research and Development Corporation for Project 2010/400.
Dwyer RG, Rathbone M, Foote DL, Bennett M, Butcher PA, Otway NM, Louden BM, Jaine FRA, Franklin CE, Kilpatrick C. 2023. Marine reserve use by a migratory coastal shark, *Carcharias taurus*. *Biol Conserv*. 283:110099.
Folpp HR, Lowry M, Gregson MA, Suthers IM. 2011. Colonisation and community development of fish assemblages associated with estuarine artificial reefs. *Brazil J Oceanog*. 59:55–67.
Folpp HR, Lowry M, Gregson MA, Suthers IM. 2013. Fish assemblages on estuarine artificial reefs: Natural rocky-reef mimics or discrete assemblages? *PloS One*. 8(6):e63505.
Folpp HR, Schilling HT, Clark GF, Lowry MB, Maslen B, Gregson M, Suthers IM. 2020. Artificial reefs increase fish abundance in habitat-limited estuaries. *J Appl Ecol*. 57(9):1752–1761.
Fowler AM, Ochwada-Doyle FA, Dowling NA, Folpp H, Hughes JM, Lowry MB, Lyle JM, Lynch TP, Miles NG, Chick RC. 2022. Integrating recreational fishing into harvest strategies: Linking data with objectives. *ICES J Mar Sci*. 79(2):285–307.

Gillies CL, Castine SA, Alleway HK, Crawford C, Fitzsimons JA, Hancock B, Koch P, McAfee D, McLeod IM, Zu Ermgassen PS. 2020. Conservation status of the oyster reef ecosystem of southern and eastern Australia. *Global Ecology and Conservation*. 22:e00988.
Grove R, Yuge J, Sonu C. 1983. *Artificial Reef Technology: A Strategy for Active Impact Mitigation*. Proceedings OCEANS '83, San Francisco, CA, USA, 29 August–1 September 1983.
Harasti D, Lee KA, Gallen C, Hughes JM, Stewart J. 2015. Movements, home range and site fidelity of Snapper (*Chrysophrys auratus*) within a temperate marine protected area. *Plos One*. 10(11):e0142454.
Henry GW, Lyle JM. 2003. *The National Recreational and Indigenous Fishing Survey*. Canberra: Australian Government Department of Agriculture, Fisheries and Forestry.
Holland MM, Becker A, Smith JA, Everett JD, Suthers IM. 2021. Fine-scale spatial and diel dynamics of zooplanktivorous fish on temperate rocky and artificial reefs. *Mar Ecol Prog Ser*. 674:221–239.
Howie AH, Bishop MJ. 2021. Contemporary oyster reef restoration: Responding to a changing world. *Front Ecol Evol*. 9.
Keller K, Smith JA, Lowry MB, Taylor MD, Suthers IM. 2017a. Multispecies presence and connectivity around a designed artificial reef. *Mar Freshw Res*. http://dx.doi.org/10.1071/MF16127.
Keller K, Steffe AS, Lowry MB, Murphy JJ, Smith JA, Suthers IM. 2017b. Estimating the recreational harvest of fish from a nearshore designed artificial reef using a pragmatic approach. *Fish Res*. 187:158–167.
Keller K, Steffe AS, Lowry MB, Murphy JJ, Suthers IM. 2016. Monitoring boat-based recreational fishing effort at a nearshore artificial reef with a shore-based camera. *Fish Res*. 181:84–92.
Layman CA, Allgeier JE. 2020. An ecosystem ecology perspective on artificial reef production. *J Appl Ecol*. 57(11):2139–2148.
Lindberg WJ. 1997. Can science resolve the attraction-production issue? *Fisheries*. 22(4):10–13.
Lorenzen K, Leber KM, Loneragan NR, Schloesser RW, Taylor MD. 2021. Developing and integrating enhancement strategies to improve and restore fisheries. *Bull Mar Sci*. 97(4):475–488.
Lowry M, Folpp H, Gregson M. 2011. Evaluation of an underwater solid state memory video system with application to fish abundance and diversity studies in southeast Australia. *Fish Res*. 110(1):10–17.
Lowry M, Folpp H, Gregson M, Suthers I. 2012. Comparison of baited remote underwater video (BRUV) and underwater visual census (UVC) for assessment of artificial reefs in estuaries. *J Exp Mar Biol Ecol*. 416:243–253.
Lowry MB, Becker A, Folpp H, McLeod J, Taylor MD. 2017. Residency and movement patterns of yellowfin bream (*Acanthopagrus australis*) released at natural and artificial reef. *Mar Freshw Res*. 68(8):1479–1488.
Lowry MB, Glasby TM, Boys CA, Folpp H, Suthers I, Gregson M. 2014. Response of fish communities to the deployment of estuarine artificial reefs for fisheries enhancement. *Fish Manag Ecol*. 21(1):42–56.
McAfee D, McLeod IM, Boström-Einarsson L, Gillies CL. 2020. The value and opportunity of restoring Australia's lost rock oyster reefs. *Restoration Ecology*. 28(2):304–314.
Moore A, Schirmer J, Magnusson A, Keller K, Galeano G, Woodhams D, Wright JC, Maloney D, Dix A. 2023. *National Social and Economic Survey of Recreational Fishers 2018–2021*. Report to the Fisheries Research and Development Corporation on Project 2018–161, Australian Bureau of Agricultural and Resource Economics and Sciences, Canberra.
Moore CH, Harvey ES, Van Niel KP. 2009. Spatial prediction of demersal fish distributions: Enhancing our understanding of species–environment relationships. *ICES J Mar Sci*. 66(9):2068–2075.
Murphy JJ, Ochwada-Doyle FA, West LD, Stark KE, Hughes JM. 2020. *Survey of Recreational Fishing in NSW, 2017/18*. Nelson Bay, Fisheries Final Report Series No. 158, NSW Department of Primary Industries.
Murphy JJ, Ochwada-Doyle FA, West LD, Stark KE, Hughes JM, Taylor MD. 2022. *Survey of Recreational Fishing in NSW, 2019/20*. Nelson Bay, Fisheries Final Report Series No. 161, NSW Department of Primary Industries.
Murphy JJ, Ochwada-Doyle FA, West LD, Stark KE, Hughes JM, Taylor MD. 2024. *Survey of Recreational Fishing in NSW, 2021/22*. Nelson Bay, Fisheries Final Report Series, NSW Department of Primary Industries.
New South Wales Recreational Fishing Trusts. 2023. *New South Wales Recreational Fishing Trusts Annual Report 2022/23*. New South Wales Government, Sydney.
Nous Group. 2019. *Social Return on Investment of NSW Artificial Reefs Program*. Report to NSW Department of Primary Industries, Nous Group, Sydney.
NSW Department of Primary Industries. 2014. *The New South Wales Marine Fish Stocking Fishery Management Strategy*. Port Stephens Fisheries Institute, Taylors Beach.
Pollard D. 1989. Artificial habitats for fisheries enhancement in the Australian region. *Mar Fish Rev*. 51(4):11–26.

Puckeridge AC, Becker A, Taylor MD, Lowry MB, McLeod J, Schilling H, Suthers IM. 2021. Foraging behaviour and movements of an ambush predator reveal benthopelagic coupling on artificial reefs. *Mar Ecol Prog Ser*. 666:171–182.

Ramm LAW, Florisson JH, Watts SL, Becker A, Tweedley JR. 2021. Artificial reefs in the Anthropocene: A review of geographical and historical trends in their design, purpose, and monitoring. *Bull Mar Sci*. 97(4):699–728.

Rees MJ, Knott NA, Neilson J, Linklater M, Osterloh I, Jordan A, Davis AR. 2018. Accounting for habitat structural complexity improves the assessment of performance in no-take marine reserves. *Biol Conserv*. 224:100–110.

Scott ME, Smith JA, Lowry MB, Taylor MD, Suthers IM. 2015. The influence of an offshore artificial reef on the abundance of fish in the surrounding pelagic environment. *Mar Freshw Res*. 66(5):429–437.

Scoulding B, Gastauer S, Taylor JC, Boswell KM, Fairclough DV, Jackson G, Sullivan P, Shertzer K, Campanella F, Bacheler N et al. 2023. Estimating abundance of fish associated with structured habitats by combining acoustics and optics. *J Appl Ecol*. 60(7):1274–1285.

Smith JA, Cornwell WK, Lowry MB, Suthers IM. 2017. Modelling the distribution of fish around an artificial reef. *Mar Freshw Res*. 68:1955–1964.

Smith JA, Lowry MB, Champion C, Suthers IM. 2016. A designed artificial reef is among the most productive marine fish habitats: New metrics to address 'production versus attraction'. *Mar Biol*. 163(9):188.

Smith JA, Lowry MB, Suthers IM. 2015. Fish attraction to artificial reefs not always harmful: A simulation study. *Ecology and Evolution*. 5(20):4590–4602.

Stelling-Wood TP, Gribben PE, Adam P, Birch G, Bishop MJ, Blount C, Booth DJ, Brown C, Bruce E, Bugnot AB et al. 2023. A deep dive into the ecology Gamay (Botany Bay): Current knowledge and future priorities for this highly modified coastal waterway. *Mar Freshw Res*. 74:1003–1025.

Taylor MD, Becker A, Lowry MB. 2018. Investigating the functional role of an artificial reef within an estuarine seascape: A case study of Yellowfin Bream (*Acanthopagrus australis*). *Estuar Coast*. 41(6):1782–1792.

Taylor MD, Palmer PJ, Fielder DS, Suthers IM. 2005. Responsible estuarine finfish stock enhancement: An Australian perspective. *J Fish Biol*. 67(2):299–331.

Taylor MD, Payne NL, Becker A, Lowry MB. 2017. Feels like home: Homing of mature large-bodied fish following translocation from a power-station canal. *ICES J Mar Sci*. 74(1):301–310.

Vaishampayan A, Miles N, Taylor MD, van Eeden L, Morrongiello JR, Ochwada-Doyle FA. 2025. *Place, Not Events, Shapes Fishing Motivations*. Social Science Research Network.

West LD, Stark KE, Murphy JJ, Lyle JM, Ochwada-Doyle FA. 2015. *Survey of Recreational Fishing in New South Wales and the ACT, 2013/14*. Nelson Bay, Fisheries Final Report Series No. 149, NSW Department of Primary Industries.

Woolley S, Becker A, Schilling HT, Suthers IM. 2026. Temporal variation in fish communities on artificial reefs is stochastic and does not follow a traditional successional trajectory. *Mar. Ecol. Prog. Ser.* 782:15099.

# 5 Understanding the Ecological Role of Artificial Structures amidst Global Change

## *A Perspective*

*Avery B. Paxton**
* Corresponding author: avery.paxton@noaa.gov

## 5.1 INTRODUCTION

Marine ecosystems are experiencing rapid global change due to cumulative human impacts (Halpern et al. 2015). One expanding human use of oceans is the installation of artificial structures. These structures include artificial reefs, offshore wind turbine foundations, port infrastructure, oil platforms, and modified shorelines (typically deployed through specific planning and permitting processes). The deployment of these structures is expanding, with projections that their global areal extent (e.g., "footprint") will grow from 32,000 km$^2$ as of 2018 to 39,400 km$^2$ in 2028 (Bugnot et al. 2020). Artificial structures can influence ecological changes, resulting in unintended consequences and, in some cases, benefits (e.g., Heery et al. 2017; Firth et al. 2024; Paxton et al. 2025b). Despite their potential to affect ecological change, artificial structures may serve important roles in the seascape, not only in terms of their physical area or footprint but also in how they function ecologically (sometimes uniquely) and how they may facilitate highly mobile predators in responding to global change scenarios.

As such, the goal of this perspective is to provide an evidence-based, personal viewpoint from research on how artificial structures function ecologically with global change (e.g., changing human uses of the ocean and environmental conditions). This perspective, which was presented in March 2025 at the 12th International Conference on Artificial Reefs and Related Aquatic Habitats, addressed four key questions:

1) How does the extent of artificial structures compare to that of natural habitats?
2) How do artificial structures function compared to natural habitats?
3) Can artificial structures help highly mobile consumers respond to global change?
4) Which traits of artificial structures predict ecological functions and services?

In addressing these questions, observations are drawn from personal experiences while conducting scientific diving, using marine robotics, and implementing fisheries echosounders, complemented by modeling and synthesis approaches. Importantly, the body of work presented here focuses on research personally directed or participated in. However, the findings shared here are truly collaborative and reflect efforts by the multitude of coauthors on the peer-reviewed publications referenced herein.

DOI: 10.1201/9781003676775-5

## 5.2 HOW DOES THE EXTENT OF ARTIFICIAL STRUCTURES COMPARE TO THAT OF NATURAL HABITATS?

To begin to understand how artificial structures may facilitate fish in experiencing global change, it is necessary to quantify how the extent of artificial structures compares to naturally occurring habitats. This question reflects an inherent disparity. Scientists (and to some degree, society) have focused on understanding the science of natural habitats yet do not fully understand how artificial structures function and provide ecosystem services. Paxton et al. (2025b) compiled existing global or regional datasets on six types of natural habitats (i.e., seagrass, mangrove, kelp, salt marsh, coral reefs, and oysters) and found that each was declining in extent. In contrast, the area of artificial structures (i.e., artificial reefs, offshore wind turbines, submarine cables, saltwater aquaculture, living shorelines, and breakwaters) increased. For example, as of 2020, the footprint (i.e., seafloor area or extent) of structures intentionally sunk to form artificial reefs in the United States covered approximately 19 $km^2$ of the seafloor, reflecting a large increase from 1970 to 2020 (Paxton et al. 2024d). Additionally, calculations highlight that the expected offshore wind footprint in the United States of all leased areas as of January 2025, if "built out" (i.e., maximally covered) over the next decade, will be approximately 18.2 $km^2$ (Runde and Paxton 2025). Despite the different trajectories in extent between natural and artificial habitats, the extent of artificial structures is orders of magnitude lower than that of natural habitats. For instance, Steward et al. (2022) found that off the Southeast United States, artificial reef seafloor coverage (~3 $km^2$) occupied 885 times less area than natural reefs (2,654 $km^2$).

## 5.3 HOW DO ARTIFICIAL STRUCTURES FUNCTION COMPARED TO NATURAL HABITATS?

While artificial structures may occupy a smaller (but increasing) extent within the seascape than naturally occurring habitats, it is possible that they may have an outsized ecological impact. Artificial and natural habitats share some ecological functions, yet other functions differ between these two habitat types. For example, a global meta-analysis revealed that artificial reefs host similar fish assemblages to those in fish communities on naturally occurring rocky reefs and coral reefs that they are often intended to mimic, with some location- and material-specific nuances (Paxton et al. 2020b). Despite the global pattern of comparable functions, nuances emerge at finer spatial and temporal scales, where differences in reef material and structural complexity can manifest as differences and similarities in fish community composition (Paxton et al. 2017). Additionally, evidence demonstrates that large-reef-associated predators (e.g., jacks, mackerels, barracudas, sharks) are over four times as abundant on artificial reefs as natural reefs (Paxton et al. 2020a) and that isolated, deep artificial structures can form hotspots (e.g., congregating areas) for groupers (Johnson et al. 2020; Paxton et al. 2021). Likewise, artificial reefs can host fish species at their species range edges to a higher degree than natural reefs (Paxton et al. 2019b). Similarly, microalgae communities can differ between artificial and natural reefs (Lim et al. 2023). Underwater soundscapes also share similar temporal patterns on artificial reefs versus natural reefs but each has a distinct spectral content (Van Hoeck et al. 2020).

## 5.4 CAN ARTIFICIAL STRUCTURES HELP HIGHLY MOBILE CONSUMERS RESPOND TO GLOBAL CHANGE?

Global changes due to increasing uses of the coastal ocean (e.g., fishing, energy extraction) and environmental changes (e.g., increasing water temperature) raise several questions. For example, how do highly mobile fauna, such as reef fishes and sharks, utilize artificial structures amidst global change? Additionally, can strategic placement and design of artificial structures help organisms respond to global changes? When evaluating the potential of artificial structures to form habitat for species predicted to be moving poleward, it was noted that these species likely use networks of artificial structures as "stepping stones" (Paxton et al. 2019b). Tests of whether artificial structures can be strategically placed to enhance the movement of reef fishes indicated that demersal species, such as Black Seabass (*Centropristis striata*) and Gag (*Mycteroperca microlepis*), exhibit a high degree of site fidelity to artificial structures, whereas water-column-associated predators, such as Greater Amberjack (*Seriola dumerili*) and Almaco Jack (*S. rivoliana*), more regularly moved greater distances within and away from artificial structure complexes (Tharp et al. 2024). Similar investigations of large predators along their seasonal migration routes have discovered that individuals of a large coastal shark species (Sand Tiger Shark; *Carcharias taurus*) returned to the same or nearby artificial structures over time (Paxton et al. 2019a). This pattern of site fidelity suggests that artificial structures can provide valuable migration corridors for large predators. Moreover, evaluations of spatial distributions of reef fishes around artificial structures (which tend to exhibit consistent patterns across trophic levels; Paxton et al. 2019c) demonstrate that newly deployed artificial structures should be installed at least 60–90 m from existing structured habitats to optimize fish habitat enhancement (Rosemond et al. 2018). Further, habitat attributes, such as spatial footprint and structural complexity, affect whether artificial structures mimic natural reefs (Paxton et al. 2017; Lemoine et al. 2019; Gilliland et al. 2023).

## 5.5 WHICH TRAITS OF ARTIFICIAL STRUCTURES PREDICT ECOLOGICAL FUNCTIONS AND SERVICES?

Field studies and synthesis research suggest that artificial structures can facilitate marine life, especially highly mobile consumers, in their response to global change. However, given that artificial structures are often installed to meet human needs (i.e., extract energy, create recreation opportunities), it is necessary to understand which attributes of these structures can allow the prediction of ecological functions and services. Such an understanding can contribute to a vision where strategic design, siting, and management of artificial structures could benefit both nature and people and minimize consequences. Perhaps paramount among the opportunities is that artificial structures are often designed to achieve a primary human-driven goal, but these structures can be designed strategically to support natural habitats or at least minimize ecological harm (Paxton et al. 2025b).

Nine concepts can help make this vision a reality: 1) preserve artificial structures' original function (i.e., designing for nature should not compromise the structure's primary purpose), 2) manage artificial structures holistically as ecosystems, 3) embrace co-benefits at project inception, 4) integrate with marine spatial planning, 5) innovate strategies in cross-disciplinary teams, 6) develop co-ownership of artificial structures across a variety of stakeholder groups (e.g., government, industry, general public), 7) increase long-term returns on investment, 8) "green" up the environmental compatibility of existing artificial structures for added benefits, and 9) evaluate artificial structures' seascape effects (Paxton et al. 2025a). There is also a need to plan for when the structure's original function is complete (e.g., decommissioning scenarios). Additionally, there are ways to incorporate ecological theory into artificial structures to optimize outcomes. This includes building upon ecological theories useful for ecosystem restoration (Silliman et al. 2024), as well as those ecological processes and functions that manifest in shipwrecks (Paxton et al. 2024b). Lastly, given that gaps remain in our understanding of how to intentionally design artificial structures for optimal ecological outcomes, we can use evidence synthesis to disentangle or synthesize existing evidence to develop lessons learned and best practices (Paxton et al. 2024a, 2024c).

## 5.6 PUTTING IT ALL TOGETHER

An opportunity exists to use the features of artificial structures to predict and maximize ecological services for people and nature. This opportunity exists because of knowledge gained about the extent of artificial structures and how they can sometimes function similarly (but sometimes differently) from natural habitats. Additionally, they can facilitate fish with their responses to global change. Increasingly, the science of artificial structures is revealing that there is, in fact, an ability to predict how artificial structures may function. This knowledge can be employed to understand how artificial structures will affect ecosystem services and ensure that these structures function effectively to facilitate marine life under scenarios of global change. This knowledge can be integrated into marine spatial planning frameworks to minimize ocean use conflicts within a crowded ocean and help optimize desired outcomes from newly installed (or decommissioned) or previously deployed structures (Paxton et al. 2022). Additionally, there may be new opportunities to use artificial reefs in combination with restoration and marine-protected areas to garner synergistic effects. However, despite the increased predictive capabilities, science still has a long way to go to fully understand how artificial structures function amidst global change, as many knowledge gaps remain that require additional scientific attention (Czaja et al. 2025; Sura et al. 2025). Importantly, the perspective here hinges upon scientific work that I have had the opportunity to lead or contribute to and is meant to complement broader perspectives and evidence syntheses (e.g., Lemasson et al. 2023, 2024; Firth et al. 2024; Watson et al. 2024) on artificial marine structures to increase environmental sustainability and reduce unintended consequences.

## 5.7 ACKNOWLEDGMENTS

I thank the coauthors of the primary peer-reviewed journal articles cited herein for their collaboration and contribution to the field of marine artificial structure ecology. I also thank J.C. Taylor, B.R. Silliman, and J. Walter for their thoughtful reviews of this perspective manuscript. The views and conclusions contained in this document are those of the author and should not be interpreted as representing the opinions or policies of the U.S. Government. Mention of trade names or commercial products does not constitute endorsement or recommendation for use.

## REFERENCES

Bugnot, A. B., M. Mayer-Pinto, L. Airoldi, E. C. Heery, E. L. Johnston, L. P. Critchley, E. M. A. Strain, R. L. Morris, L. H. L. Loke, M. J. Bishop, E. V. Sheehan, R. A. Coleman, and K. A. Dafforn. 2020. Current and projected global extent of marine built structures. *Nature Sustainability* **4**:33–41. DOI. 10.1038/s41893-020-00595-1.

Czaja, R. E., S. A. Sura, W. F. Patterson, D. D. Chagaris, J. F. Walter, S. R. Sagarese, A. B. Paxton, W. D. Heyman, and H. E. Harris. 2025. Opportunity knocks: Leveraging offshore wind development as a natural experiment to address the ecological function of artificial reefs. *Fish and Fisheries* **26**:688–698. DOI. 10.1111/faf.12906.

Firth, L. B., J. Bone, A. Bartholomew, M. J. Bishop, A. Bugnot, F. Bulleri, S.-Y. Chee, L. Claassens, K. A. Dafforn, T. P. Fairchild, A. E. Hall, M. E. Hanley, V. Komyakova, A. J. Lemasson, L. H. L. Loke, M. Mayer-Pinto, R. Morris, L. Naylor, M. J. Perkins, S. Pioch, F. Porri, K. A. O'Shaughnessy, N. Schaefer, E. A. Strain, J. D. Toft, N. Waltham, M. Aguilera, L. Airoldi, F. Bauer, P. Brooks, J. Burt, C. Clubley, J. R. Cordell, F. Espinosa, A. J. Evans, V. Farrugia-Drakard, W. Froneman, J. Griffin, S. J. Hawkins, E. Heery, R. J. H. Herbert, E. Jones, K. M. Y. Leung, P. Moore, J. Sempere-Valverde, D. Sengupta, M. Sheaves, S. Swearer, R. C. Thompson, P. Todd, and A. M. Knights. 2024. Coastal greening of grey infrastructure: An update on the state-of-the-art. *Proceedings of the Institution of Civil Engineers—Maritime Engineering*: 1–69. DOI. 10.1680/jmaen.2023.003.

Gilliland, V. A., A. E. Fessler, A. B. Paxton, E. F. Ebert, R. M. Tharp, B. J. Runde, N. M. Bacheler, J. A. Buckel, and J. C. Taylor. 2023. Spatial extent and isolation of marine artificial structures mediate fish density. *Frontiers in Marine Science* **10**. DOI. 10.3389/fmars.2023.1240344.

Halpern, B. S., M. Frazier, J. Potapenko, K. S. Casey, K. Koenig, C. Longo, J. S. Lowndes, R. C. Rockwood, E. R. Selig, K. A. Selkoe, and S. Walbridge. 2015. Spatial and temporal changes in cumulative human impacts on the world's ocean. *Nature Communications* **6**:7615. DOI. 10.1038/ncomms8615.

Heery, E. C., M. J. Bishop, L. P. Critchley, A. B. Bugnot, L. Airoldi, M. Mayer-Pinto, E. V. Sheehan, R. A. Coleman, L. H. L. Loke, E. L. Johnston, V. Komyakova, R. L. Morris, E. M. A. Strain, L. A. Naylor, and

K. A. Dafforn. 2017. Identifying the consequences of ocean sprawl for sedimentary habitats. *Journal of Experimental Marine Biology and Ecology* **492**:31–48. DOI. 10.1016/j.jembe.2017.01.020.
Johnson, K. H., A. B. Paxton, J. C. Taylor, J. Hoyt, J. McCord, and W. Hoffman. 2020. Extracting ecological metrics from archaeological surveys of shipwrecks using submersible video and laser-line scanning. *Ecosphere* **11**:e03210. DOI.
Lemasson, A. J., P. J. Somerfield, M. Schratzberger, and A. M. Knights. 2023. Challenges of evidence-informed offshore decommissioning: An environmental perspective. *Trends in Ecology & Evolution* **38**:688–692. DOI. 10.1016/j.tree.2023.04.003.
Lemasson, A. J., P. J. Somerfield, M. Schratzberger, M. S. A. Thompson, L. B. Firth, E. Couce, C. L. McNeill, J. Nunes, C. Pascoe, S. C. L. Watson, and A. M. Knights. 2024. A global meta-analysis of ecological effects from offshore marine artificial structures. *Nature Sustainability* **7**:485–495. DOI. 10.1038/s41893-024-01311-z.
Lemoine, H. R., A. B. Paxton, S. C. Anisfeld, R. C. Rosemond, and C. H. Peterson. 2019. Selecting the optimal artificial reefs to achieve fish habitat enhancement goals. *Biological Conservation* **238**. DOI. 10.1016/j.biocon.2019.108200.
Lim, P., A. B. Paxton, J. C. Taylor, and N. S. Hall. 2023. Survey of epiphytic microalgae to evaluate risk of ciguatera fish poisoning across natural and artificial reefs in North Carolina. *Frontiers in Marine Science* **10**. DOI. 10.3389/fmars.2023.1232524.
Paxton, A. B., E. Blair, C. Blawas, M. H. Fatzinger, M. Marens, J. Holmberg, C. Kingen, T. Houppermans, M. Keusenkothen, J. McCord, B. R. Silliman, and L. M. Penfold. 2019a. Citizen science reveals female sand tiger sharks (*Carcharias taurus*) exhibit signs of site fidelity on shipwrecks. *Ecology* **100**:e02687. DOI.
Paxton, A. B., I. R. Foxfoot, C. Cutshaw, D. N. Steward, L. Poussard, T. N. Riley, T. M. Swannack, C. D. Piercy, S. Altman, B. J. Puckett, C. D. Storlazzi, and T. S. Viehman. 2024a. Evidence on the ecological and physical effects of built structures in shallow, tropical coral reefs: A systematic map. *Environmental Evidence* **13**:12. DOI. 10.1186/s13750-024-00336-3.
Paxton, A. B., S. L. Harter, S. W. Ross, C. M. Schobernd, B. J. Runde, P. J. Rudershausen, K. H. Johnson, K. W. Shertzer, N. M. Bacheler, J. A. Buckel, G. T. Kellison, and J. C. Taylor. 2021. Four decades of reef observations illuminate deep-water grouper hotspots. *Fish and Fisheries* **22**:749–761. DOI. 10.1111/faf.12548.
Paxton, A. B., S. E. Lester, C. S. Smith, S. Narayan, C. Angelini, B. J. Runde, M. I. Saunders, R. K. Gittman, J. Allgeier, M. L. Vozzo, D. a. N. Steward, H. R. Lemoine, S. R. Valdez, R. L. Morris, D. P. Nowacek, W. Seaman, P. N. Halpin, and B. R. Silliman. 2025a. Recommendations for built marine infrastructure that supports natural habitats. *Frontiers in Ecology and the Environment*:e2840. DOI. 10.1002/fee.2840.
Paxton, A. B., C. McGonigle, M. Damour, G. Holly, A. Caporaso, P. B. Campbell, K. S. Meyer-Kaiser, L. J. Hamdan, C. H. Mires, and J. C. Taylor. 2024b. Shipwreck ecology: Understanding the function and processes from microbes to megafauna. *BioScience* **74**:12–24. DOI. 10.1093/biosci/biad084.
Paxton, A. B., E. A. Newton, A. M. Adler, R. V. Van Hoeck, E. S. Iversen, Jr., J. C. Taylor, C. H. Peterson, and B. R. Silliman. 2020a. Artificial habitats host elevated densities of large reef-associated predators. *PLoS One* **15**:e0237374. DOI. 10.1371/journal.pone.0237374.
Paxton, A. B., C. H. Peterson, J. C. Taylor, A. M. Adler, E. A. Pickering, and B. R. Silliman. 2019b. Artificial reefs facilitate tropical fish at their range edge. *Communications Biology* **2**:168. DOI. 10.1038/s42003-019-0398-2.
Paxton, A. B., E. A. Pickering, A. M. Adler, J. C. Taylor, and C. H. Peterson. 2017. Flat and complex temperate reefs provide similar support for fish: Evidence for a unimodal species-habitat relationship. *PLoS One* **12**:e0183906. DOI. 10.1371/journal.pone.0183906.
Paxton, A. B., T. N. Riley, C. L. Steenrod, B. J. Puckett, I. J. Alemu, S. T. Paliotti, A. M. Adler, L. Exar, J. E. T. McLean, J. Kelley, Y. S. Zhang, C. S. Smith, R. K. Gittman, and B. R. Silliman. 2024c. Evidence on the performance of nature-based solutions interventions for coastal protection in biogenic, shallow ecosystems: A systematic map. *Environmental Evidence* **13**:28. DOI. 10.1186/s13750-024-00350-5.
Paxton, A. B., B. J. Runde, C. S. Smith, S. E. Lester, M. L. Vozzo, M. I. Saunders, D. a. N. Steward, H. R. Lemoine, S. R. Valdez, R. K. Gittman, S. Narayan, J. E. Allgeier, R. L. Morris, D. P. Nowacek, W. Seaman, P. N. Halpin, C. Angelini, and B. R. Silliman. 2025b. Leveraging built marine structures to benefit and minimize impacts on natural habitats. *BioScience* **75**:172–183. DOI. 10.1093/biosci/biae135.
Paxton, A. B., K. W. Shertzer, N. M. Bacheler, G. T. Kellison, K. L. Riley, and J. C. Taylor. 2020b. Meta-analysis reveals artificial reefs can be effective tools for fish community enhancement but are not one-size-fits-all. *Frontiers in Marine Science* **7**:282. DOI. 10.3389/fmars.2020.00282.
Paxton, A. B., D. a. N. Steward, Z. H. Harrison, and J. C. Taylor. 2022. Fitting ecological principles of artificial reefs into the ocean planning puzzle. *Ecosphere* **13**:e3924. DOI. 10.1002/ecs2.3924.

Paxton, A. B., D. a. N. Steward, K. J. Mille, J. Renchen, Z. H. Harrison, J. S. Byrum, C. Brinton, A. Nelson, E. Simpson, P. J. Clarke, C. LaPorta, P. D. Barrett, M. Rousseau, D. C. Newton, R. B. Rigby, D. T. Williams, J. B. Shipley, P. Murakawa, B. J. Runde, K. L. Riley, N. M. Bacheler, G. T. Kellison, and J. C. Taylor. 2024d. Artificial reef footprint in the United States ocean. *Nature Sustainability* **7**:140–147. DOI. 10.1038/s41893-023-01258-7.

Paxton, A. B., J. C. Taylor, C. H. Peterson, S. R. Fegley, and J. H. Rosman. 2019c. Consistent spatial patterns in multiple trophic levels occur around artificial habitats. *Marine Ecology Progress Series* **611**:189–202. DOI.

Rosemond, R. C., A. B. Paxton, H. R. Lemoine, S. R. Fegley, and C. H. Peterson. 2018. Fish use of reef structures and adjacent sand flats: Implications for selecting minimum buffer zones between artificial reefs and existing reefs. *Marine Ecology Progress Series* **587**:187–199. Doi.

Runde, B. J., and A. B. Paxton. 2025. The seafloor footprint of offshore wind infrastructure in the United States Ocean. *Discover Oceans* **2**. DOI. 10.1007/s44289-025-00097-y.

Silliman, B. R., M. J. S. Hensel, J. P. Gibert, P. Daleo, C. S. Smith, D. J. Wieczynski, C. Angelini, A. B. Paxton, A. M. Adler, Y. S. Zhang, A. H. Altieri, T. M. Palmer, H. P. Jones, R. K. Gittman, J. N. Griffin, M. I. O'Connor, J. van de Koppel, J. R. Poulsen, M. Rietkerk, Q. He, M. D. Bertness, T. van der Heide, and S. R. Valdez. 2024. Harnessing ecological theory to enhance ecosystem restoration. *Current Biology* **34**:R418–R434. DOI. 10.1016/j.cub.2024.03.043.

Steward, D. a. N., A. B. Paxton, N. M. Bacheler, C. M. Schobernd, K. Mille, J. Renchen, Z. Harrison, J. Byrum, R. Martore, C. Brinton, K. L. Riley, J. C. Taylor, and G. T. Kellison. 2022. Quantifying spatial extents of artificial versus natural reefs in the seascape. *Frontiers in Marine Science* **9**. DOI. 10.3389/fmars.2022.980384.

Sura, S. A., R. E. Czaja, N. Brugnone, S. L. Gibbs, J. R. Hendon, W. Klajbor, A. B. Paxton, R. R. Rindone, Skyler R. Sagarese, K. Wing, L. Bosarge, D. D. Chagaris, William D. Heyman, M. A. Johnston, J. A. Morris, William F. Patterson, J. M. Tolan, J. F. Walter, and H. E. Harris. 2025. Science priorities to evaluate the effects of offshore wind energy development on fish and fisheries in the Gulf of America. *Marine and Coastal Fisheries* **17**. DOI. 10.1093/mcfafs/vtaf009.

Tharp, R. M., N. J. Hostetter, A. B. Paxton, J. C. Taylor, and J. A. Buckel. 2024. Artificial structure selection by economically important reef fishes at North Carolina artificial reefs. *Frontiers in Marine Science* **11**. DOI. 10.3389/fmars.2024.1373494.

Van Hoeck, R. V., A. B. Paxton, D. R. Bohnenstiehl, J. C. Taylor, F. J. Fodrie, D. P. Nowacek, C. M. Voss, and C. H. Peterson. 2020. Soundscapes of natural and artificial temperate reefs: Similar temporal patterns but distinct spectral content. *Marine Ecology Progress Series* **649**:35–51. DOI. 10.3354/meps13434.

Watson, S. C. L., P. J. Somerfield, A. J. Lemasson, A. M. Knights, A. Edwards-Jones, J. Nunes, C. Pascoe, C. L. McNeill, M. Schratzberger, M. S. A. Thompson, E. Couce, C. L. Szostek, H. Baxter, and N. J. Beaumont. 2024. The global impact of offshore wind farms on ecosystem services. *Ocean & Coastal Management* **249**. DOI. 10.1016/j.ocecoaman.2024.107023.

# 6 Enhancing Nektonic Communities and Juvenile Fish Biodiversity with Reef cubes® Deployments in Torbay, United Kingdom

*Samuel Hickling*, James Murphy, and Ferrante Grasselli**
* Corresponding author: sam.hickling@arcmarine.co.uk, ferrante@arcmarine.co.uk

## 6.1 INTRODUCTION

The application of coastal and marine engineering has long played a central role in the development of modern coastal societies (Palmer and Tritton Limited Development and Engineering Consultants 1996). Today, more than three billion people are estimated to live in coastal areas (Airoldi et al. 2021), and the total global footprint of marine infrastructure, estimated at 32,000 km² in 2018, is projected to grow substantially by 2028 (Bugnot et al. 2020). Alongside other global pressures, the rapid proliferation of artificial structures, a phenomenon known as 'ocean sprawl' (Firth et al. 2016), and the intensification of human activity in the marine environment have contributed to an ongoing biodiversity and ecosystem functioning crisis (Bishop et al. 2017; Heery et al. 2017), particularly severe in coastal waters (Todd et al. 2019). The spatial concentration of multiple, often competing, commercial interests in coastal areas presents a major challenge for implementing strategies aimed at conserving, restoring, or enhancing ecological conditions (Todd et al. 2019). In this context, minimising space-use conflicts among sea users has become an additional key objective that marine spatial planning efforts must address when advancing sustainable development (Douvere 2008; Schupp et al. 2019).

In response to these growing demands, increasing efforts are being directed toward the design of multifunctional structures that can simultaneously deliver environmental, social, and economic benefits while occupying a minimal spatial footprint (Chapman and Underwood 2011, Dafforn et al. 2015). The Reef Enhancement of Fisheries and Aquaculture project (REFAS project, hereafter), funded by the UK (United Kingdom) Government's Seafood Innovation Fund through the UK Centre for Environment, Fisheries and Aquaculture Science, was conceived with a multifunctional approach. The goal of the REFAS project was to develop an artificial reef that could promote the development of biotic marine assemblages by acting as an artificial reef, while providing mooring support for adjacent aquaculture operations and ensuring that all stakeholders benefit from a growing blue economy (Seafood Innovation Fund 2020). To achieve these objectives, the REFAS project deployed Reef cubes®, a patented artificial reef modular unit developed by ARC marine Ltd. (Torquay, UK). Each cube features an internal spherical chamber and six lateral passageways (Birbeck and Doddrell 2020), with a design expected to outperform traditional substrates in ecological performance and habitat functionality (Hunter and Sayer 2009; Rouse et al. 2020). They are

DOI: 10.1201/9781003676775-6

produced using a proprietary low-carbon concrete (Hickling et al. 2022), specifically developed to reduce carbon emissions by fully replacing Portland cement (Meyer 2009; Provis 2018).

Indeed, conventional artificial substrates, such as plain solid concrete blocks, have often proved poor surrogates for natural reef habitats (Burt et al. 2009; Sánchez-Caballero et al. 2021). Apparently, it is not the colonisation time but rather the intrinsic simplicity and lack of habitat niches and refuges from stressors, predators, and competitors that have been repeatedly linked to a reduction in biodiversity associated with some artificial substrates (Moschella et al. 2005; Perkol-Finkel et al. 2006; Bulleri and Chapman 2010). By contrast, nature-inclusive design seeks to intentionally incorporate ecological features into artificial structures to promote biodiversity and ecosystem functioning. Defined as 'a design approach that integrates habitat features within man-made infrastructure to support native species and ecological processes' (Hermans et al. 2020), nature-inclusive design offers evidence-based tools to enhance ecological performance while maintaining engineering function. Structures developed embedding these nature-positive approaches aim to replicate natural complexity and provide conditions that support marine species throughout their life stages (Naylor et al. 2017; Airoldi et al. 2021; Evans et al. 2021). Thus, the deployed habitat is designed to facilitate natural habitat utilisation across diel cycles (Taira et al. 2024), which is particularly important for juvenile growth and survival (Scyphers et al. 2015; Mercader et al. 2017). However, the novelty of these technologies currently limits our ability to generalise about their effectiveness, and caution is warranted to avoid inadvertently contributing to further 'ocean sprawl' through their uncritical application (Firth et al. 2020, 2024). Continued field-based research is therefore essential to build a robust evidence base on the ecological performance of such eco-engineered structures (Bortone 2006).

By implementing Reef cubes®, the REFAS project established a long-term research site to assess the ecological performance of a multifunctional artificial reef system designed according to nature-inclusive principles in contributing to biodiversity. Building on previous surveys at the site (Hickling and Murphy 2022; Hickling et al. 2023), this study represents a short-term investigation aimed at extending the temporal scale of nektonic monitoring with an additional focus on juvenile fish and diel cycles. The goal is to provide preliminary insights into how an artificial reef composed of nature-inclusive design structures supports essential habitat functioning, particularly during critical life stages of fish. By doing so, the research explores the potential for nature-inclusive designs, specifically Reef cubes®, as an effective management option for marine spatial planning and for mitigating competition among coastal sea users.

## 6.2 MATERIALS AND METHODS

### 6.2.1 Study Site

This study was conducted in Torbay, Devon, United Kingdom. In the northern part of the Bay facing the English Channel, the REFAS2 site (Reef Enhancement of Fisheries and Aquaculture Sites) was established in 2021 at 50°26.97′N, 003°29.59′W on a sandy mud bottom at a depth of 11 m. At this site, 11-reef Reef cube® modules were deployed in June 2021 (each no more than 50 m from each other), forming a reef group (*sensu* Grove and Sonu 1983) and extending over 0.78 hectares. Each reef module had a volume ranging from 3.4 $m^3$ to 4.0 $m^3$ and was composed of three different Reef cube® formations, covering a total footprint of 28 $m^2$: (i) a cluster of four 1-meter-sided cubes, (ii) a cluster of eight 0.75-meter-sided cubes, and (iii) a single 1.5-meter-sided cube (Figure 6.1). All Reef cubes® were made using Marine Crete®, a low-carbon concrete incorporating alkali-activated cementitious material and recycled components.

### 6.2.2 Survey Methods

To assess the performance of the artificial reef on the nektonic community, the REFAS2 project area was surveyed during three periods: Spring 2021, three months prior to reef deployment (9–18

FIGURE 6.1 Photographs of Reef cube® formations prior to deployment. From left to right: (i) a cluster of four 1-meter-sided cubes, (ii) a cluster of eight 0.75-meter-sided cubes, and (iii) a single 1.5-meter-sided cube.

March); Autumn 2021, four months post-deployment (18 September–9 October); and Spring 2022, ten months post-deployment (23 March–14 April). Surveys were conducted during both daytime and nighttime, defined respectively as hours of full daylight and hours of full darkness, excluding civil twilight. A total of 26 surveys were conducted for each time across three sampling phases, resulting in a total sample size of 156. Of these, 11 surveys were conducted within the reef group (i.e., one survey per reef module), while the remaining 15 surveys were conducted at increasing distances (15 m, 40 m, and 130 m) from the reef group (Figure 6.2). For the pre-deployment surveys, the reef group was surveyed at the estimated future positions of the modules.

A modified Remotely Operated Vehicle (ROV) BlueROV2 (Blue Robotics, Torrance, California, USA) was used to conduct the surveys as a Baited Remote Underwater Video (BRUV) (Cappo et al. 2004). An ROV was used to conduct surveys at the reef modules to record video samples, landing approximately at two metres from each module, and the position of the modules was considered for the position of the replicate. For each video sample recorded during the pre-deployment surveys or at the progressive distance radiating from the reef group, the ROV was aligned with the surface-tending vessel during the ROV descent, and the vessel's GPS was considered as the sample location. The ROV was equipped with a bait arm mounted at the front and two parallel lasers to assist with following fish for measurement. The bait consisted of 200 g of chopped Atlantic Salmon (*Salmo salar*). Once the ROV was guided into the target position, it was gently landed on the seafloor, and 30 minutes of video were recorded.

### 6.2.3 Data Analyses

To assess the effect of the artificial reef system on the nekton assemblage, survey videos were analysed by identifying organisms to the lowest possible taxonomic resolution and quantifying the abundance of each taxon as the maximum number observed within a single video frame (Max N) (Ellis and DeMartini 1995). Two parallel lasers spaced 10 cm apart, along with a 14 cm wide baited cage, were used as visual references to estimate fish lengths. Individuals were assigned to one of six size classes: <2.5 cm, 2.5–5 cm, 5–10 cm, 10–20 cm, 20–40 cm, and >40 cm. Length at maturity for each taxon was obtained from FishBase (Froese and Pauly 2025). For each replicate, the following community metrics were derived: taxa richness, total nekton abundance (defined as the sum of MaxN across all taxa), multivariate community composition, and juvenile abundance (defined as the total MaxN of individuals belonging to size classes smaller than the size class containing the

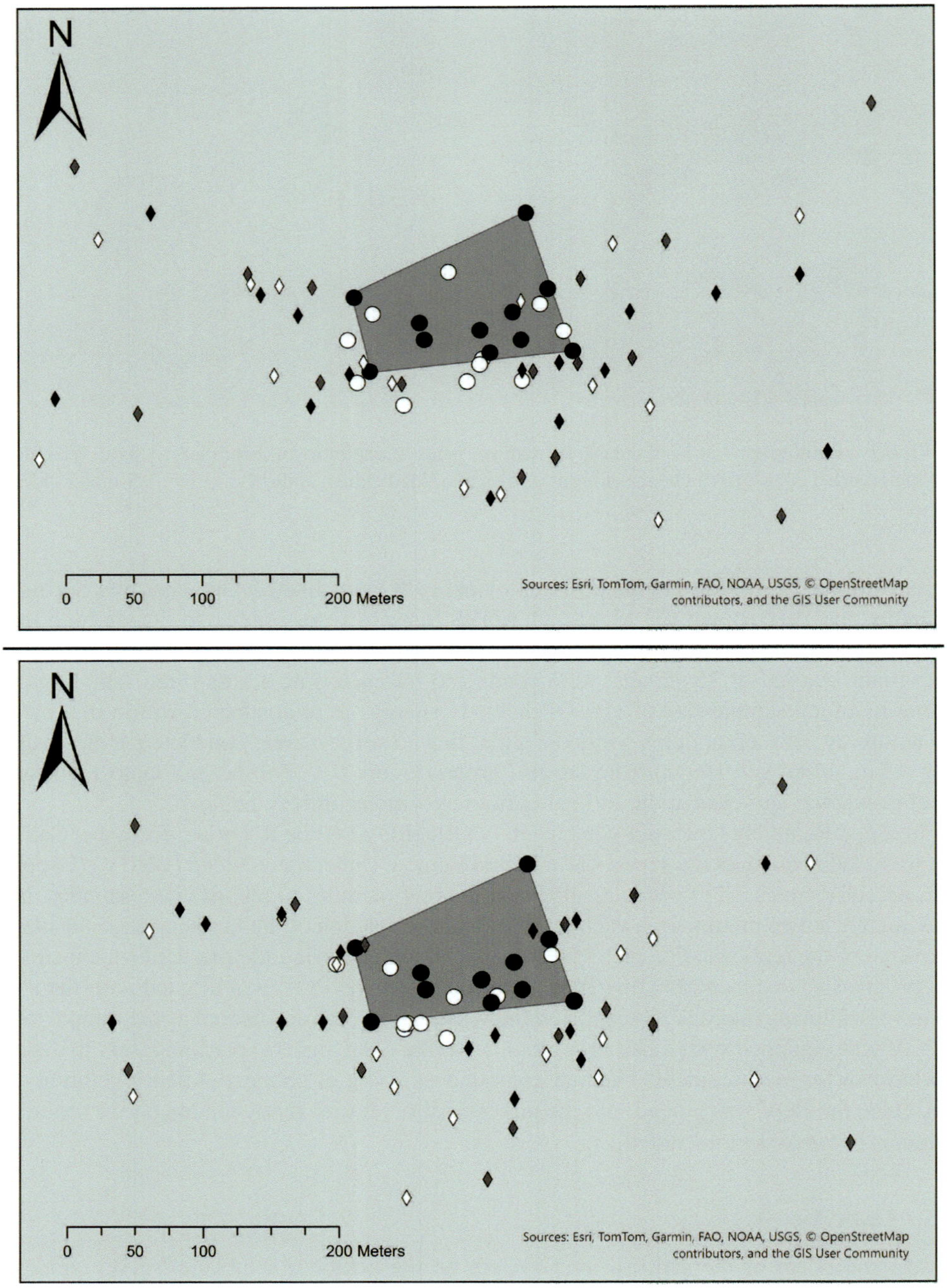


**FIGURE 6.2** Maps of surveys at the REFAS2 site with BRUVs surveys conducted during the daytime (top) and nighttime (bottom). The grey polygon represents the REFAS2 reef group site extension. Symbol colour indicates the survey period: white for before surveys; grey for three months after surveys; black for ten months after surveys. Symbol shape indicates survey position: circles for surveys at the reef group site, diamonds for surveys at non-reef sites. Note: Overlapping symbols (e.g., grey and black circles) may result in only the uppermost symbol being visible. Source: Esri, TomTom, Garmin, FAO, NOAA, USGS, © OpenStreetMap contributors, and the GIS User Community.

species-specific length at maturity). These metrics were modelled using a Before-After Gradient (BAG) design (Ellis and Schneider 1997), which was preferred over a more canonical Before-After Control-Impact (BACI) design (Methratta 2020).

The BAG design was tested using generalized linear models (GLMs). The full model included two fully crossed, fixed categorical factors—that is, the survey period (Before vs. After) and the time-of-day (Day vs. Night)—and their interactions with a continuous covariate: the natural logarithmic (ln)-transformed distance from the reef modules. In addition, the survey event (e.g., Before 1, After 1, After 2) was included as a random effect nested within the survey period. Backward stepwise selection was performed on the full model, aided by the Akaike Information Criteria (AIC), to progressively remove non-informative terms for model parsimony. Diagnostics were then performed on the reduced model to assess overdispersion (McCullagh and Nelder 2019). If overdispersion was detected, model selection was repeated using a GLM with a negative binomial error distribution instead of the Poisson family (Hilbe 2011). Multivariate nekton assemblage composition was directly modelled using a GLM with a negative binomial distribution. Post hoc, simple-slopes analyses were finally performed to elucidate any significant interaction term effects. Additionally, the effect of predictors on individual taxa was investigated on the basis of estimated coefficients from the multivariate GLM. Significance levels were then indicated as follows: * for $p < 0.05$, ** for $p < 0.01$, and *** for $p < 0.001$.

All data analyses were performed in R (version 4.4.1), using several packages for statistical modelling and data visualisation. Specifically, we used: 'AER' for regression diagnostic (Kleiber and Zeileis 2008); 'interactions' to explore and visualise interaction effects in regression models (Long 2024); 'MASS' for fitting GLMs with negative binomial distribution (Venables and Ripley 2009); 'mvabund' for multivariate abundance analyses (Wang et al. 2022); and 'sjPlot' to visualise model outputs (Lüdecke 2013).

## 6.3 RESULTS

### 6.3.1. Nekton Assemblage

A total of 806 individuals spanning 29 different taxa were observed across the surveys, with the dominance of Poor Cod (*Trisopterus minutus*), with 164 individuals observed across 26 surveys, and Sand Goby (*Pomatoschistus minutus*), with 153 individuals across 64 samples (Table 6.1). The nekton assemblage was predominantly comprised of fish species (94%), with a minor representation (6%) of free-swimming invertebrates, including Common Cuttlefish (*Sepia officinalis*) and unidentified squid species (Loliginidae).

Taxa richness (S) and total abundance (MaxN) varied considerably across all the investigated factors. After the reef deployment, the total number of taxa increased in the survey area from 9 to 26. The highest average taxa richness was observed at reef group sites (S = 3.0 ± 0.3; Figure 6.3a), primarily driven by particularly high values recorded during nighttime in autumn 2021 (S = 3.9 ± 0.6). Although lower richness levels were found than at non-reef sites, these values still exceeded those observed before reef deployment. Among all low richness values observed during the pre-deployment phase, the minimum occurred at the 40-metre distance from the reef group (S = 0.6 ± 0.2; Figure 6.3a). Similarly, abundance was lower during the pre-deployment period, with an average of one individual per survey. Following the deployment, the average abundance (MaxN = 10.8, ±2.4; Figure 6.3b) was highest at the reef group. In 2022, the total number of individuals observed during the 'After 2' period was lower (213) than in autumn 2021, four months after reef deployment (540 individuals). Additionally, their distribution across distance gradients was more even in 2022 compared to the previous period.

These tendencies were reflected in the GLM results. Taxa richness was significantly higher after reef deployment (Before = −1.16***) and increased with proximity to the reef group (ln-distance = −0.16**).

**TABLE 6.1**

**List of fish taxa recorded during the study, by survey period. For each taxon, the total number of individuals observed is shown, with the number of individuals classified as juveniles (i.e., below the length at maturity when available, otherwise NA, from FishBase) reported in parentheses.**

| Common name (Scientific name) [length at maturity in cm] | Before (spring 2021) | After 1 (autumn 2021) | After 2 (spring 2022) |
|---|---|---|---|
| Actinopterygii spp. [NA] | 3 | 51 | 62 |
| Argentine *(Argentina sphyraena)* [12.0] | 0 | 11 (11) | 0 |
| Dragonet *(Callionymus lyra)* [17.4] | 0 | 5 (5) | 0 |
| Clupeidae spp. [10.1] | 0 | 10 (10) | 0 |
| European Seabass *(Dicentrarchus labrax)* [36.1] | 0 | 4 | 59 |
| Grey Gurnard *(Eutrigla gurnardus)* [18.8] | 7 (1) | 6 | 0 |
| Gadidae spp. [13.4] | 0 | 15 (14) | 4 |
| Atlantic Cod *(Gadus morhua)* [65.4] | 0 | 1 (1) | 0 |
| Common Dab *(Limanda limanda)* [21.5] | 6 (6) | 0 | 0 |
| Loliginidae spp. [NA] | 0 | 8 | 36 |
| Haddock *(Melanogrammus aeglefinus)* [34.9] | 0 | 1 (1) | 0 |
| Whiting *(Merlangius merlangus)* [28.2] | 6 (6) | 75 (63) | 4 (3) |
| Lemon Sole *(Microstomus kitt)* [29.3] | 1 | 0 | 0 |
| Ling *(Molva molva)* [92.5] | 0 | 0 | 1 |
| *Moronidae* spp. [NA] | 0 | 2 | 0 |
| Surmullet *(Mullus surmuletus)* [16.1] | 0 | 2 | 0 |
| Pleuronectiforme spp. [NA] | 2 | 2 | 1 |
| Sand Goby *(Pomatoschistus minutus)* [6.3] | 23 (16) | 106 (54) | 24 (15) |
| *Scombridae* spp. [NA] | 0 | 6 | 0 |
| Lesser Spotted Dogfish *(Scyliorhinus canicula)* [54.4] | 1 (1) | 3 (3) | 0 |
| Nursehound *(Scyliorhinus stellaris)* [78.0] | 0 | 1 (1) | 0 |
| Common Cuttlefish *(Sepia officinalis)* [NA] | 0 | 2 | 1 |
| Sepiidae spp. [NA] | 0 | 0 | 1 |
| Sparidae spp. [NA] | 0 | 0 | 1 |
| European Sprat *(Sprattus sprattus)* [10.1] | 4 | 0 | 0 |
| Atlantic Horse Mackerel *(Trachurus trachurus)* [24.3] | 0 | 36 (36) | 3 (2) |
| Triglidae spp. [NA] | 0 | 4 | 0 |
| Pouting *(Trisopterus luscus)* [22.1] | 0 | 33 (33) | 8 (7) |
| Poor Cod *(Trisopterus minutus)* [13.4] | 0 | 156 (98) | 8 (7) |
| TOTAL | 53 (30) | 540 (330) | 213 (38) |

Additionally, richness in spring 2022 (After 2) was significantly lower than that in autumn 2021 (After 2 = −0.55***). Total abundance (MaxN) was significantly higher after reef deployment (Before = −2.00***). Proximity to the reef also had a significant effect, although its magnitude varied with the time of day as MaxN increased more sharply with proximity to the reef group during the daytime (ln-distance × day = −0.40***) compared to nighttime (ln-distance × night = −0.12***). Once again, there was a significant difference between the two post-deployment periods, with lower abundance recorded in 2022 compared to 2021 (After 2 = −0.93***).

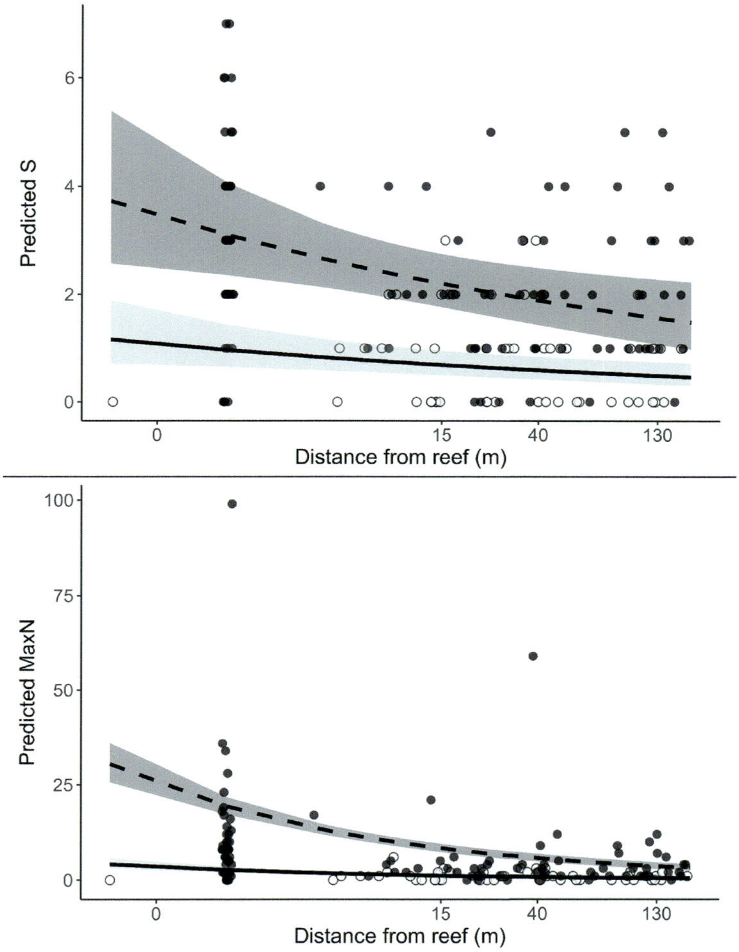


**FIGURE 6.3** Predicted response of the nektonic assemblage in terms of: a) species richness (S) and b) abundance (MaxN), as a function of distance from the artificial reef. Predictions are based on a GLM with Poisson error distribution and shown separately for the two survey periods ('Before' and 'After'). Solid and dashed lines represent the predicted means for the Before and After periods, respectively, with shaded ribbons indicating 95% confidence intervals. Observed values are shown as white (Before) and black (After) circles. For improved visualisation, observed points were jittered to reduce overlap. The x-axis reports the original (untransformed) sampling distances from the reef module group: 0, 15, 40, and 130 metres.

The multivariate GLM revealed significant main effects of reef deployment, proximity to the reef group, and diel period on multivariate community composition. However, these effects were independent of one another, as no interaction terms were statistically significant. Community composition also differed notably between the two post-installation periods.

Among all taxa, *T. minutus* exhibited the strongest model response and thus contributed the most to the observed variation in community composition. This species increased in abundance following reef installation and with greater proximity to the reef group, although it was less abundant at night and in 2022 ('After 2'; Figure 6.4). European seabass (*Dicentrarchus labrax*) was the second most influential species in the model. It increased after installation, especially during the 'After 2' period, but showed no significant response to reef proximity. Several species contributed strongly to the multivariate pattern, including Sand Goby *P. minutus*, Pouting *T. luscus*, and Clupeidae spp. during the daytime, and Atlantic Horse Mackerel *Trachurus trachurus*, Whiting *Merlangius merlangus*,

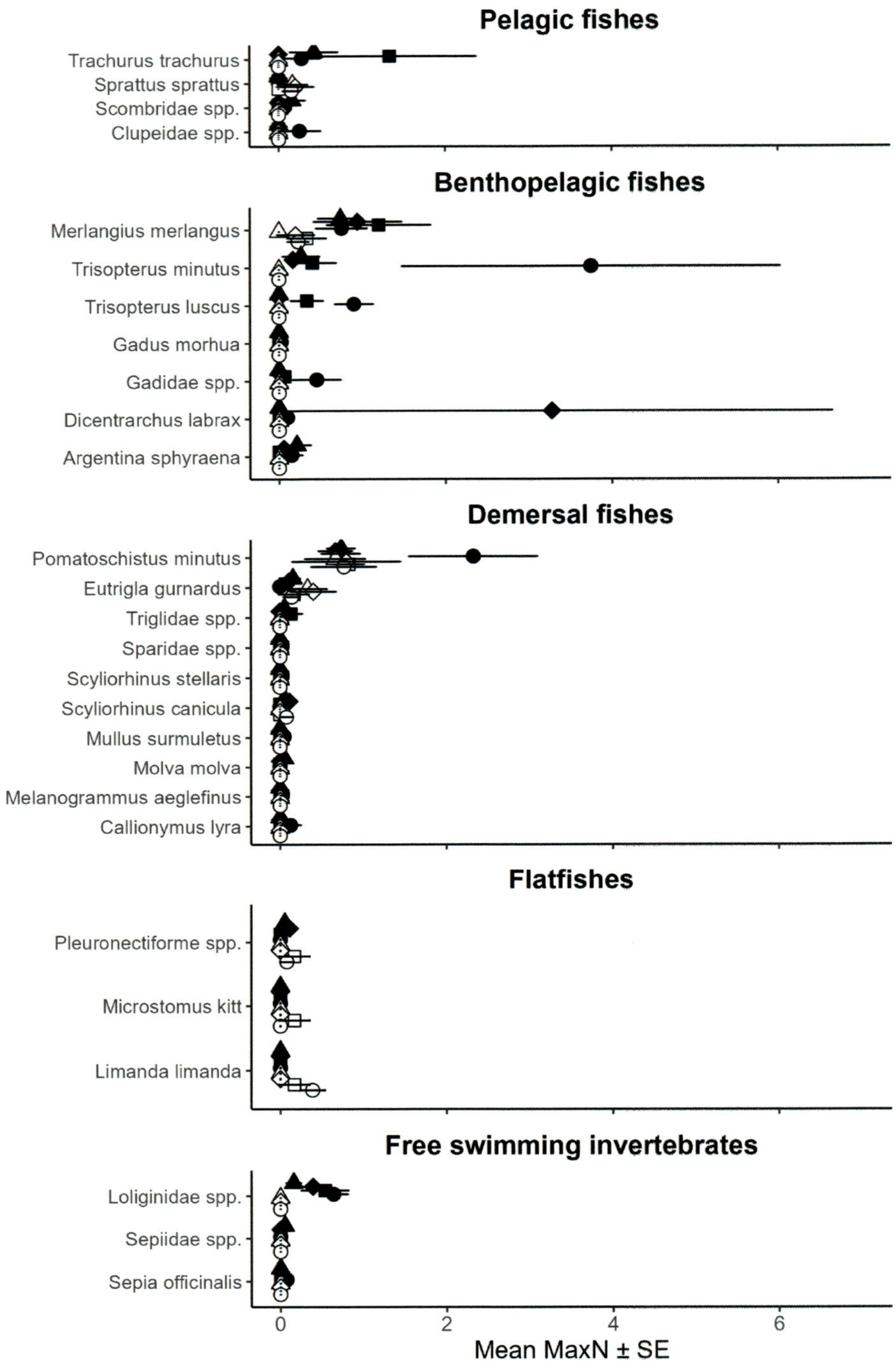


**FIGURE 6.4** Mean (±SE) abundance of taxa across survey periods and positions relative to the reef group, grouped according to ecological association as defined in FishBase (Froese and Pauly 2025). Symbol colour indicates the survey period: white for before-installation surveys, black for after-installation surveys. Symbol shape indicates the distance from the reef group: circles = within the reef group; squares, diamonds, and triangles = approximately 15 m, 40 m, and 130 m away, respectively.

and Argentine *Argentina sphyraena* during nighttime. These species also responded positively to reef installation, with more moderate responses to reef proximity (Figure 6.4).

### 6.3.2 Juvenile Fish Abundance

A total of 398 individuals representing 15 fish taxa were recorded throughout the surveys. Among them, *T. minutus*, with 105 individuals, and *P. minutus*, with 85 individuals, were the most abundant. The community was primarily composed of bony fish, while five juvenile Chondrichthyes were also observed, represented by Lesser Spotted Catshark *Scyliorhinus canicula* (four individuals) and Nursehound *S. stellaris* (one individual). Juvenile abundance was highest at the reef group site, with an average MaxN of 6.9 (±1.6) (Figure 6.5). During this first post-deployment survey, juvenile abundance was ten times higher (330 individuals) than before the deployment of reefs (30 individuals), especially concentrated within the reef group and at the nearest distance (15 m). In contrast, juvenile abundance ten months after deployment (spring 2022) returned to levels comparable to the pre-deployment baseline (38 individuals).

Based on model results, juvenile abundance was significantly higher closer to the reef group (ln-distance = –0.59**) and increased after the reefs were deployed (Before = –0.73***). However, these responses depended on the time of day as the effect of distance was greater during the day, while the positive effect of reef deployment was more pronounced at night. Nevertheless, juvenile abundance in 2022 (After 2) was significantly lower than that in autumn 2021 (After 2 = –1.60***).

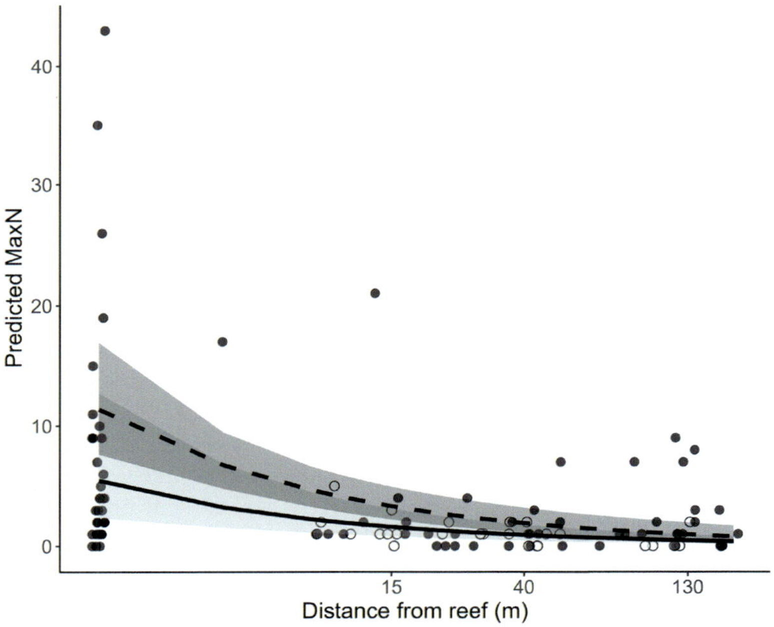


**FIGURE 6.5** Predicted response of juvenile fish abundance (MaxN) as a function of distance from the artificial reef. Predictions are based on a GLM with Poisson error distribution and are shown separately for the two survey periods ('Before' and 'After'). Solid and dashed lines represent the predicted means for the Before and After periods, respectively, with shaded ribbons indicating 95% confidence intervals. Observed values are shown as white (Before) and black (After) circles. For improved visualisation, observed points were jittered to reduce overlap. The x-axis reports the original (untransformed) sampling distances from the reef module group: 0, 15, 40, and 130 metres.

## 6.4 DISCUSSION

This study revealed marked differences in the nektonic assemblage following the deployment of artificial reefs with regard to taxon richness and total abundance, tripling and increasing tenfold, respectively, in proximity to the reef modules compared to pre-deployment conditions. Approximately 50% of the observed individuals were in the juvenile life stage of their respective species, and the ecological enhancement effect of the reefs was further confirmed by a tenfold higher abundance of juveniles after reef deployment. The generally positive community response to reef presence involved not only species typically associated with artificial reefs, such as *M. merlangus* and *Trisopterus* spp. (Fowler 1999, Reubens et al. 2013), but also more transient species (e.g., *T. trachurus* and *D. labrax*) and species commonly linked to sandy habitats (e.g., *P. minutus*). A few expected exceptions to these observations were noted for other species, such as the Common Dab *Limanda limanda* (Sørensen and Pedersen 2021). However, the assemblage enhancement owing to the reef group appeared to be greatest in the first survey period, immediately after reef deployments, with a marked decline observed ten months post-deployment.

These results align with previous findings reporting enhanced abundance and diversity around artificial reef structures compared to surrounding soft-bottom habitats (Bohnsack et al. 1994; Hylkema et al. 2020), particularly with reefs designed according to nature-inclusive principles, incorporating internal voids and higher structural complexity (Fabi 2002; Hunter and Sayer 2009; Fabi et al. 2015). Nevertheless, the discrepancy observed between the two post-deployment periods does not resolve the ongoing debate between attraction and production mechanisms at artificial reefs (Bohnsack and Sutherland 1985; Bortone 1998), highlighting the need for continued long-term monitoring to accurately evaluate artificial reef performance (Becker et al. 2022).

The 'attraction' of fishes is often presented in the literature as a hypothesis for heightened fish abundance around artificial reefs, in a dichotomy against the 'production' of biomass and fish abundance. The argument is that artificial reefs simply redistribute fish populations from natural habitats to artificial ones with no benefit to regional populations. Although secondary production enhancement is commonly cited as one of the primary goals of artificial reefs, the REFAS2 research site was not specifically designed to test this, and no dedicated procedures were included in this study to address this issue (Cresson et al. 2014; Roa-Ureta et al. 2019). Nevertheless, the high proportion of juveniles observed during our surveys, along with anecdotal evidence of Reef cubes® being used as a substrate for egg-layers such as squid, catsharks, and cuttlefish (pers. observ.), offers arguments that challenge the attraction-only hypothesis of artificial reef function (Folpp et al. 2020).

The actual contribution of these juvenile fishes to adult populations, in terms of survival and growth, was not assessed. Therefore, it remains unclear whether the juveniles observed around Reef cubes® actively navigated to the modules, guided by environmental cues such as sound, chemical signals, and visual stimuli (Montgomery et al. 2001, 2006; Atema et al. 2002), or whether they were instead passively recruited. The latter scenario appears more likely for species with limited home ranges (e.g., *P. minutus*) (Andersson et al. 2009). As a result, our findings should be viewed as preliminary indications of a potential nursery habitat role for Reef cubes® (Beck et al. 2001; Dahlgren et al. 2006). Regardless of the underlying mechanism, the higher presence of juveniles strongly suggests that Reef cubes® can benefit critical life stages of some fish species, whether as temporary shelter or as a more stable 'home reef' offering refuge from adverse marine conditions (Fowler 1999).

The performance of Reef cubes® in supporting juvenile fishes, and more broadly the nekton community, can be attributed to the physical characteristics of the nature-inclusive design purposely adopted. This approach aligns with several recognised optimal features for ecological enhancement (Blount et al. 2021; Pondella et al. 2022), and is consistent with other studies demonstrating the ecological benefits of incorporating nature-inclusive design in marine and coastal infrastructure (Lemoine et al. 2019; Sella et al. 2022; Berges et al. 2024), even when assessed within a relatively short time frame of one year (Mercader et al. 2017; Selfati et al. 2018; Hylkema et al. 2020).

Among these attributes, substrate complexity stands out as a key factor. Substrate complexity (as a reef design attribute) may play a key role in shaping nekton community dynamics (Johnson 2006; Rogers et al. 2018) and improving survival during fish early life stages (Scharf et al. 2006; Coker et al. 2012), as complexity was significantly higher in the sites with Reef cubes® compared to the pre-existing seabed. Structural complexity has been shown to reduce predation rates (Rogers et al. 2018), increase dietary input via richer epibenthic colonisation and enhanced plankton flux (Rilov and Benayahu 2002; Cresson et al. 2014; Rouse et al. 2020), or simply lower energy expenditure by providing hydrodynamically sheltered areas (Fowler 1999; Schwartzbach et al. 2020). In some of the Reef cube® modules, structural complexity was further enhanced by embedding smaller cubes within larger ones, with the explicit aim of increasing juvenile fish abundance (Hylkema et al. 2020), as several studies suggest a positive link between internal void complexity and fish diversity (Charbonnel et al. 2002; Taira et al. 2024).

Diel variation also influenced the response of both the nekton community and juvenile fishes to the artificial reef, with higher abundances generally recorded during daylight hours, although this trend varied by species. Notably, the highest juvenile fish counts were often recorded in the morning, particularly between 08:00 and 11:00 GMT (pers. observ.). These patterns likely reflect species-specific diel behaviours and predator–prey dynamics (Azzurro et al. 2007). During the day, many diurnal species (e.g., *T. minutus*) are more active and forage in the water column (Nickell and Sayer 1998), while others (e.g., *T. luscus*) may aggregate near the reef to conserve energy in response to strong tidal currents (Fowler 1999). At night, these diurnal species likely retreat into micro-habitats offering shelter and reduced visibility to predators, decreasing their detectability on video surveys (Taira et al. 2024). Conversely, nocturnal taxa such as *Scyliorhinus* spp., juvenile whiting, and squids were more frequently observed at nighttime, as their foraging activity primarily occurs under low-light conditions (Mergardt 1997; Sims et al. 2005). Moreover, the high squid detections at night may have been further amplified by light attraction to the BRUV unit (Martins et al. 2006). Interestingly, the Sand Goby, typically considered nocturnal, was more frequently recorded during the day, possibly due to stationary behaviour at nesting sites (Lehtonen and Lindström 2004). Similarly, European Seabass (*D. labrax*), a species known to feed primarily at night (Boujard et al. 1996), was more often observed during daylight, potentially due to comparable detectability factors. Overall, the lower fish detections at night may also reflect inherent limitations of BRUVs, which can underestimate nocturnal activity compared to more active techniques such as diver-based visual census.

The nekton community's response to the newly deployed reef habitat varied among species, reflecting differences in lifestyle and habitat use in association with distinct habitat types (Fabi et al. 2015). Among the most responsive taxa to reef proximity were gadoid fishes such as *M. merlangus*, *T. minutus*, and *T. luscus*. These species may have been foraging on epifaunal prey (Reubens et al. 2014) or using the reef structures as resting sites during strong tidal flows, especially at the juvenile stage (Fowler 1999). Their response is consistent with previous findings reporting increased abundance of gadoid species around novel reef structures in the Southwest UK (Bridger et al. 2022). A similar pattern was observed for more demersal species such as *P. minutus, Scyliorhinus* spp., the Dragonet *Callionymus lyra*, and *Sepia officinalis*, likely reflecting similar uses of the reefs for shelter and refuge (Sims et al. 2005), foraging (Secco et al. 2022), improved quantity and quality of nest sites (Lehtonen and Lindström 2004), recruitment (Andersson et al. 2009), and egg deposition (Guerra et al. 2016). Other more transient species, such as *D. labrax*, *T. trachurus*, and members of the Scombridae and Loliginidae families, were less closely associated with the Reef cube® modules but still appeared to be attracted to their presence, likely using them opportunistically as feeding grounds (Mavraki et al. 2021).

Despite the encouraging results, caution is warranted in interpreting these patterns given the short temporal scale and limited replication of this study. This work provides preliminary yet promising evidence that incorporating nature-inclusive design into artificial reefs can enhance nekton richness and juvenile abundance. Similar short-term benefits have been reported in other studies (e.g., Mercader et al. 2017; Hylkema et al. 2020), although conclusions remain constrained by the novelty of the approach. Nevertheless, to move beyond early-stage observations, clarify attraction versus production effects, and more robustly assess specific habitat functions (e.g., nursery roles), future research will require extended monitoring, broader spatial replication, and complementary methodologies.

## 6.5 CONCLUSION

The findings of this study highlight the potential of Reef cubes® to act as key habitats during critical life stages for some species, as suggested by the significant presence of juvenile fishes and the observed enhancement in nekton richness and abundance. The positive response of a wide range of taxa, including both resident and transient species, suggests that these nature-inclusive designed structures provide multifunctional habitat features that cater to diverse ecological needs, such as shelter, nesting, feeding, or reproduction. Despite the encouraging results, further investigation is required to strengthen the evidence base for Reef cubes® as functional habitat, particularly in terms of long-term demographic contributions to fish populations. Future research should delve more into the attraction versus production dilemma, assess ecological performance over broader temporal scales, and explore species-specific mechanisms driving habitat use, potentially through tagging, larval tracking, or analysis of growth and survival. Ultimately, this study contributes to the growing body of evidence supporting nature-inclusive design as a tool to enhance essential habitat functioning in human-dominated seascapes. By delivering ecological benefits with minimal seafloor footprint, these structures represent a promising approach for integrating biodiversity support within coastal development and marine spatial planning, while minimising spatial conflict among sea users.

## 6.6 ACKNOWLEDGEMENTS

We thank the staff at the Marine Business Technology Centre for their assistance with bathymetry surveys of the experimental site and the funders at the Centre for Environment, Fisheries and Aquaculture Science for investing in this research.

## 6.7 FUNDING

This work was funded by the UK government's Seafood Innovation Fund, administered by the Centre for Environment, Fisheries and Aquaculture Science—project code RD046—and by ARC marine Ltd, Truro, England TR1 1UT.

## REFERENCES

Airoldi L, Beck MW, Firth LB, Bugnot AB, Steinberg PD, Dafforn KA. 2021. Emerging solutions to return nature to the urban ocean. *Annual Review of Marine Science* 13: 445–477.

Andersson MH, Berggren M, Wilhelmsson D, Öhman MC. 2009. Epibenthic colonization of concrete and steel pilings in a cold-temperate embayment: A field experiment. *Helgoland Marine Research* 63: 249–260.

Atema J, Kingsford M, Gerlach G. 2002. Larval reef fish could use odour for detection, retention and orientation to reefs. *Marine Ecology Progress Series* 241: 151–160.

Azzurro E, Pais A, Consoli P, Andaloro F. 2007. Evaluating day–night changes in shallow Mediterranean rocky reef fish assemblages by visual census. *Marine Biology* 151: 2245–2253.

Beck MW, Heck KL, Able KW, Childers DL, Eggleston DB, Gillanders BM, Halpern B, Hays CG, Hoshino K, Minello TJ, Orth RJ, Sheridan PF, Weinstein MP. 2001. The identification, conservation, and management of estuarine and marine nurseries for fish and invertebrates. *BioScience* 51: 633.

Becker A, Taylor M, Folpp H, Lowry M. 2022. Revisiting an artificial reef after 10 years: What has changed and what remains the same? *Fisheries Research* 249: 106261.

Berges BJP, Knaap I van der, van Keeken OA, Reubens J, Winter HV. 2024. Strong site fidelity, residency and local behaviour of Atlantic cod (*Gadus morhua*) at two types of artificial reefs in an offshore wind farm. *Royal Society Open Science* 11: 240339.

Birbeck T, Doddrell J. 2020. Apparatus for an artificial reef and method (U.K Patent No. GB2569716B). United Kingdom Intellectual Property Office.
Bishop MJ, Mayer-Pinto M, Airoldi L, Firth LB, Morris RL, Loke LHL, Hawkins SJ, Naylor LA, Coleman RA, Chee SY, Dafforn KA. 2017. Effects of ocean sprawl on ecological connectivity: Impacts and solutions. *Journal of Experimental Marine Biology and Ecology* 492: 7–30.
Blount C, Komyakova V, Barnes L, Smith ML, Zhang D, Reeds K, McPhee D, Taylor MD, Macbeth W, Needham E. 2021. Using ecological evidence to refine approaches to deploying offshore artificial reefs for recreational fisheries. *Bulletin of Marine Science* 97: 665–698.
Bohnsack J, Harper D, McClellan D, Hulsbeck M. 1994. Effects of reef size on colonization and assemblage structure of fishes at artificial reefs off southeastern Florida, U.S.A. *Bulletin of Marine Science* 55: 796–823.
Bohnsack JA, Sutherland DL. 1985. Artificial reef research: A review with recommendations for future priorities. *Bulletin of Marine Science* 37: 11–39.
Bortone S. 2006. A perspective of artificial reef research: The past, present, and future. *Bulletin of Marine Science* 78: 1–8.
Bortone SA. 1998. Resolving the attraction-production dilemma in artificial reef research: Some yeas and nays. *Fisheries* 23: 6–10.
Boujard T, Jourdan M, Kentouri M, Divanach P. 1996. Diel feeding activity and the effect of time-restricted self-feeding on growth and feed conversion in European sea bass. *Aquaculture* 139: 117–127.
Bridger D, Attrill MJ, Davies BFR, Holmes LA, Cartwright A, Rees SE, Cabre LM, Sheehan EV. 2022. The restoration potential of offshore mussel farming on degraded seabed habitat. *Aquaculture, Fish and Fisheries* 2: 437–449.
Bugnot AB, Mayer-Pinto M, Airoldi L, Heery EC, Johnston EL, Critchley LP, Strain EMA, Morris RL, Loke LHL, Bishop MJ, Sheehan EV, Coleman RA, Dafforn KA. 2020. Current and projected global extent of marine built structures. *Nature Sustainability* 4: 33–41.
Bulleri F, Chapman MG. 2010. The introduction of coastal infrastructure as a driver of change in marine environments. *Journal of Applied Ecology* 47: 26–35.
Burt J, Bartholomew A, Usseglio P, Bauman A, Sale PF. 2009. Are artificial reefs surrogates of natural habitats for corals and fish in Dubai, United Arab Emirates? *Coral Reefs* 28: 663–675.
Cappo M, Speare P, De'ath G. 2004. Comparison of baited remote underwater video stations (BRUVS) and prawn (shrimp) trawls for assessments of fish biodiversity in inter-reefal areas of the Great Barrier Reef Marine Park. *Journal of Experimental Marine Biology and Ecology* 302: 123–152.
Chapman MG, Underwood AJ. 2011. Evaluation of ecological engineering of "armoured" shorelines to improve their value as habitat. *Journal of Experimental Marine Biology and Ecology* 400: 302–313.
Charbonnel E, Serre C, Ruitton S, Harmelin J-G, Jensen A. 2002. Effects of increased habitat complexity on fish assemblages associated with large artificial reef units (French Mediterranean coast). *ICES Journal of Marine Science* 59: S208–S213.
Coker DJ, Graham NAJ, Pratchett MS. 2012. Interactive effects of live coral and structural complexity on the recruitment of reef fishes. *Coral Reefs* 31: 919–927.
Cresson P, Ruitton S, Harmelin-Vivien M. 2014. Artificial reefs do increase secondary biomass production: Mechanisms evidenced by stable isotopes. *Marine Ecology Progress Series* 509: 15–26.
Dafforn KA, Glasby TM, Airoldi L, Rivero NK, Mayer-Pinto M, Johnston EL. 2015. Marine urbanization: An ecological framework for designing multifunctional artificial structures. *Frontiers in Ecology and the Environment* 13: 82–90.
Dahlgren C, Kellison G, Adams A, Gillanders B, Kendall M, Layman C, Ley J, Nagelkerken I, Serafy J. 2006. Marine nurseries and effective juvenile habitats: Concepts and applications. *Marine Ecology Progress Series* 312: 291–295.
Douvere F. 2008. The importance of marine spatial planning in advancing ecosystem-based sea use management. *Marine Policy* 32: 762–771.
Ellis DM, DeMartini E. 1995. Evaluation of a video camera technique for indexing abundances of juvenile pink snapper, *Pristipomoides filamentosus*, and other Hawaiian insular shelf fishes. *Fishery Bulletin* 93: 67–77.
Ellis JI, Schneider DC. 1997. Evaluation of a gradient sampling design for environmental impact assessment. *Environmental Monitoring and Assessment* 48: 157–172.
Evans AJ, Lawrence PJ, Natanzi AS, Moore PJ, Davies AJ, Crowe TP, McNally C, Thompson B, Dozier AE, Brooks PR. 2021. Replicating natural topography on marine artificial structures—A novel approach to eco-engineering. *Ecological Engineering* 160: 106144.
Fabi G. 2002. An assessment of biomass and diel activity of fish at an artificial reef (Adriatic sea) using a stationary hydroacoustic technique. *ICES Journal of Marine Science* 59: 411–420.

Fabi G, Scarcella G, Bortone S, Charbonnel E, Goutayer J, Hadded N, Lok A, Trommelen M, Spagnolo A. 2015. *Practical guidelines for the use of artificial reefs in the Mediterranean and the Black Sea*. Rome: Food and Agriculture Organisation of the United Nations—General Fisheries Commission for the Mediterranean.

Firth LB, Airoldi L, Bulleri F, Challinor S, Chee S-Y, Evans AJ, Hanley ME, Knights AM, O'Shaughnessy K, Thompson RC, Hawkins SJ. 2020. Greening of grey infrastructure should not be used as a Trojan horse to facilitate coastal development. *Journal of Applied Ecology* 57: 1762–1768.

Firth LB, Bone J, Bartholomew A, Bishop MJ, Bugnot A, Bulleri F, Chee S-Y, Claassens L, Dafforn KA, Fairchild TP, Hall AE, Hanley ME, Komyakova V, Lemasson AJ, Loke LHL, Mayer-Pinto M, Morris R, Naylor L, Perkins MJ, Pioch S, Porri F, O'Shaughnessy KA, Schaefer N, Strain EA, Toft JD, Waltham N, Aguilera M, Airoldi L, Bauer F, Brooks P, Burt J, Clubley C, Cordell JR, Espinosa F, Evans AJ, Farrugia-Drakard V, Froneman W, Griffin J, Hawkins SJ, Heery E, Herbert RJH, Jones E, Leung KMY, Moore P, Sempere-Valverde J, Sengupta D, Sheaves M, Swearer S, Thompson RC, Todd P, Knights AM. 2024. Coastal greening of grey infrastructure: An update on the state-of-the-art. *Proceedings of the Institution of Civil Engineers—Maritime Engineering* 1–69.

Firth WLB, Knights AM, Bridger D, Evans AJ, Mieszkowska N, Moore PJ, O'Connor NE, Sheehan EV, Hawkins RCT& SJ. 2016. *Ocean Sprawl: Challenges and opportunities for biodiversity management in a changing world*. Oceanography and Marine Biology CRC Press.

Folpp HR, Schilling HT, Clark GF, Lowry MB, Maslen B, Gregson M, Suthers IM. 2020. Artificial reefs increase fish abundance in habitat-limited estuaries. *Journal of Applied Ecology* 57: 1752–1761.

Fowler A. 1999. Age structure and diel activity of pouting on the Poole Bay artificial reef. *Journal of Fish Biology* 54: 944–954.

Froese R, Pauly D. 2025. FishBase. World Wide Web electronic publication. Available from: www.fishbase.org (December 2025).

Grove RS, Sonu CJ. 1983. Review of Japanese fisheries reef technology. Report 83-RD137. Southern California Edison Company, Rosemead, California.

Guerra Á, Hernández-Urcera J, Garci ME, Sestelo M, Regueira M, Gilcoto M, González ÁF. 2016. Spawning habitat selection by the common cuttlefish *Sepia officinalis* in the Cíes Islands (Northwest Spain). *Fisheries Research* 183: 44–54.

Heery EC, Bishop MJ, Critchley LP, Bugnot AB, Airoldi L, Mayer-Pinto M, Sheehan EV, Coleman RA, Loke LHL, Johnston EL, Komyakova V, Morris RL, Strain EMA, Naylor LA, Dafforn KA. 2017. Identifying the consequences of ocean sprawl for sedimentary habitats. *Journal of Experimental Marine Biology and Ecology* 492: 31–48.

Hermans A, Bos OG, Prusina I. 2020. *Nature-inclusive Design: A catalogue for offshore wind infrastructure: Technical report*. Witteveen+ Bos.

Hickling S, Matthews J, Murphy J. 2022. The suitability of alkali activated slag as a substrate for sessile epibenthos in Reef Cubes®. *Ecological Engineering* 174: 106471.

Hickling S, Murphy J, Cox C, Mynott S, Birbeck T, Wright S. 2023. Benthic invertebrate biodiversity enhancement with reef cubes®, evidenced by environmental DNA analysis of sediment samples. *Ecological Engineering* 195: 107064.

Hickling SJS, Murphy JAJ. 2022. Preliminary evidence of early Reef Cube ® enhancements to mobile benthic ecosystems. *IOP Conference Series: Materials Science and Engineering* 1245: 012015.

Hilbe JM. 2011. *Negative binomial regression*. Cambridge University Press.

Hunter WR, Sayer MDJ. 2009. The comparative effects of habitat complexity on faunal assemblages of northern temperate artificial and natural reefs. *ICES Journal of Marine Science* 66: 691–698.

Hylkema A, Debrot AO, Osinga R, Bron PS, Heesink DB, Izioka AK, Reid CB, Rippen JC, Treibitz T, Yuval M, Murk AJ. 2020. Fish assemblages of three common artificial reef designs during early colonization. *Ecological Engineering* 157: 105994.

Johnson DW. 2006. Predation, habitat complexity, and variation in density-dependent mortality of temperate reef fishes. *Ecology* 87: 1179–1188.

Kleiber C, Zeileis A. 2008. *Applied Econometrics with R*. Springer-Verlag. Available from https://doi.org/10.1007/978-0-387-77318-6. R package available at https://CRAN.R-project.org/package=AER

Lehtonen T, Lindström K. 2004. Changes in sexual selection resulting from novel habitat use in the sand goby. *Oikos* 104: 327–335.

Lemoine HR, Paxton AB, Anisfeld SC, Rosemond RC, Peterson CH. 2019. Selecting the optimal artificial reefs to achieve fish habitat enhancement goals. *Biological Conservation* 238: 108200.

Long JA. 2024. Interactions: Comprehensive, user-friendly toolkit for probing interactions (R package version 1.2.0). Available from: https://doi.org/10.32614/CRAN.package.interactions

Lüdecke D. 2013. sjPlot: Data visualization for statistics in social science (R package version 2.8.17). Available from: https://CRAN.R-project.org/package=sjPlot.
Martins R, Perez J, Perez A. 2006. Cephalopods and fish attracted by night lights in coastal shallow-waters, off Southern Brazil, with the description of squid and fish behavior. *Revista de Etologia* 8.
Mavraki N, Degraer S, Vanaverbeke J. 2021. Offshore wind farms and the attraction–production hypothesis: Insights from a combination of stomach content and stable isotope analyses. *Hydrobiologia* 848: 1639–1657.
McCullagh P, Nelder JA. 2019. *Generalized linear models*. Routledge.
Mercader M, Mercière A, Saragoni G, Cheminée A, Crec'hriou R, Pastor J, Rider M, Dubas R, Lecaillon G, Boissery P, Lenfant P. 2017. Small artificial habitats to enhance the nursery function for juvenile fish in a large commercial port of the Mediterranean. *Ecological Engineering* 105: 78–86.
Mergardt N. 1997. Diel pattern of food intake in whiting (Merlangius merlangus) investigated from the weight of partly digested food particles in the stomach and laboratory determined particle decay functions. *ICES Journal of Marine Science* 54: 226–242.
Methratta ET. 2020. Monitoring fisheries resources at offshore wind farms: BACI vs. BAG designs. *ICES Journal of Marine Science* 77: 890–900.
Meyer C. 2009. The greening of the concrete industry. *Cement and Concrete Composites* 31: 601–605.
Montgomery JC, Jeffs A, Simpson SD, Meekan M, Tindle C. 2006. *Sound as an orientation cue for the pelagic larvae of reef fishes and decapod crustaceans*. Advances in Marine Biology Elsevier. p. 143–196.
Montgomery JC, Tolimieri N, Haine OS. 2001. Active habitat selection by pre-settlement reef fishes. *Fish and Fisheries* 2: 261–277.
Moschella PS, Abbiati M, Åberg P, Airoldi L, Anderson JM, Bacchiocchi F, Bulleri F, Dinesen GE, Frost M, Gacia E, Granhag L, Jonsson PR, Satta MP, Sundelöf A, Thompson RC, Hawkins SJ. 2005. Low-crested coastal defence structures as artificial habitats for marine life: Using ecological criteria in design. *Coastal Engineering* 52: 1053–1071.
Naylor LA, Kippen H, Coombes MA, Horton B, MacArthur M, Jackson N. 2017. Greening the grey: A framework for Integrated Green Grey Infrastructure (IGGI). Available from: from http://eprints.gla.ac.uk/15067 2
Nickell LA, Sayer MDJ. 1998. Occurrence and activity of mobile macrofauna on a sublittoral reef: Diel and seasonal variation. *Journal of the Marine Biological Association of the United Kingdom* 78: 1061–1082.
Palmer R, Tritton Limited Development and Engineering Consultants. 1996. *History of coastal engineering in Great Britain*. 25th International Coastal Engineering Conference. p. 214–274.
Perkol-Finkel S, Shashar N, Benayahu Y. 2006. Can artificial reefs mimic natural reef communities? The roles of structural features and age. *Marine Environmental Research* 61: 121–135.
Pondella DJ, Claisse JT, Williams CM. 2022. Theory, practice, and design criteria for utilizing artificial reefs to increase production of marine fishes. *Frontiers in Marine Science* 9: 983253.
Provis JL. 2018. Alkali-activated materials. *Cement and Concrete Research* 114: 40–48.
Reubens JT, Braeckman U, Vanaverbeke J, Van Colen C, Degraer S, Vincx M. 2013. Aggregation at windmill artificial reefs: CPUE of Atlantic cod (*Gadus morhua*) and pouting (*Trisopterus luscus*) at different habitats in the Belgian part of the North Sea. *Fisheries Research* 139: 28–34.
Reubens JT, Degraer S, Vincx M. 2014. The ecology of benthopelagic fishes at offshore wind farms: A synthesis of 4 years of research. *Hydrobiologia* 727: 121–136.
Rilov G, Benayahu Y. 2002. Rehabilitation of coral reef-fish communities: The importance of artificial-reef relief to recruitment rates. *Bulletin of Marine Science* 70: 185–197.
Roa-Ureta RH, Santos MN, Leitão F. 2019. Modelling long-term fisheries data to resolve the attraction versus production dilemma of artificial reefs. *Ecological Modelling* 407: 108727.
Rogers A, Blanchard JL, Newman SP, Dryden CS, Mumby PJ. 2018. High refuge availability on coral reefs increases the vulnerability of reef-associated predators to overexploitation. *Ecology* 99: 450–463.
Rouse S, Porter JS, Wilding TA. 2020. Artificial reef design affects benthic secondary productivity and provision of functional habitat. *Ecology and Evolution* 10: 2122–2130.
Sánchez-Caballero CA, Borges-Souza JM, Abelson A. 2021. Can wrecks serve as exploitable surrogate habitats for degraded natural reefs? *Marine Environmental Research* 169: 105399.
Scharf FS, Manderson JP, Fabrizio MC. 2006. The effects of seafloor habitat complexity on survival of juvenile fishes: Species-specific interactions with structural refuge. *Journal of Experimental Marine Biology and Ecology* 335: 167–176.
Schupp MF, Bocci M, Depellegrin D, Kafas A, Kyriazi Z, Lukic I, Schultz-Zehden A, Krause G, Onyango V, Buck BH. 2019. Toward a common understanding of ocean multi-use. *Frontiers in Marine Science* 6: 165.

Schwartzbach A, Behrens J, Svendsen J. 2020. Atlantic cod *Gadus morhua* save energy on stone reefs: Implications for the attraction versus production debate in relation to reefs. *Marine Ecology Progress Series* 635: 81–87.

Scyphers S, Powers S, Heck K. 2015. Ecological value of submerged breakwaters for habitat enhancement on a residential scale. *Environmental Management* 55: 383–391.

Seafood Innovation Fund. 2020. [Internet]. Available from: https://www.seafoodinnovation.fund/projects/reef-enhancement-for-aquaculture-sites-refas-phase-2-rd046/

Secco S, Scalici M, Lomiri S, Nonnis O, Ausili A, Tomassetti P. 2022. Effect of artificial reefs on the alimentary strategies of two Mediterranean Sea teleosts. *Sustainability* 15: 11.

Selfati M, El Ouamari N, Lenfant P, Fontcuberta A, Lecaillon G, Mesfioui A, Boissery P, Bazairi H. 2018. Promoting restoration of fish communities using artificial habitats in coastal marinas. *Biological Conservation* 219: 89–95.

Sella I, Hadary T, Rella AJ, Riegl B, Swack D, Perkol-Finkel S. 2022. Design, production, and validation of the biological and structural performance of an ecologically engineered concrete block mattress: A Nature-Inclusive Design for shoreline and offshore construction. *Integrated Environmental Assessment and Management* 18: 148–162.

Sims DW, Southall EJ, Wearmouth VJ, Hutchinson N, Budd GC, Morritt D. 2005. Refuging behaviour in the nursehound *Scyliorhinus stellaris* (chondrichthyes: elasmobranchii): Preliminary evidence from acoustic telemetry. *Journal of the Marine Biological Association of the United Kingdom* 85: 1137–1140.

Sørensen OJR, Pedersen T. 2021. Effects of season, bottom substrate and population dynamics on fish communities in shallow subarctic northeast Atlantic waters. *Journal of Sea Research* 178: 102136.

Taira D, Mark RYY, Hsiung AR, Jaafar Z, Todd PA. 2024. Fish responses to manipulated microhabitat complexity in urbanised shorelines. *Journal of Applied Ecology* 61: 1095–1108.

Todd PA, Heery EC, Loke LHL, Thurstan RH, Kotze DJ, Swan C. 2019. Towards an urban marine ecology: Characterizing the drivers, patterns and processes of marine ecosystems in coastal cities. *Oikos* 128: 1215–1242.

Venables WN, Ripley BD. 2002. *Modern applied statistics with S*. 4th edn. Springer. ISBN 0-387-95457-0.

Wang Y, Naumann U, Eddelbuettel D, Wilshire J, Warton D. 2022. mvabund: Statistical methods for analysing multivariate abundance data (R package version 4.2.1). Available from: https://CRAN.R-project.org/package=mvabund.

# 7 New Insights into the Use of Baited Video and Baited Traps for Assessing the Structure and Function of Fish and Mobile Invertebrate Communities Associated with Artificial Reefs

*Zhongxin Wu*, Liwei Si, James R. Tweedley, Neil R. Loneragan, Yanchao Zhang, Zheyang Ou, and Tao Tian*
* Corresponding author: wuzhongxin@dlou.edu.cn

## 7.1 INTRODUCTION

The data in this chapter were originally published by Zhang et al. (2024) in the *Journal of Experimental Marine Biology and Ecology* and were subsequently presented at the 12th CARAH (International Conference on Artificial Reefs and Related Aquatic Habitats). Conversations at that conference stimulated new ideas among the authors and ways of reconsidering the findings of that data. As a result of these reflections, new statistical analyses have been conducted, all figures redrawn, and the text has been extensively modified from the original publication. In particular, the new analyses shed clearer light on the different sampling methods (i.e., Baited Remote Underwater Video (BRUV) and baited traps) and the functional trait modalities of the species they recorded.

The depletion of fishery resources and degradation of coastal habitats have become global challenges that need to be urgently addressed, particularly in the context of United Nations Sustainable Development Goal 14 (UN, 2015), which calls for the conservation and sustainable use of aquatic resources (Ye and Gutierrez, 2017; Mao et al., 2019). In China, marine ranching, which integrates fisheries, aquaculture, habitat enhancement, and other marine resource management strategies, was proposed as a potential solution to restore depleted fisheries (Yu and Zhang, 2020; Lorenzen et al., 2021). As it has in South Korea (Lee and Rahman, 2018) and Japan (Kitada, 2020), marine ranching has expanded in China, with 189 national-level demonstration zones built or under construction by 2024 (Yuan et al., 2024). These ranches produce important ecological and economic benefits, estimated at 94 and 178 billion Yuan (approximately USD 13.2 billion and 25.0 billion, respectively) in 2022 (Zhang et al., 2022; Yuan et al., 2024).

One of the core components of marine ranching systems is habitat enhancement, with an emphasis on deploying artificial reefs (Wu et al., 2019; Lorenzen et al., 2021; Dirhamsyah et al., 2024). Globally, artificial reefs are presumed to restore the productivity of degraded ecosystems (Becker et al., 2018; Lima et al., 2019; Vivier et al., 2021). In South Korea, artificial reefs were incorporated into a comprehensive Fisheries Rebuilding Plan initiated in 2005, and by 2015, the area of these reefs exceeded 2,070 km$^2$ (Lee and Rahman, 2018). China is among the earliest countries to develop artificial reef engineering, with historical records indicating the use of reefs in freshwater

DOI: 10.1201/9781003676775-7 

systems as early as the Spring and Autumn and the Warring States periods from 770 to 221 BC (Pan et al., 2022). However, the large-scale construction and systematic development of artificial reefs in marine environments progressed slowly over the following centuries, with the first formally designed and government-sponsored artificial reef deployed in 1979. Since the implementation of the "China Outline of Action Plan for the Conservation of Aquatic Biological Resources" in 2006 (MOA, 2006), artificial reef construction and deployment have entered a rapid stage of development. By 2017, artificial reefs in China's coastal waters covered an area of 853 $km^2$, with a total volume of about 50 million $m^3$ (Zhang et al., 2023).

Monitoring and evaluating the ecological effects of artificial reefs is crucial for guiding their responsible construction and deployment (Seaman, 2000; Becker et al., 2018). Most studies have focused on the roles these structures might play in either attracting or enhancing the production of fish and macroinvertebrates by comparing the faunal assemblages on the artificial reefs to those in nearby natural habitats (Pickering and Whitmarsh, 1997; Brickhill et al., 2005; Streich et al., 2017; Wu et al., 2019; Hammond et al., 2020). In addition to surveying community structure using abundance and/or biomass, functional diversity has emerged as a promising method. It has been used on anthropogenically-derived structures, such as oil and gas pipelines (Madgett et al., 2023) and is particularly useful for assessing the effect of artificial reefs. The addition of these structures to "unvegetated" habitats can increase the number of niches in the ecosystem and, consequently, alter the functional ecology of the local environment (Schroeter et al., 2015).

To understand any ecosystem and establish baseline measures of ecosystem state, an effective sampling method and design are needed. A wide range of methods have been used to monitor fish assemblages associated with artificial reefs around the world, including: (i) traditional fishing gears, for example, nets, traps, and longline; (ii) underwater visual census (UVC) and video equipment, such as BRUVs; and (iii) acoustics (Lima et al., 2019; Ramm et al., 2021; Duarte et al., 2024). These sampling methods can be divided into two main categories: destructive and non-destructive (or observational) techniques (Murphy and Jenkins, 2010; Lima et al., 2019).

Destructive fishing gears typically retain a sample for biological measurements, but they may have other consequences. For example, gill nets, which are widely used in studies of artificial reefs (Lima et al., 2019), are prone to becoming entangled on and in the structures, preventing their removal but continuing to catch fish (i.e., ghost fishing; Matsuoka et al., 2005). Similarly, bottom trawls may also become snagged on artificial structures, so they are typically operated at a set distance away from the reef, thus limiting their effectiveness in obtaining representative catches of resident fish.

Among the observational, non-destructive methods used for studying the faunas associated with artificial reefs, UVC has been one of the most utilized sampling techniques due to its adaptability to varied habitats, including those with complex topography. Surveys using UVC also allow for replicate sampling to be repeated over short time intervals and at low cost (Lima et al., 2019; Ramm et al., 2021). However, UVC surveys are limited by water depth and may be biased due to the reef obstructing the diver's view and the behavioral responses of fish to divers (Lowry et al., 2012; Ghazilou et al., 2019; La Manna et al., 2021). Alternative observational methods to UVC include various video systems (Patterson et al., 2009; Bernard and Gotz, 2012; Ajemian et al., 2015; Asher et al., 2019; Goetze et al., 2019). The benefits of BRUVs include their relatively low cost, ease of replication, permanent and reviewable records, and suitability for a variety of habitat types and water depths. With improvements in image quality, storage, and battery capacity (Mallet and Pelletier, 2014), BRUVs have been increasingly used to study fish assemblages (Lowry et al., 2012; Barker and Cowan, 2017; Becker et al., 2017; Florisson et al., 2018; Balaji et al., 2019; Bicknell et al., 2019) and behavior (Krispyn et al., 2025) in a variety of artificial habitats.

In China, methods commonly used to investigate fish and invertebrate assemblages on artificial reefs include trammel nets, beam trawls, fish traps, and UVC (Wang et al., 2015; Jiao et al., 2016; Wu et al., 2019). Fish traps have been effectively used globally to estimate the relative abundance of reef-associated fish species in various habitats (Harvey et al., 2012; Bacheler et al., 2013, 2017; Wakefield et al., 2013; Bosch et al., 2017; Finucci et al., 2019). Their simple structure, ease of operation, low cost, and efficient handling of catch make traps a commonly used fishing gear in the coastal waters of the northern Yellow Sea (Liu et al., 2016).

This study describes the first application of BRUVs in China to determine the fish and mobile invertebrates associated with artificial reefs and an adjacent natural habitat (i.e., bare substratum, which serves as a control or reference site) off Dachangshan Island in the Northern Yellow Sea (see Figure 7.1a). We hypothesize that artificial reef sites will support a higher abundance and diversity of fauna than the control area due to the provision of habitat, and that BRUVs will record a broader range of species than traps, particularly those that are pelagic. The data obtained from BRUVs sampling have been compared with those obtained from traditional baited fish traps (hereafter traps) deployed in the same habitats at the same time. Aspects of community structure were determined using traditional taxonomic metrics, such as the number of species, abundance (number of individuals), Shannon index (H′), and faunal composition, as well as functional metrics of diversity, including richness, evenness, and divergence, and community weighted means (CWMs) based on functional traits that take into account the use of space, food resources, and swimming abilities of fish. These metrics were compared between sampling methods to determine whether BRUVs are an

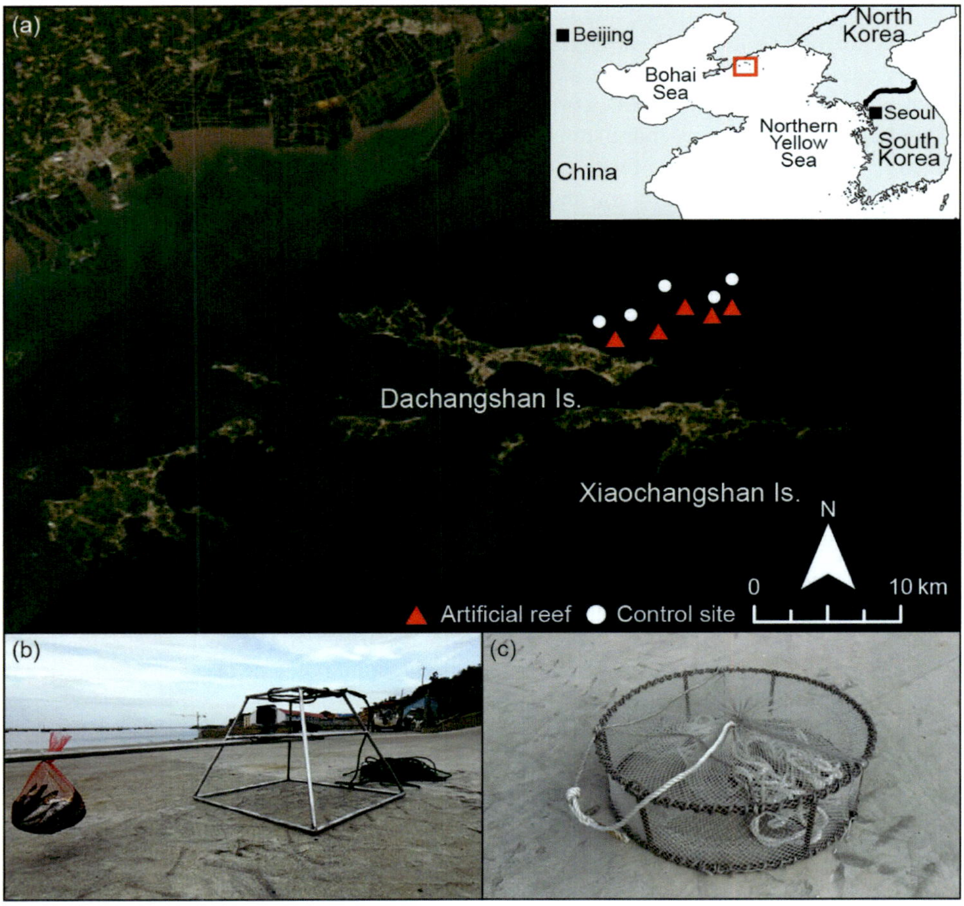


FIGURE 7.1 (a) Map showing the location of the study area (red square in inset) where (b) BRUVs (base = 50 × 50 cm, total height = 40 cm, bait arm = 60 cm) and (c) fish traps (diameter = 60 cm, height = 16 cm, mesh = 15 mm) were deployed to sample fish and mobile invertebrates on artificial reefs and control sites (bare substratum) in the waters off Dachangshan Island, Northern Yellow Sea, China. World Imagery basemap in (a) provided by Earthstar Geographics.

effective method for evaluating the mobile fauna of artificial reefs. Biomass data were also used to estimate these metrics for the data from traps, which were not possible to collect for BRUVs.

## 7.2 MATERIAL AND METHODS

### 7.2.1 Study Area

Sampling was conducted in the coastal waters near Dachangshan Island (39°16′ N, 122°34′ E; Figure 7.1a), the largest island in the Changshan Archipelago off the Liaodong Peninsula (Liaoning Province, China) in the Northern Yellow Sea. The Changshan Archipelago Sea covers an area of approximately 4,030 $km^2$, with water depths ranging from 5 to 20 m in the study area. The sediment consists of sand and mud, with seagrasses (mainly *Zostera marina*, *Zostera japonica*, and *Zostera caespitosa*) present in shallow subtidal waters (2–5 m deep). Tides are semidiurnal, with a tidal range of 2–4 m and a current velocity of approximately 1.5–2.5 m $s^{-1}$ (Xu et al., 2021). Water quality around the island is generally good, but it may decline in August due to rainfall and the resulting input of sediments, nutrients, and heavy metals from the surrounding terrestrial environments, leading to increased phytoplankton and macroalgae production (Zhang et al., 2018). The study area is a core area of the Haiyang Island Fishing Ground, which is well-known for producing high-quality seafood (Wang et al., 2006). Seafood production in the region reached 679,220 in 2022 and mainly comprises scallops, finfish, sea cucumbers, sea urchins, and abalone. Song et al. (2010) used a Pressure-State-Response framework to model the biomass of fish species in the area between 1987/8 and 2006/7. They found that fishing pressure increased during this time, and the biomass of long-lived fish decreased, leading them to conclude that the system was in an unhealthy state.

To help ameliorate over-exploitation and enhance fishery stocks, the Dachangshan artificial habitat demonstration zone was constructed in November 2013 (Zhang et al., 2018). A total of 3,000 concrete reef modules (cubic modules, each with 1.2 m sides) were deployed. Additionally, 1,700 units of shell reefs were also placed into 19 reef groups over an area of 3.64 $km^2$ within the artificial reef zone. The shell reefs were low-profile, polygonal concrete structures with an external diameter of 2.4 m, a height of 0.75 m, and a weight of 2.5 t. They were designed to provide habitat for the sea cucumber *Apostichopus japonicus* and the sea urchin *Strongylocentrotus nudus* (Zhao et al., 2019).

### 7.2.2 Sampling Methodology

Five sites in the artificial reef zone and five adjacent control sites were sampled seven times (each time referred to as a "sampling occasion") between July and November 2020 (i.e., 12 July, 30 August, 25 September, 10 and 25 October, and 10 and 24 November) in waters between 7.2 and 24.2 m deep off Dachangshan Island (Figure 7.1a). BRUVs and baited fish traps were deployed at each site and on each sampling occasion. Traps were deployed 30 minutes after BRUVs, as the capture and extraction of fauna from the traps could potentially reduce the abundance of organisms recorded on the BRUVs. All sampling was conducted on the slack tide during daylight when water movement was minimal and visibility was optimal.

Each BRUV unit comprised a single GoPro Hero 7 Black action camera with an ultra-wide-angle lens mounted on the stainless-steel frame (bottom: 50 × 50 cm, camera platform: 25 × 25 cm, total height: 40 cm, camera platform height: 25 cm) (Figure 7.1b). The camera was equipped with a waterproof housing rated to 40 m depth and recorded video footage using a resolution of 1080p at 60 fps. A bait arm extended 60 cm from the camera and had a mesh bait bag with 500 g of crushed Japanese Anchovy, *Engraulis japonicus*. Each BRUV was deployed for 60 mins, a duration commonly used to sample fauna present on artificial reefs in temperate waters elsewhere (Florisson et al., 2018; Ou et al., 2021).

Each BRUV video was reviewed, and the MaxN (i.e., the maximum number of individuals of a given species seen in any one video frame) was recorded for each species. Note that while MaxN is the most widely used method of estimating abundance from underwater video (Whitmarsh et al., 2017; Ou et al., 2021), it is a conservative index of relative abundance (Schobernd et al., 2013). All taxa were identified to the taxonomic level of species.

The traps were 60 cm in diameter, 16 cm high, and constructed of 15 mm mesh (Figure 7.1c). Four entrance funnels (26 cm × 11 cm) allowed fauna to enter but inhibited exit from the trap. Each trap was baited with 500 g of crushed *E. japonicus* (i.e., the same type and quantity of bait as used in the BRUV) and deployed for 3 hours. All organisms captured in the traps were euthanized in an ice slurry, placed in labeled plastic bags, and transported to the laboratory, where they were identified to species, counted, measured to the closest 1 mm in length, and weighed to the nearest 1 g. Six environmental variables were recorded at each site on each sampling occasion. Of these, four were measured using a YSI Pro Plus multiparameter meter: water temperature (°C), dissolved oxygen concentration (mg/L), salinity, and pH. Transparency (cm) was measured using a Secchi disk, and water depth (m) with an SM-5 A depth sounder.

### 7.2.3 Statistical Analyses

#### 7.2.3.1 Taxonomic Diversity and Composition

The abundances of each species, recorded in each of the 140 samples, were combined into a single data matrix and subjected to the DIVERSE routine to calculate the number of species, total number of individuals per deployment: that is, total MaxN for BRUVs (1 hour) and total catch rate for traps (for 3 hours), and the Shannon index (i.e., $H' = -\sum Pi \ln(Pi)$). The total biomass in each sample was calculated only for the trap data.

The data for the four variables were used to construct a draftsman plot to visually assess skewness and determine which type of transformation would best mitigate it (Clarke and Gorley, 2015). This plot demonstrated that biomass required a fourth-root transformation. The values for each variable were used to construct a Euclidean distance matrix and subjected to three-way Permutational Analysis of Variance (PERMANOVA; Anderson et al., 2008) to determine if the values for a variable differed significantly between method (two levels; BRUVs and traps), habitat (two levels; artificial reef and control) and sampling occasion (seven levels; July, August, September, early October, late October, early November and late November). All factors were treated as fixed in these and all subsequent tests. The null hypothesis of no significant difference among *a priori* groups was rejected if the significance level ($P$) was $\leq 0.05$. The relative influence of each term in the model was determined by calculating the percentage mean squares relative to the total mean squares for each analysis. Note that biomass was subjected to a two-way analysis (habitat and sampling occasion) as this variable could only be calculated from the catches in traps. Unless otherwise stated, PRIMER v7 (Clarke and Gorley, 2015) with the PERMANOVA+ add-on (Anderson et al. 2008) was used to conduct all statistical analyses.

#### 7.2.3.2 Multivariate Analyses

The abundance of each species in each sample was converted to percentage composition data and square-root transformed to balance the contributions of species with high and low values. The biomass of each species in each sample was dispersion-weighted and square-root transformed. The transformed abundance and biomass data were then used to construct separate Bray–Curtis resemblance matrices. Each matrix was subjected to a three-way PERMANOVA comparing method, habitat, and sampling occasion. Trends among (i) method and/or habitat and (ii) sampling occasion were visualized using metric multidimensional scaling (mMDS) ordination plots produced by the Bootstrap Averages routine (Clarke et al., 2014a). Superimposed on each plot were the group averages (i.e., the average of the bootstrapped averages) and the associated, smoothed, and marginally bias-corrected 95% confidence regions. The effect of interactions involving sampling occasion was simplified and displayed using centroid non-metric multidimensional scaling (nMDS) ordination plots. These plots were produced using a distance-among-centroids matrix, which creates averages in the "Bray–Curtis space" from points representing the levels of each factor in the interaction (Lek et al., 2011). Superimposed onto the centroid nMDS plot are vectors for species whose contribution changed in a linear direction: that is, Pearson correlation > 0.6 (abundance data) and 0.5 (biomass data) relative to the nMDS axes.

Two shade plots were constructed and used to visualize the trends exhibited by the transformed average assemblage composition data (i) percentage contribution by abundance (BRUVs and traps) and (ii) total biomass (traps only) of each species in each combination of method, habitat, and sampling occasion. On the plots, white space indicates that a species was not recorded, while the depth of shading from gray to black is linearly proportional to the contribution/biomass of that species (Clarke et al., 2014b). Species (y-axis) were seriated using an Index of Association resemblance matrix. Method, habitat, and sampling occasion combinations (x-axis) were arranged in sequential monthly order within each method and habitat combination.

### 7.2.3.3 Functional Diversity

Nine functional traits, comprising 29 trait modalities and two continuous variables, cover two major categories of function: Feeding (with four traits)—that is, diet, mouth position, mouth size, and trophic level; and Locomotion (with five traits)—that is, habitat type, residency, relationship with the reef (Nakamura, 1985), body shape, and maximum recorded total length. Each fish species was classified into trait modalities (Table 7.1) based on data from Froese and Pauly (2022) and Zhang et al. (2024).

Trait modalities are as follows: Diet: Benthivore (B); Benthivore/Piscivore (B/Pi); Piscivore (Pi); Planktivore/Benthivore (Pl/B). Mouth position: Ventral (V); Inferior (I); Terminal (T); Superior (S). Relative mouth size: Ends before eye (Be); Ends at the anterior edge of the eye (A); Ends at the center of the eye (C); Ends at the posterior of the eye (P); Ends behind the eye (Bh). Trophic level: Continuous variable (range = 3.1 to 4.5). Habitat type: Continental shelf demersal fish (CD); Continental shelf benthopelagic fish (CBD); Continental shelf pelagic-neritic fish (CPN); Continental shelf reef-associated fish (CRA); Oceanic bathydemersal fish (OMP). Residency: Resident; Migrant. Relationship with reefs: "In the reef"—fish closely associated with reefs and their structure, occupying holes, crevasses, and narrow openings; "On the reef"—fish that swim around the reefs while remaining near the bottom; "Above the reef"—fish that hover above the reef while remaining in the middle and upper parts of the water column. Body shape: Fusiform; Compressiform; Depressiform; Anguilliform; Asymmetric; Band-shaped. Max. length (cm): Continuous variable (range = 11 to 150 cm).

Three univariate measures of functional diversity were evaluated by calculating using the "FD" package (Laliberté et al., 2014) in R (version 3.6.1): (i) Functional richness—that is, the amount of niche space occupied by the species within a community (Mason et al., 2005); (ii) Functional divergence—that is, the degree to which the abundance distribution in niche space maximizes divergence in functional characters within the community. High values indicate a high degree of niche differentiation and therefore low resource competition (Cronin-O'Reilly et al., 2025); and (iii) Functional evenness—that is, the extent to which the individuals in a community are distributed across niche space. To provide a more robust measure of functional diversity, these metrics were calculated using the abundance of each species for each combination of method, habitat, and sampling occasion. Differences in the values of the three functional measures were tested using a two-way PERMANOVA comparing method and habitat, using the seven sampling occasions as replicates. A draftsman's plot indicated that no transformations were required.

The FD package was also used to calculate the CWMs of each trait modality. The CWM represents the proportion of individuals expressing each trait modality, or for the two continuous variables (i.e., trophic level and maximum length), the weighted average of trait values based on the relative abundances of each species. The continuous data were standardized against the maximum possible value (i.e., the greatest trophic level and largest maximum size) to place all values on the same scale:

**TABLE 7.1**
**Generalized body shape and trait modalities of 17 fish species recorded on BRUVs and traps in waters off Dachangshan Island, Northern Yellow Sea, China. The number in parentheses indicates the number of traits in each category and modalities in each trait.**

| | Categories | Feeding (4) | | | | Locomotion (5) | | | | |
|---|---|---|---|---|---|---|---|---|---|---|
| **Species and generalized body shape** | **Functional traits** | **Diet (4)** | **Mouth position (4)** | **Relative mouth size (5)** | **Trophic level** | **Habitat type (5)** | **Residency (2)** | **Relationship with reefs (3)** | **Body shape (6)** | **Max. length (cm)** |
| *Hexagrammos otakii* |  | Pi/B | T | Be | 3.8 | CRA | Resident | In the reef | Compressiform | 57 |
| *Sebastes schlegelii* |  | B/Pi | T | A | 3.8 | CRA | Resident | In the reef | Compressiform | 65 |
| *Scomberomorus niphonius* |  | Pi | T | A | 4.5 | CPN | Migrant | Above the reef | Fusiform | 113 |
| *Platichthys bicoloratus* |  | B | T | Be | 3.7 | CD | Migrant | On the reef | Asymmetric | 50 |
| *Pennahia argentatus* |  | B/Pi | T | C | 4.1 | CBD | Migrant | On the reef | Compressiform | 40 |
| *Conger myriaster* |  | B/Pi | I | P | 4.0 | CD | Migrant | In the reef | Anguilliform | 100 |
| *Acanthogobius flavimanus* |  | B | I | A | 3.4 | CD | Resident | On the reef | Band shape | 30 |
| *Sillago sihama* |  | B | T | Be | 3.3 | CD | Migrant | On the reef | Compressiform | 31 |
| *Hexagrammos agrammus* |  | B | T | Be | 3.3 | CRA | Resident | In the reef | Compressiform | 30 |
| *Pterogobius zacalles* |  | B | T | Bf | 3.1 | CD | Resident | On the reef | Band shape | 16 |

*(Continued)*

**TABLE 7.1** *(Continued)*

**Generalized body shape and trait modalities of 17 fish species recorded on BRUVs and traps in waters off Dachangshan Island, Northern Yellow Sea, China. The number in parentheses indicates the number of traits in each category and modalities in each trait.**

| Categories | | Feeding (4) | | | | Locomotion (5) | | | | |
|---|---|---|---|---|---|---|---|---|---|---|
| Species and generalized body shape | Functional traits | Diet (4) | Mouth position (4) | Relative mouth size (5) | Trophic level | Habitat type (5) | Residency (2) | Relationship with reefs (3) | Body shape (6) | Max. length (cm) |
| *Okamejei kenojei* |  | B/Pi | V | Be | 3.4 | CD | Migrant | On the reef | Depressiform | 57 |
| *Thamnaconus modestus* |  | B | T | Be | 3.4 | CD | Migrant | On the reef | Compressiform | 36 |
| *Tridentiger trigonocephalus* |  | B | T | P | 3.4 | CD | Resident | In the reef | Anguilliform | 11 |
| *Chaeturichthys stigmatias* |  | B | T | Be | 3.8 | CD | Resident | On the reef | Anguilliform | 28 |
| *Amblychaeturichthys hexanema* |  | B | T | Be | 3.4 | CD | Resident | On the reef | Anguilliform | 17 |
| *Lophius litulon* |  | Pi | S | Bh | 4.1 | OMP | Migrant | On the reef | Depressiform | 150 |
| *Enedrias nebulosus* |  | B | T | C | 3.2 | CD | Resident | In the reef | Band shape | 30 |

0 (not recorded) to 100 (all individuals possess a particular trait modality for a functional trait). The CWM values for all trait modalities were used to create a Bray–Curtis similarity matrix. This was, in turn, subjected to a two-way PERMANOVA and used to produce a bootstrapped mMDS plot and a Canonical Analysis of Principal Coordinates (CAP; Anderson et al., 2008) plot. Superimposed on the latter plot were vectors for trait modalities with a Pearson correlation > 0.5 relative to the CAP axes. Finally, the CWM data were used to produce a shade plot. Trait modalities (y-axis) were seriated using an index of association resemblance matrix. Method, habitat, and sampling occasion combinations (x-axis) were arranged in sequential monthly order within each method and habitat combination.

## 7.3 RESULTS

### 7.3.1 Environmental Conditions

Water temperatures and salinities were almost identical at the artificial reefs and control sites (Figure 7.2a, b). Mean temperature varied by ~15 °C over the seven sampling occasions, reaching

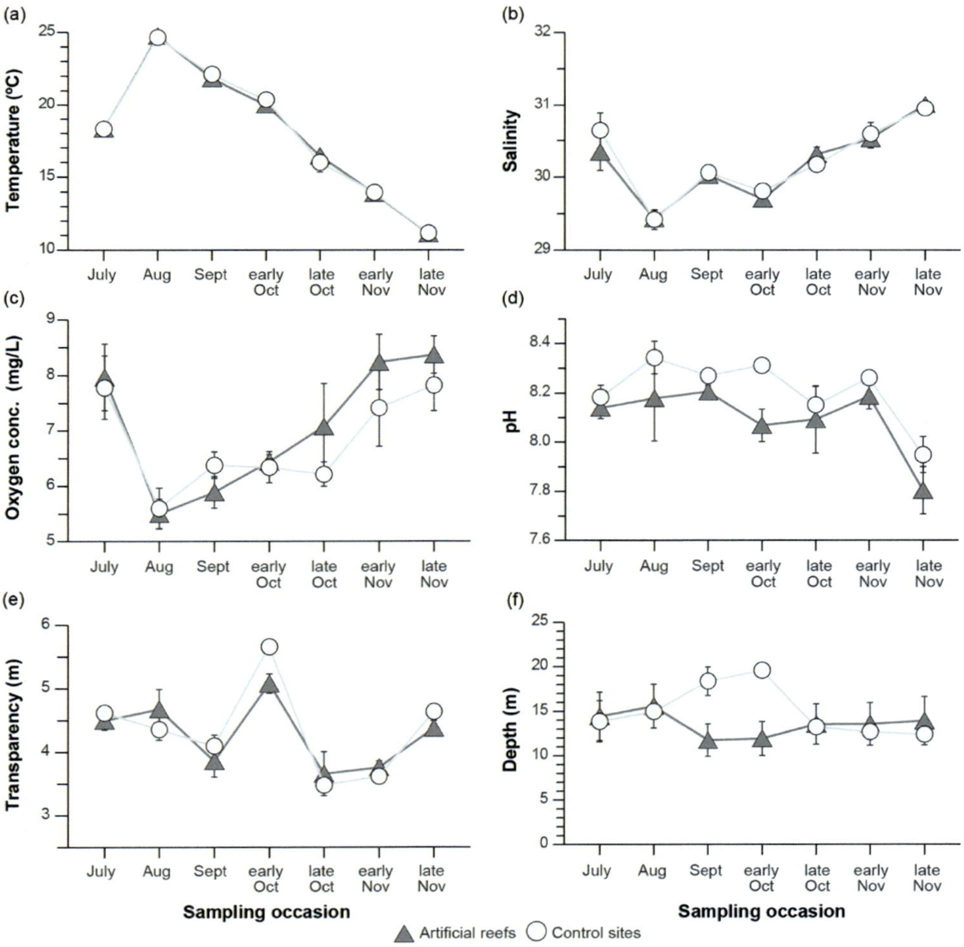


FIGURE 7.2 Mean values (± 1 SE) for (a) temperature (°C), (b) salinity (ppt), (c) dissolved oxygen concentration (mg/L), (d) pH, (e) transparency (m), and (f) depth (m) among sampling occasions at artificial reefs and control sites in waters off Dachangshan Island, Northern Yellow Sea, China.

a maximum of 24.7 °C in August and then declining sequentially to a minimum of 11.1 °C in late November. In contrast, salinity was relatively consistent, ranging from 29.4 ppt in August to 31.0 ppt in early November. Mean dissolved oxygen was ~8 mg $L^{-1}$ in July, before declining to a minimum of 5.5 mg $L^{-1}$ in August and then increasing to over 7 and 8 mg $L^{-1}$ in control sites and artificial reefs, respectively, in November (Figure 7.2c). The mean pH was typically around 8.2, except in late November when it was below 8 (i.e., 7.80 for artificial reefs and 7.95 for control sites, Figure 7.2d). Mean pH was consistently higher in the control sites than in artificial reefs, but only by less than 0.2. Water transparency was typically around 3–4 m, but reached 5 m in early October when it was ~0.5 m greater in the control sites (5.2 m) than in the artificial reef (4.6 m, Figure 7.2e). The mean depth over the entire study area for all sampling occasions was 14.3 m, with means ranging from 11.7 m (artificial reefs in September) to 19.6 m (control sites in early October; Figure 7.2f).

### 7.3.2 The Fauna

A total of 18 species, representing 13 families, were recorded using BRUVs, of which 15 species were teleost fishes, two were crustaceans, and one was an elasmobranch (Table 7.2). In comparison, 15 species (12 families), including 11 teleosts, three crustaceans, and one mollusk species were collected in traps (Table 7.2). No species were recorded in five of the 70 samples using BRUVs and five of the 70 using traps. A single species was recorded on eight video recordings (11.4% of BRUVs) and caught in 33 traps (47.1% of traps). The four most abundant species, the portunid crab *Charybdis japonica* and benthic fishes *Hexagrammos otakii* and *Sebastes schlegelii*, were ranked in the top four species by abundance for each method and habitat combination. Six species were recorded only using BRUVs, including the pelagic Japanese Spanish Mackerel (*Scomberomorus niphonius*), which was the third most abundant species at the control site and fifth most abundant on the artificial reef (Table 7.2). The five other species only recorded using BRUVs were the gobiid fishes *Pterogobius zacalles* and *Chaeturichthys stigmatias*, the rajid *Okamejei kenojei*, the monacanthid *Thamnaconus modestus*, and the lophiid *Lophius litulon*. Three species, each among the least abundant species recorded, were caught only using traps, namely the gastropod *Neptunea cumingi*, the decapod crab *Pugettia quadridens*, and the pholid teleost *Enedrias nebulosus* (Table 7.2).

BRUVs recorded 16 species associated with artificial reefs and 13 at the control sites, with 11 of those 13 (i.e., all except *C. stigmatias* and *L. litulon*) also occurring on the reefs (Table 7.2). The mean total MaxN was greater on the artificial reefs than the control sites (7.89 vs. 4.03), with the abundance of almost all species, and particularly *H. otakii*, showing the same pattern. Like the BRUVs, a greater number of species were collected in traps deployed on artificial reefs than on the control sites (13 vs. 8), and most species were recorded in greater abundance in traps on artificial reefs (Table 7.2). The mean total abundance in traps was ~50% greater on the artificial reefs than the control site (4.70 vs. 3.18 $h^{-1}$), but the mean biomass in traps on the artificial reefs was ~80% greater than in the control sites (13,747 vs. 7,744 g; Table 7.2).

### 7.3.3 Taxonomic Diversity Measures

The mean number of species, number of individuals, and Shannon index differed significantly between methods, habitats, and among sampling occasions, with the method × habitat interaction also significant for the number of species and Shannon index (Table 7.3). The factor method contributed the greatest percentage to the total mean squares in these analyses: 70% for the number of species; 77% for the Shannon index, except for the number of individuals (~20%), where habitat had the greatest contribution (48%).

Overall, a greater mean number of species were recorded using BRUVs than traps (2.9 vs. 1.6) and on the artificial reefs than control sites (2.5 vs. 2.0). Among the sampling occasions, the mean number of species was always highest in BRUVs on artificial reefs (≥3.2 on occasions except July; Figure 7.3a). Conversely, on all sampling occasions, the lowest values were recorded using traps

**TABLE 7.2**

**Relative abundances (mean ± 1 standard error [SE]) of all species recorded using BRUVs (MaxN) and traps (Count), together with their percentage contributions to the total abundance from the artificial reefs and control sites in waters off Dachangshan Island, Northern Yellow Sea, China. Species are listed by total abundance.**

| | | BRUV | | | | | | Trap | | | | | |
|---|---|---|---|---|---|---|---|---|---|---|---|---|---|
| | | Artificial reefs | | | Control sites | | | Artificial reefs | | | Control sites | | |
| **Species** | **Family** | **MaxN** | **SE** | **%** | **MaxN** | **SE** | **%** | **Count** | **SE** | **%** | **Count** | **SE** | **%** |
| *Charybdis japonica* | Portunidae | 0.97 | 0.10 | 12.29 | 0.83 | 0.14 | 20.60 | 2.51 | 0.43 | 53.40 | 2.17 | 0.35 | 68.24 |
| *Hexagrammos otakii* | Hexagrammidae | 2.65 | 0.45 | 33.59 | 0.86 | 0.19 | 21.34 | 1.49 | 0.52 | 31.70 | 0.63 | 0.22 | 19.81 |
| *Sebastes schlegelii* | Scorpaenidae | 0.63 | 0.17 | 7.98 | 0.43 | 0.21 | 10.67 | 0.14 | 0.07 | 2.98 | 0.11 | 0.05 | 3.46 |
| *Scomberomorus niphonius* | Scombridae | 0.49 | 0.32 | 6.21 | 0.49 | 0.32 | 12.16 | | | | | | |
| *Platichthys bicoloratus* | Pleuronectidae | 0.57 | 0.04 | 7.22 | 0.11 | 0.05 | 2.73 | 0.03 | 0.03 | 0.64 | 0.03 | 0.03 | 0.94 |
| *Pennahia argentatus* | Sciaenidae | 0.37 | 0.15 | 4.69 | 0.31 | 0.11 | 7.69 | 0.03 | 0.03 | 0.64 | | | |
| *Conger myriaster* | Congridae | 0.29 | 0.03 | 3.68 | | | | 0.20 | 0.12 | 4.26 | 0.09 | 0.06 | 2.83 |
| *Acanthogobius flavimanus* | Gobiidae | 0.40 | 0.15 | 5.07 | 0.11 | 0.07 | 2.73 | 0.03 | 0.03 | 0.64 | | | |
| *Sillago sihama* | Sillaginidae | 0.06 | 0.06 | 0.76 | 0.42 | 0.26 | 10.42 | 0.03 | 0.03 | 0.64 | | | |
| *Hexagrammos agrammus* | Hexagrammidae | 0.34 | 0.12 | 4.31 | 0.09 | 0.05 | 2.23 | 0.03 | 0.03 | 0.64 | 0.03 | 0.03 | 0.94 |
| *Pterogobius zacalles* | Gobiidae | 0.37 | 0.15 | 4.69 | | | | | | | | | |
| *Oratosquilla oratoria* | Squillidae | 0.09 | 0.05 | 1.14 | 0.14 | 0.07 | 3.47 | | | | 0.09 | 0.06 | 2.83 |
| *Okamejei kenojei* | Rajidae | 0.23 | 0.07 | 2.92 | 0.09 | 0.05 | 2.23 | | | | | | |
| *Thamnaconus modestus* | Monacanthidae | 0.23 | 0.18 | 2.92 | | | | | | | | | |
| *Tridentiger trigonocephalus* | Gobiidae | 0.14 | 0.09 | 1.77 | | | | 0.06 | 0.04 | 1.28 | | | |
| *Chaeturichthys stigmatias* | Gobiidae | | | | 0.09 | 0.06 | 2.23 | | | | | | |
| *Amblychaeturichthys hexanema* | Gobiidae | 0.06 | 0.06 | 0.76 | | | | 0.03 | 0.03 | 0.64 | | | |
| *Neptunea cumingi* | Buccinidae | | | | | | | 0.03 | 0.03 | 0.64 | 0.03 | 0.03 | 0.94 |
| *Lophius litulon* | Lophiidae | | | | 0.06 | 0.04 | 1.49 | | | | | | |
| *Pugettia quadridens* | Majidae | | | | | | | 0.06 | 0.04 | 1.28 | | | |
| *Enedrias nebulosus* | Pholidae | | | | | | | 0.03 | 0.03 | 0.64 | | | |
| **Number of samples** | | | **35** | | | **35** | | | **35** | | | **35** | |
| **Number of species** | | | **16** | | | **13** | | | **14** | | | **8** | |
| **Mean total abundance** | | | **7.89** | | | **4.03** | | | **4.7** | | | **3.18** | |
| **Total biomass (g)** | | | | | | | | | **13,747** | | | **7,741** | |

## TABLE 7.3

**Mean squares (MS), percentage contribution of mean squares (%MS), pseudo-F (pF), and significance levels (*P*) derived using three- and two-way PERMANOVAs on taxonomic and functional metrics of diversity and composition of the fish and macroinvertebrates derived from samples obtained from BRUVs and traps deployed on artificial reefs and control sites in waters off Dachangshan Island, Northern Yellow Sea, China. df = degrees of freedom. Significant results ($P \leq 0.05$) are highlighted in bold.**

| | | Number of species | | | | Total number of individuals | | | |
|---|---|---|---|---|---|---|---|---|---|
| **Source** | **df** | **MS** | **%MS** | **pF** | ***P*** | **MS** | **%MS** | **pF** | ***P*** |
| Method | 1 | **64.46** | **70.08** | **47.88** | **0.001** | **48.03** | **19.65** | **5.05** | **0.030** |
| Habitat | 1 | **12.01** | **13.05** | **8.92** | **0.004** | **117.03** | **47.89** | **12.30** | **0.001** |
| Sampling occasion | 6 | **3.10** | **3.37** | **2.30** | **0.045** | **28.50** | **11.66** | **3.00** | **0.013** |
| Method × habitat | 1 | **6.01** | **6.53** | **4.46** | **0.044** | 4.11 | 1.68 | 0.43 | 0.529 |
| Method × sampling occasion | 6 | 2.51 | 2.73 | 1.87 | 0.085 | 13.15 | 5.38 | 1.38 | 0.207 |
| Habitat × sampling occasion | 6 | 1.89 | 2.06 | 1.40 | 0.215 | 21.38 | 8.75 | 2.25 | 0.054 |
| Method × habitat × sampling occasion | 6 | 0.66 | 0.71 | 0.49 | 0.816 | 2.66 | 1.09 | 0.28 | 0.951 |
| Residual | 112 | 1.35 | 1.46 | | | 9.51 | 3.89 | | |

| | | Shannon index | | | | Faunal composition: abundance | | | |
|---|---|---|---|---|---|---|---|---|---|
| **Source** | **df** | **MS** | **%MS** | **pF** | ***P*** | **MS** | **%MS** | **pF** | ***P*** |
| Method | 1 | **10.86** | **77.25** | **59.45** | **0.001** | **30,869** | **49.69** | **13.86** | **0.001** |
| Habitat | 1 | **1.00** | **7.08** | **5.45** | **0.029** | **5,390** | **8.68** | **2.42** | **0.025** |
| Sampling occasion | 6 | **0.48** | **3.43** | **2.64** | **0.027** | **13,096** | **21.08** | **5.88** | **0.001** |
| Method × habitat | 1 | **0.88** | **6.26** | **4.81** | **0.026** | 1,825 | 2.94 | 0.82 | 0.548 |
| Method × sampling occasion | 6 | 0.27 | 1.90 | 1.46 | 0.188 | **5,152** | **8.29** | **2.31** | **0.001** |
| Habitat × sampling occasion | 6 | 0.31 | 2.20 | 1.69 | 0.132 | 1,662 | 2.68 | 0.75 | 0.887 |
| Method × habitat × sampling occasion | 6 | 0.08 | 0.59 | 0.46 | 0.841 | 1,900 | 3.06 | 0.85 | 0.742 |
| Residual | 112 | 0.18 | 1.30 | | | 2,227 | 3.58 | | |

*(Continued)*

**TABLE 7.3 *(Continued)***

**Mean squares (MS), percentage contribution of mean squares (%MS), pseudo-F (pF), and significance levels (*P*) derived using three- and two-way PERMANOVAs on taxonomic and functional metrics of diversity and composition of the fish and macroinvertebrates derived from samples obtained from BRUVs and traps deployed on artificial reefs and control sites in waters off Dachangshan Island, Northern Yellow Sea, China. df = degrees of freedom. Significant results ($P \leq 0.05$) are highlighted in bold.**

| | | Total biomass | | | | Faunal composition: biomass | | | |
|---|---|---|---|---|---|---|---|---|---|
| Source | df | MS | %MS | pF | *P* | MS | %MS | pF | *P* |
| Habitat | 1 | 4.22 | 34.04 | 2.26 | 0.148 | 2,747 | 20.62 | 1.64 | 0.155 |
| Sampling occasion | 6 | **4.41** | **35.50** | **2.36** | **0.042** | **7,303** | **54.82** | **4.35** | **0.001** |
| Method × sampling occasion | 6 | 1.91 | 15.38 | 1.02 | 0.416 | 1,595 | 11.97 | 0.95 | 0.545 |
| Residual | 24 | 1.87 | 15.06 | | | 1,678 | 12.59 | | |

| | | Functional richness | | | | Functional divergence | | | |
|---|---|---|---|---|---|---|---|---|---|
| Source | df | MS | %MS | pF | *P* | MS | %MS | pF | *P* |
| Method | 1 | **0.14** | **81.81** | **16.81** | **0.001** | **0.97** | **50.50** | **7.46** | **0.014** |
| Habitat | 1 | 0.02 | 13.18 | 2.71 | 0.111 | 0.46 | 24.11 | 3.56 | 0.068 |
| Method × habitat | 1 | 0.00 | 0.14 | 0.03 | 0.870 | 0.36 | 18.62 | 2.75 | 0.112 |
| Residual | 24 | 0.01 | 4.87 | | | 0.13 | 6.77 | | |

| | | Functional evenness | | | | Community-weighted means | | | |
|---|---|---|---|---|---|---|---|---|---|
| Source | df | MS | %MS | pF | *P* | MS | %MS | pF | *P* |
| Method | 1 | **0.62** | **45.98** | **6.21** | **0.022** | **3,918** | **57.93** | **3.29** | **0.016** |
| Habitat | 1 | 0.38 | 28.57 | 3.86 | 0.063 | 731 | 10.80 | 0.61 | 0.707 |
| Method × habitat | 1 | 0.24 | 18.05 | 2.44 | 0.136 | 925 | 13.68 | 0.78 | 0.564 |
| Residual | 24 | 0.10 | 7.41 | | | 1,190 | 13.68 | | |

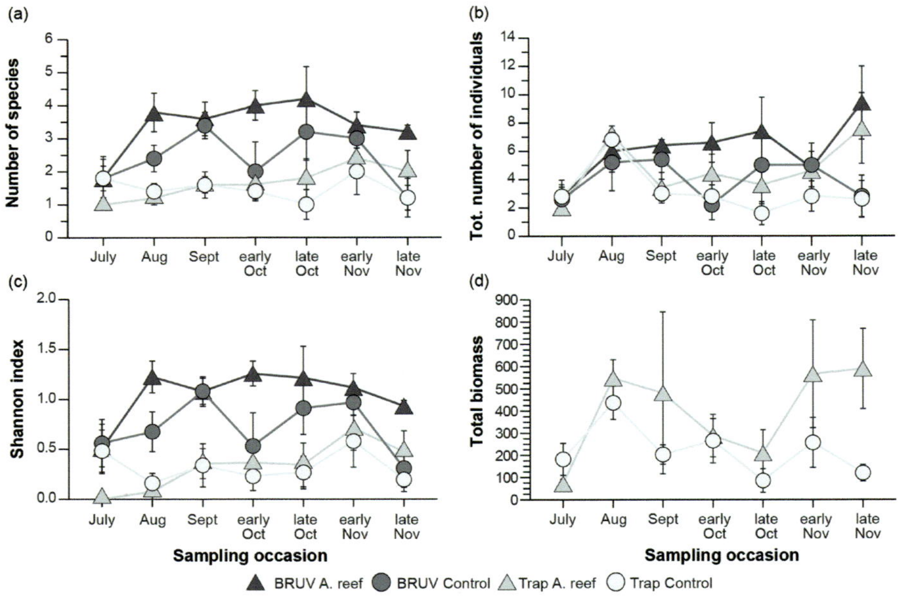


**FIGURE 7.3** Mean values (±1 SE) for mobile macroinvertebrate fauna recorded in BRUV surveys and fish traps for (a) number of species, (b) total number of individuals, (c) Shannon index, and (d) total biomass of the fish in traps only. The surveys were conducted on artificial reefs and control sites in waters off Dachangshan Island, Northern Yellow Sea, China on each sampling occasion.

deployed on either artificial reefs or control sites (1.7 and 1.5, respectively). No clear temporal trend was evident in the number of species, except for the lower number of species recorded using BRUVs on both habitats in July (Figure 7.3a). The mean total number of individuals followed a similar pattern between methods to that for the number of species, with overall greater values recorded using BRUVs (MaxN: 1 $h^{-1}$) than traps (abundance: 3 $h^{-1}$) (5.1 vs. 3.9) and on the artificial reefs than control sites (5.4 vs. 3.6). The mean number of individuals was greater when using BRUVs on artificial reefs (2.8 to 9.4) than with all other treatments, except in July, August, and early November. In July, the mean value recorded using BRUVs on artificial reefs was the same as that using traps at control sites (both 2.8). In August, the value recorded using BRUVs on artificial reefs (6.0) was lower than that using BRUVs and traps on control sites (7.2 and 6.8), while in early November (4.8), it was slightly lower than that using BRUVs on control sites (5.0; Figure 7.3b). Trends among sampling occasions were mainly due to lower numbers of individuals in all combinations of methods and habitats in July and higher numbers of individuals on artificial reefs (both methods) in late November.

The mean Shannon index values followed a similar pattern to those of the number of species, with higher values obtained using BRUVs (overall mean = 0.88, range = 0.3 to 1.3) compared to traps (overall mean = 0.32, range = 0.0 to 1.1). In most sampling occasions, the highest values were recorded using BRUVs on artificial reefs, followed by BRUVs on control sites (Figure 7.3c). The mean biomass (only available for traps) was greater on the artificial reefs on each sampling occasion except July (overall mean = 392 g, range = 70 to 589 g) than the control sites (overall mean = 221 g, range = 85 to 437 g; Figure 7.3d).

### 7.3.4 Faunal Composition

The composition of the fauna based on abundance differed significantly among all main effects, and the method × sampling occasion interaction was also significant (Table 7.3). Method explained the greatest proportion of the total mean squares (50%), followed by sampling occasion (21%), habitat

(9%), and the interaction (8%). The large difference in faunal composition between the BRUVs and traps is clearly shown on the mMDS ordination plot. The bootstrapped averages and their associated 95% confidence regions for the two methods are clearly separated. All points representing BRUV samples are on the left-hand side of the plot, and all those from traps are on the right, with no overlap between BRUV and trap samples (Figure 7.4a). The faunal composition for each method

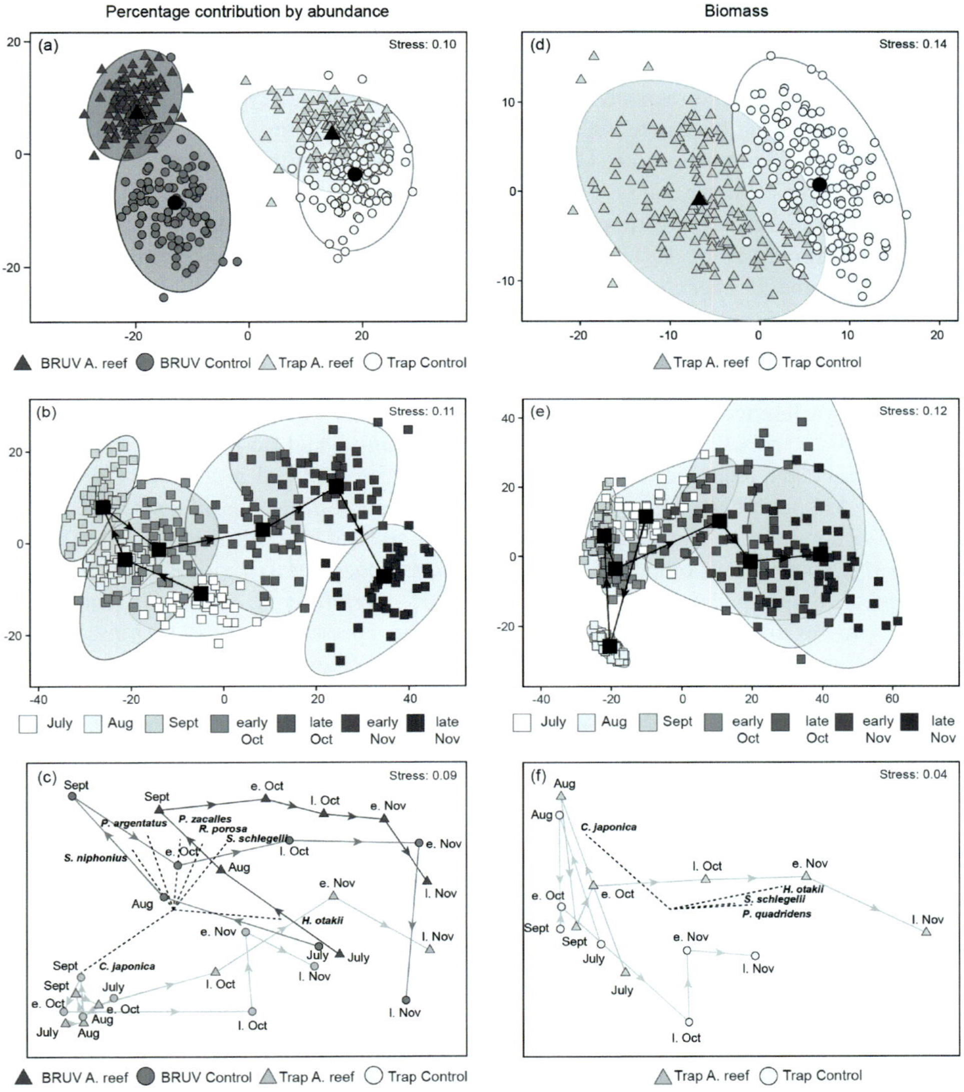


**FIGURE 7.4** mMDS ordination plots constructed from bootstrap averages for (a) method and habitat and (b) sampling occasion based on percentage contribution by abundance and (d) habitat and (e) sampling occasion based on biomass. Group averages (larger black symbols) and approximate 95% region estimates (shaded areas) are fitted to the bootstrap averages. Centroid non-metric nMDS plots, derived from a distance among centroid matrix constructed from Bray–Curtis matrix of the (c) percentage contribution by abundance and (f) biomass of the fish and mobile macroinvertebrates in waters off Dachangshan Island, Northern Yellow Sea, China. Arrows in (b, c) and (e, f) show sequential sampling occasions. Vectors shown in (c) and (f) are for species whose contribution by abundance/biomass changes in a linear direction (Pearson correlation > 0.6) relative to the nMDS axes.

also differed between the artificial reef and control area, with less overlap of samples from BRUVs than traps (Figure 7.4a). The mMDS of sampling occasions for each method showed a clear cyclical pattern, with substantial overlap in confidence regions between sequential months (e.g., August, September, and early October) but separation between those sampling occasions that were further apart (e.g., August vs. early and late November; Figure 7.4b). The centroid nMDS plot demonstrated that the sequential pattern of change among sampling occasions differed markedly between the two methods and, to a lesser extent, between habitats (Figure 7.4c).

The composition of the fauna sampled in traps based on biomass differed significantly with sampling occasion, but not habitat, and the habitat × sampling occasion interaction was not significant (Table 7.3). The lack of a habitat effect is demonstrated by the substantial overlap in 95% confidence regions between the artificial reefs and control sites (Figure 7.4d). A cyclical pattern of change was evident among sampling occasions on the mMDS ordination, with the assemblage in July being particularly distinct. This pattern of change with sampling occasions was similar on the artificial reefs and control sites (Figure 7.4f).

Differences in faunal composition between sampling methods were due to a greater variety of species being recorded using BRUVs. Fish species such as *S. schlegelii*, *Hexagrammos agrammus*, *O. kenojei*, and *P. argentatus*, as well as the mantis shrimp *Oratosquilla oratoria* had the highest frequency of occurrence (Figure 7.5a). Few species were more abundant or recorded more frequently in traps than with BRUVs, except for the Whitespotted Conger, *Conger myriaster*, a species that was recorded on five sampling occasions in traps but only once with BRUVs (Figure 7.5a). In terms of the difference between artificial reefs and control areas, the gobiids (i.e., *Tridentiger trigonocephalus*, *P. zacalles*, and *Acanthogobius flavimanus*) and, to a lesser extent, *O. kenojei*, were more frequently recorded and more abundant on the artificial reefs than the control sites (Figure 7.5a). Although both *Scomboromorus niphonius* and *Sillago sihama* were more abundant on the BRUVs deployed on control sites, they were also present on the artificial reefs. The fauna in July was fairly depauperate, comprised mainly of *C. japonica*, *H. otakii*, and *O. oratoria*. By August and September, these species were associated with *P. argentatus*, *S. niphonius*, and *S. sihama*, which were present until the end of October. At this time, species such as *Platichthys bicoloratus*, *H. agrammus*, and *S. schlegelli* had increased in abundance (Figure 7.5a).

A similar suite of species was found in traps deployed on artificial reefs and control sites (Figure 7.5b). The portunid *C. japonica* dominated the assemblages in both habitats during all sampling occasions and comprised the vast majority of the biomass in July and August. Minor contributions to the fauna were also made by *H. otakii* in October and November.

### 7.3.5 Functional Diversity

Values for all three univariate functional diversity indices differed significantly between methods and were far greater using BRUV than traps (Table 7.3; Figure 7.6). Although not significant, functional diversity (FDiv) and evenness (FEve) also tended to differ between habitats ($P = 0.063$ and 0.068, respectively, Table 7.3), due to the particularly low values for traps deployed on control sites (Figure 7.6).

The CWMs of the trait modalities differed significantly between methods, but not between habitats, nor was the method × habitat interaction significant (Table 7.3). On the mMDS plot, the points representing BRUV samples are on the left of the plot, essentially discrete from those representing the traps (Figure 7.7a). The lack of a significant difference in habitats reflects the substantial overlap in 95% confidence regions between the two habitats sampled with either method. Similarly, the samples representing each method are separated along axis 1 of the CAP plot, but within which the points for artificial reefs and control sites intermingle (Figure 7.7b). The vectors indicate that fish caught in the traps mainly resided "in the reef", whereas those recorded using BRUVs were mainly "on the reef" and on one sampling occasion also "above the reef". Other trait modalities

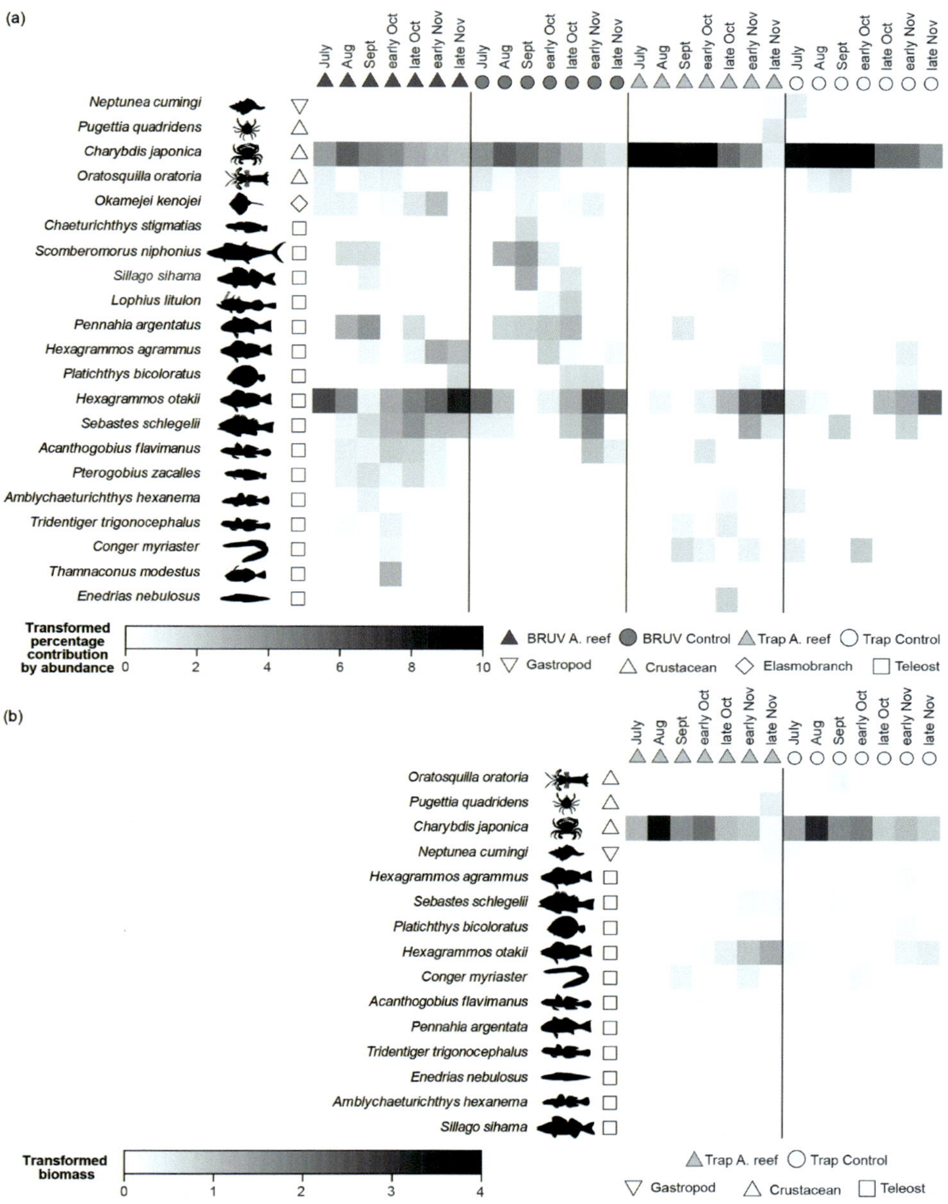


**FIGURE 7.5** Shade plot of (a) square-root transformed percentage contribution by abundance and (b) dispersion-weighted and square-root transformed biomass of each fish and mobile macroinvertebrate species recorded using BRUVs and traps at the artificial reefs and control sites in waters off Dachangshan Island, Northern Yellow Sea, China. Note that biomass data were only available for traps.

were associated with fish recorded using BRUVs, most notably those with a depressiform or band-shaped morphology.

The shade plot of the CWM data clearly shows that the fish recorded using BRUVs possessed a larger range of trait modalities (Figure 7.8). Fish caught in the traps tended to be resident, continental shelf benthopelagic fish that were compressiform and living within the reef. Moreover, they had small, terminal mouths and fed on plankton and benthic invertebrates. Although these modalities

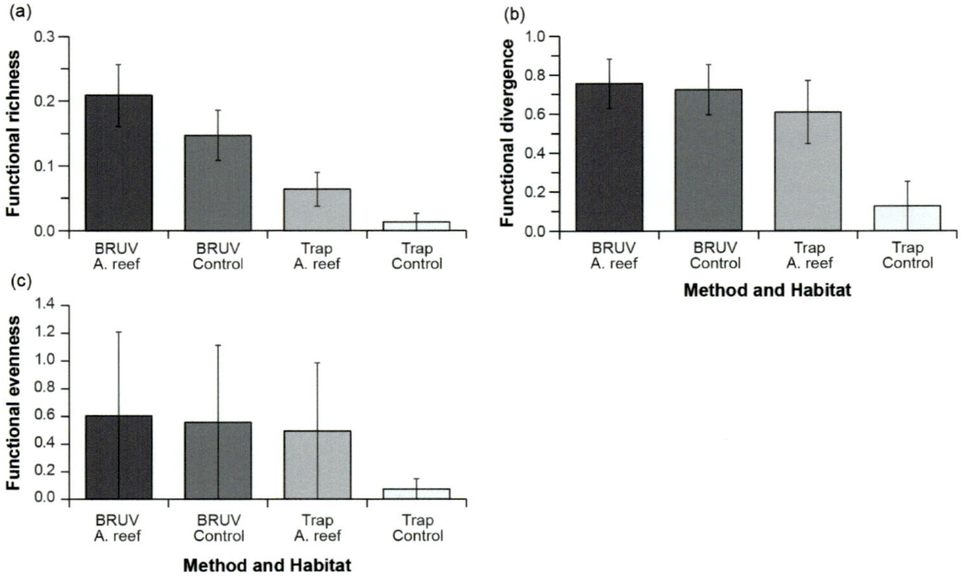


**FIGURE 7.6** Mean values (± 1 SE) for functional diversity indices calculated from the fish and mobile macroinvertebrate fauna recorded using BRUVs and traps on artificial reefs and control sites in waters off Dachangshan Island, Northern Yellow Sea, China. Note that data were pooled across the seven sampling occasions.

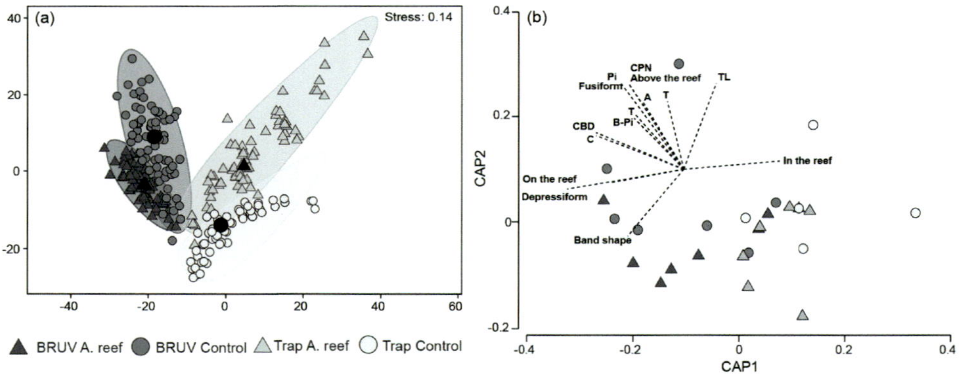


**FIGURE 7.7** (a) mMDS ordination plot constructed from bootstrap averages of the trait modalities derived from the fish species recorded using BRUVs and traps on the artificial reefs and control sites in waters off Dachangshan Island, Northern Yellow Sea, China. Group averages (larger black symbols) and approximate 95% confidence region estimates (shaded areas) are fitted to the bootstrap averages. (b) CAP plot of the same trait modalities with vectors overlaid for functional traits whose contribution changes in a linear direction (Pearson correlation > 0.5) relative to the CAP axes. See Table 7.1 for full names of trait modalities.

were also found in fish recorded using BRUVs, they were less dominant due to the presence of other modalities. For example, migratory fish and those that reside above, particularly on the reef, were abundant on some sampling occasions. Moreover, fish that are oceanic, bathydemersal, or continental shelf, pelagic-neritic piscivores, and fusiform were only recorded using BRUVs. Within each method, the trait modalities were similar, although the relative contribution varied among sampling occasions. For example, band-shaped fish were proportionally more abundant on BRUV samples

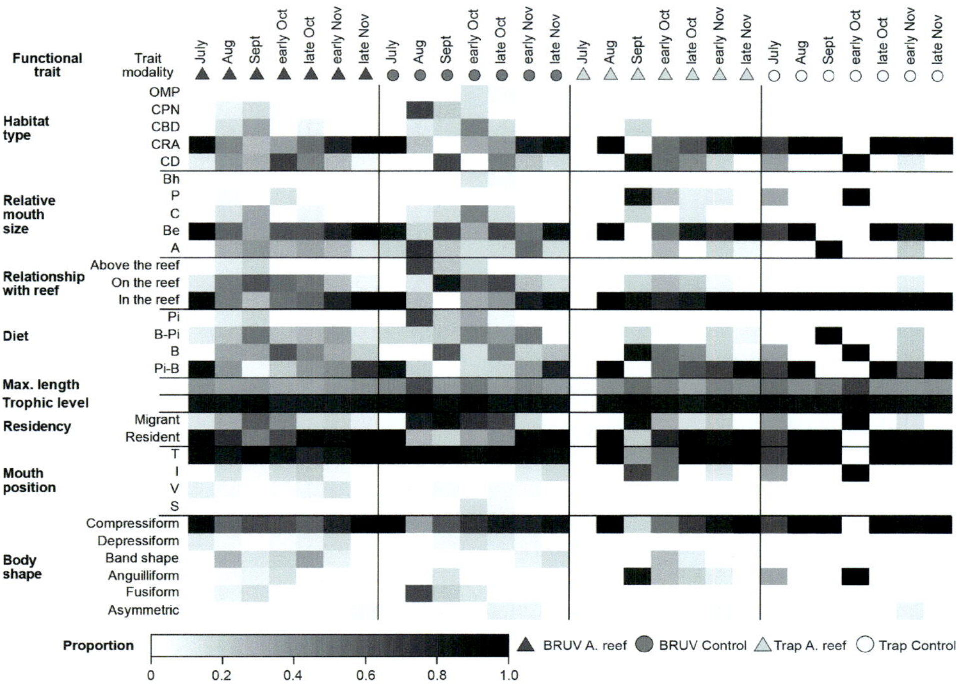


**FIGURE 7.8** Shade plot of the community-weighted means (CWMs) of trait modalities from each functional trait derived from the fish species recorded using BRUVs and traps on the artificial reefs and control sites in waters off Dachangshan Island, Northern Yellow Sea, China. See Table 7.1 for the full names of trait modalities.

from artificial reefs than on control sites, whereas the reverse was true for asymmetric fish. In trap samples, fish residing on the reef and those that were band-shaped made a greater contribution to the fauna on artificial reefs.

## 7.4 DISCUSSION

This study evaluates the first use of BRUVs in Chinese coastal waters to investigate the spatial and temporal variation in the composition and diversity of the fish and mobile invertebrate community on artificial reefs. It is also the first study to estimate the functional diversity of the mobile fauna on artificial reefs. The data and metrics derived from BRUV samples have been compared with those from commonly used commercial baited fish traps (traps). In addition to traditional diversity indices and species composition, functional diversity traits, which define species in terms of their ecological roles, that is, how they interact with the environment and with other species, were compared. A greater number of species and higher relative abundance were recorded using BRUVs than traps in the same location, indicating that BRUVs are a more effective fishery-independent monitoring approach than traps for detecting changes in reef-associated species on the artificial reefs over time.

### 7.4.1 Effectiveness of the Two Methods

The faunal assemblage recorded using BRUVs contained a greater total number of species and relative abundance than that caught in traps. Similarly, six species were detected only using BRUVs. These species included four demersal fish (i.e., *O. kenojei*, *L. litulon*, *P. zacalles*, and *C. stigmatias*) and two seasonal migratory fish (i.e., *S. niphonius* and *T. modestus*). Multivariate analyses also confirmed significant differences in the composition of the mobile fauna between the two methods. These general findings were consistent with previous studies that compared the sampling effectiveness of BRUVs and traps in estuaries and tropical continental shelf areas (i.e., the number of fish species and abundances obtained by sampling with BRUVs was higher than traps), providing greater statistical power for monitoring differences in fish assemblages (Harvey et al., 2012; Wakefield et al., 2013; Bacheler et al., 2017).

BRUVs can detect faunal elements at a maximum range (field of vision underwater) of 10 m in high visibility water conditions (Boutros et al., 2015). Providing fish enter the field of view, regardless of whether they are attracted by bait or swimming through, they will be recorded by the camera. In this study, other than the predominant demersal resident fish *H. otakii* and *S. schlegelli*, the pelagic *S. niphonius* was also observed using benthic BRUVs. This is possibly related to the dispersal and attraction of the bait/scent plume of crushed *E. japonicus* (the preferred prey of *S. niphonius*; Zhang et al., 2022), the only fish categorized in the "above the reef" orientation (Wu et al., 2016). This common species has rarely been documented in fish monitoring programs around artificial reefs in the northern coastal waters of China (Wu et al., 2019). This may be due to the limited capability of traditional benthic sampling gear (e.g., traps) in capturing pelagic species from artificial reefs. In the future, sampling *in situ* with BRUVs, including both demersal and pelagic BRUVs, is a promising method for monitoring artificial reefs to comprehensively understand fish community structure and function (Scott et al., 2015). In addition, BRUVs in this study detected the Spiny Skate *O. kenojei*, which is listed as vulnerable by the IUCN (Rigby et al., 2021). Sampling *in situ* with BRUVs also has the advantage of not removing species from the environment compared to traps, a feature particularly important for conservation-listed species. Moreover, fish traps also sample a more limited size range of fauna than BRUVs, as the size of individuals caught is limited by the size of the entrance funnel to the trap. The limited space in the traps and the interaction between species within the traps may hinder the entry of other species and individuals, particularly after the dominant species, for example, *C. japonica* and *H. otakii*, enter traps (High and Ellis, 1973).

The metric MaxN for BRUVs is likely to have provided a conservative estimate of the relative abundance of species (Cappo et al., 2006; Schobernd et al., 2013). However, the MaxN derived from BRUVs deployed for one hour was greater than the actual catches in traps deployed for over three hours. The high selectivity and limited capture capacity of traps for both size and range of species resulted in many organisms not being recorded and underestimated the number of species and abundance, especially species that are not attracted by bait but are visually detected by BRUVs (Harvey et al., 2012; Bacheler et al., 2017). All sampling was conducted during the day on slack tides to optimize visibility for BRUV recordings. The state of the tide and time of day are likely to influence the catchability of both methodologies (Taylor et al., 2013; Rieucau et al., 2015). Generally, based on their relative abundances, samples using BRUVs recorded more fish than traps.

When comparing the application of the two methods, the time required to conduct sampling is an important consideration. In this study, BRUVs had a soak time of one hour, compared to three hours for the traps. Although the soak time was shorter for a BRUV than a trap, more species and a greater total number of individuals were recorded in BRUVs. Therefore, in terms of sampling efficiency, a one-hour BRUV soak time is preferable to the three hours traditionally used for the traps.

The performance of BRUVs is, however, dependent on the prevailing light conditions and, particularly, water clarity. This is important because even a moderate decrease in visibility can drastically inhibit one's ability to identify and count fish in the field of view in deeper waters. These technologies have primarily been tried and tested in high-visibility and biodiverse environments, such as those found on reef systems in Australia (e.g., French et al., 2021). It has been reported that spring tides can drive fish away from the seafloor and decrease visibility due to greater water movement and sediment resuspension, thereby reducing the ability to identify and count fish accurately (Lecrivain et al., 2021). Improving the efficacy of BRUVs in turbid waters has been a recent focus of research. For example, a clear liquid optical chamber has been experimentally added to BRUV systems to enhance visibility and increase the capability of observing fauna in turbid waters (Jones et al., 2019).

The extensive time required for manual video processing is a disadvantage of using BRUVs, as noted by Bernard and Gotz (2012) and Goetze et al. (2019), who found that an average of 5.7 hours was required to extract the MaxN values from a 1-hour video. In this study, while the exact time for determining the MaxN from each video was not recorded, it was estimated to be approximately 3 hours, depending on the number of species and individuals recorded. In contrast, traps only required 0.5 hours of processing to identify, count, and measure all individuals. Reducing the time needed for BRUV video processing can help reduce operating costs. Additionally, adopting "deep learning", an artificial intelligence (AI) technique that enables computers to identify patterns in images automatically, is increasingly being used to identify fish from underwater video (Sheehan et al., 2019; Ditria et al., 2020, 2022; Connolly et al., 2022).

The benthic BRUV used in this study documented the fish species characteristic of the study area, such as the pelagic teleost *S. niphonius* and vulnerable elasmobranch *O. kenojei*. However, non-aggressive or lower trophic-level species were underrepresented when sampled using BRUVs. For example, the pholid *Enedrias nebulosus* (a small benthic prey fish) was only caught in traps on the artificial reefs and was not recorded by the BRUVs. This species was also collected by long fishing traps in the Lidao artificial reef zone and Qianshan Island in the Yellow Sea (Zhang et al., 2016; Wu et al., 2019). In addition, the Whitespotted Conger (*C. myriaster*), an eel species widely distributed in the coastal to deep waters around China, the Korean Peninsula, and Japan (Tang and Wu, 1988), was recorded in lower numbers and less frequently from BRUVs than traps. This is possibly related to the behavior of *C. myriaster*, as this species is often found in caves (Tang and Wu, 1988) and thus unlikely to be recorded on horizontally facing BRUVs unless they face directly into a cave. *Conger myriaster* has been found to have a close relationship with artificial reefs in Korea and China from studies using trammel nets, gill nets, and traps (Noh et al., 2017; Wu et al., 2019). Fish traps, with a specialized design entrance funnel, similar to the structure of a cave, are particularly effective for capturing *C. myriaster*. This has been evidenced by the dominance of the trait "in the reef" in the CWM for traps. The application of the BRUVs in this study had a horizontal-view angle for sampling, which led to a decreased sensitivity to detect *C. japonica* and *C. myriaster*. In the future, supplementing horizontal-view BRUVs with downward-facing cameras is likely to facilitate recording these species and other cryptic, reef-dwelling species (Hewitt et al., 2015).

The location of the baited fish trap on the substratum and the characteristics of this gear make it ineffective for capturing pelagic fish. As a result, the catches from traps lacked pelagic fish species. Although benthic BRUVs successfully recorded pelagic fish, it is possible that a considerable portion of the pelagic fish community was missed due to the deployment location. This limitation could be addressed through the addition of pelagic BRUVs (e.g., Bouchet et al., 2018). A study using benthic and pelagic BRUVs on temperate seagrass, reef, and sand habitats showed that each method of orientation detected a different fish fauna (Clarke et al., 2019). The use of both visual fields, i.e., downward and outward facing, conjointly could also facilitate an understanding of benthic–pelagic coupling.

### 7.4.2 Spatial and Temporal Variation in Faunal Composition

The simultaneous application of both BRUVs and traps demonstrated that artificial reefs yielded a higher abundance of mobile fauna than the adjacent natural habitat. These findings are supported by other surveys that utilized BRUVs to compare artificial reefs with adjacent unstructured areas (Balaji et al., 2019; Folpp et al., 2020; Maus et al., 2024). The reef structure influences local hydrology and provides shelter and food for larger organisms, which may affect the spatial and temporal variation in faunal composition of the associated reef assemblage. In this study, differences in topography and benthic relief between artificial reefs and natural areas might explain the observed differences in the reef-associated species. The shell reef deployed in this study had a regular, hexagonal contour that has been observed to attract and enhance the abundance of sea cucumbers, sea urchins, and other reef-associated species (Zhao et al., 2019). This observation can be further confirmed by studies where the complex, topographic structure of shell reefs was found to provide greater structural complexity and increased habitats, thus enhancing species richness and diversity (Charton et al., 2000; Davis and Smith, 2017). Additionally, the CWM analysis of the current study indicated that the trait "in the reef" is more dominant in artificial reef sites than in the control sites.

A total of 17 fish species were detected by the 140 BRUVs and traps deployed on seven occasions, spanning five months from summer to autumn in this study. This level of species richness approaches the number of fish species found in the coastal artificial reef zone waters of Qianshan Island (22 species) and Lidao (25 species) in the Yellow Sea using trammel nets and long fishing traps deployed over 12 months (Zhang et al., 2016; Wu et al., 2019). The number of species/taxa in this study is, however, far fewer than the 76 species recorded in the coastal waters of the northern Yellow Sea in 2006 (Liu et al., 2014). Moreover, a total of 44 fish species were recorded from 24 videos taken from the two artificial reefs in Geographe Bay in temperate Western Australia (Florisson et al., 2018). The Geographe Bay study recorded total MaxN values of 118 in one area (Dunsborough) and 26 in another (Bunbury), while 18 species and the greatest total MaxN value of 7.89 were recorded in this study. The far fewer number of species and individuals recorded in our study may be attributed to the depletion of fishery resources, particularly large predatory fishes, in the coastal waters of northern China (Liu et al., 2014). This study included a relatively limited spatial and temporal scale. This may explain why some common species in this region, such as the small Yellow Croaker, *Larimichthys polyactis*, and Red Tonguesole, *Cynoglossus joyneri* (Wan et al., 2020; Gao et al., 2023), were not recorded.

Temporal changes in the composition of the reef-associated community were detected in this study, with fewer species recorded in July and an increase in the number of species and a shift in faunal composition from August to October. According to previous studies, water temperature was a primary environmental factor that influenced the distributional shift in fish species in the northern Yellow Sea (Wu et al., 2019). Seasonal migratory fishes, including *S. niphonius* and *T. modestus*, were only recorded from August to early October, indicating that these two species migrated to the coastal waters of Dachangshan Island during this period. Local fishers anecdotally report that fewer individuals of these two species are caught from July to August (Wu, unpublished discussions with fishers), which is consistent with findings from the data recorded using BRUVs reported here. Therefore, the low number of species and abundance observed in early July was more likely related to species' life history features, such as spawning times, growth, and movement patterns of these migratory fish, than environmental factors during the five months of this study.

In this study, employing BRUVs demonstrated several advantages over traditional baited traps for assessing mobile fauna on artificial reefs. They recorded higher species richness and abundance, captured a broader range of functional traits, and provided a non-destructive and replicable method for long-term monitoring. However, their effectiveness is limited for cryptic benthic species, such as *C. japonica* and *C. myriaster*, likely due to the horizontal camera orientation, and their observational capacity is constrained by water transparency. The substantial time required for manual video analysis further adds to the methodological challenges, though these could be alleviated with deep learning or AI-based automated processing. Baited traps, on the other hand, are cost-effective and particularly effective at sampling benthic and cryptic species, but they generally capture fewer species and individuals and may be biased toward certain sizes or behaviors. Together, these findings suggest that while BRUVs offer a more comprehensive assessment of reef-associated mobile fauna, traps can complement BRUV surveys by targeting species that are less detectable on video. Future research could enhance sampling efficacy by integrating BRUVs with complementary methods, adjusting camera orientation, extending deployment durations, and incorporating technological advancements to overcome the current limitations.

## 7.5 CONCLUSION

This study confirms that employing fishery-independent BRUVs to sample faunas *in situ* is a more effective monitoring method than baited, fishery-dependent fish traps for detecting changes in reef-associated species in an artificial reef zone over time. BRUVs recorded a significantly greater number of species, higher relative abundance, and higher functional diversity metrics in terms of feeding and space use than traps. These findings indicate that data obtained via BRUVs can provide a more comprehensive representation of the mobile fauna for establishing baseline fish community conditions than traps and are a more effective monitoring tool for marine ranching operations in China.

## 7.6 ACKNOWLEDGMENTS

We thank the staff from Zhengda Marine Biology Company Limited for assisting in the field sampling. We are grateful to the following funding providers for supporting the research: National Natural Science Foundation of China (41906125), National Key Research & Development Program of China (2023YFD2401102, 2019YFD0901304), General Program of Liaoning Provincial Department of Education (LJKMZ20221120), Dalian Science and Technology Innovation Fund (2021JJ11CG001, 2022JJ13FG112). The silhouette images used were provided by PhyloPic under a Attribution-ShareAlike 4.0 International license. See https://www.phylopic.org/permalinks/cfdb647215219f54faf36fb45daba76f62e00ff5884f11a8881718cd3c19e823 for a full list of attributions. We gratefully acknowledge the support and critical feedback provided by the editors of the International Conference on Artificial Reefs and Related Aquatic Habitats.

## REFERENCES

Ajemian, M.J., Wetz, J.J., Shipley-Lozano, B., Shively, J.D., Stunz, G.W., 2015. An analysis of artificial reef fish community structure along the northwestern Gulf of Mexico shelf: Potential impacts of "rigs-to-reefs" programs. *PLoS One* 10 (5), e0126354. https://doi.org/10.1371/journal.pone.0126354.

Anderson, M.J., Gorley, R.N., Clarke, K.R., 2008. *PERMANOVA + for PRIMER: Guide to Software and Statistical Methods*. PRIMER-E, Plymouth, UK. https://cir.nii.ac.jp/crid/1370572092711312918.

Asher, J., Williams, I.D., Harvey, E.S., 2019. Is seeing believing? Diver and video-based censuses reveal inconsistencies in roving predator estimates between regions. *Mar. Ecol. Prog. Ser.* 630, 115–136. https://doi.org/10.3354/meps.

Bacheler, N.M., Geraldi, N.R., Burton, M.L., Munoz, R.C., Kellison, G.T., 2017. Comparing relative abundance, lengths, and habitat of temperate reef fishes using simultaneous underwater visual census, video, and trap sampling. *Mar. Ecol. Prog. Ser.* 574, 141–155. https://doi.org/10.3354/meps.

Bacheler, N.M., Schobernd, Z.H., Berrane, D.J., Schobernd, C.M., Mitchell, W.A., Geraldi, N.R., 2013. When a trap is not a trap: Converging entry and exit rates and their effect on trap saturation of black sea bass (*Centropristis striata*). *ICES J. Mar. Sci.* 70 (4), 873–882. https://doi.org/10.1093/icesjms/fst062.

Balaji, V., Sekar, V., Masilamani Selvam, M., 2019. Fish assemblages associated with natural and artificial habitats in Palk Bay, southeast coast of India: Comparison using baited remote underwater video. *J. Mar. Biol. Assoc. India* 61 (1), 54–60. https://doi.org/10.1016/j.ocecoaman.2021.105901.

Barker, V.A., Cowan, J.H., 2017. The effect of artificial light on the community structure of reef-associated fishes at oil and gas platforms in the northern Gulf of Mexico. *Environ. Biol. Fish* 101 (1), 153–166. https://doi.org/10.1007/s10641-017-0688-9.

Becker, A., Taylor, M.D., Folpp, H., Lowry, M.B., 2018. Managing the development of artificial reef systems: The need for quantitative goals. *Fish Fish.* 19 (4), 740–752. https://doi.org/10.1111/faf.12288.

Becker, A., Taylor, M.D., Lowry, M.B., Grabowski, J., 2017. Monitoring of reef associated and pelagic fish communities on Australia's first purpose built offshore artificial reef. *ICES J. Mar. Sci.* 74 (1), 277–285. https://doi.org/10.1093/icesjms/fsw133.

Bernard, A.T.F., Gotz, ¨ A., 2012. Bait increases the precision in count data from remote underwater video for most subtidal reef fish in the warm-temperate Agulhas bioregion. *Mar. Ecol. Prog. Ser.* 471, 235–252. https://doi.org/10.3354/meps.
Bicknell, A.W.J., Sheehan, E.V., Godley, B.J., Doherty, P.D., Witt, M.J., 2019. Assessing the impact of introduced infrastructure at sea with cameras: A case study for spatial scale, time and statistical power. *Mar. Environ. Res.* 147, 126–137. https://doi.org/10.1016/j.marenvres.2019.04.007.
Bosch, N.E., Gonçalves, J.M.S., Tuya, F., Erzini, K., 2017. Marinas as habitats for nearshore fish assemblages: Comparative analysis of underwater visual census, baited cameras and fish traps. *Sci. Mar.* 81 (2), 159–169. http://dx.doi.org/10.3989/scimar.04540.20A.
Bouchet, P.M., Huveneers, C.J., Langlois, T., Letessier, T., Lowry, M., Rees, M., Santana Garcon, J., Scott, M., Taylo, M., Thompson, C., Vigliola, L., Whitmarsh, S., 2018. *Field Manuals for Marine Sampling to Monitor Australian Waters, Version 1.* NESP Marine Biodiversity Hub Canberra, Australia. https://marine-sampling-field-manual.github.io.
Boutros, N., Shortis, M.R., Harvey, E.S., 2015. A comparison of calibration methods and system configurations of underwater stereo-video systems for applications in marine ecology. *Limnol. Oceanogr. Methods* 13 (5), 213–265. https://doi.org/10.1002/lom 3.10020.
Brickhill, M., Lee, S.Y., Connolly, R.M., 2005. Fishes associated with artificial reefs: Attributing changes to attraction or production using novel approaches. *J. Fish Biol.* 67, 53–71. https://doi.org/10.1111/j.0022-1112.2005.00915.x.
Cappo, M., Harvey, E., Shortis, M., 2006. Counting and measuring fish with baited video techniques-an overview. *In*: *Proceedings of the 2006 Australian Society of Fish Biology Conference and Workshop Proceedings*, Australia, pp. 101–114. https://www.academia.edu/47756818/Cutting_edge_technologies_in_fish_and_fisheries_science#page=109.
Charton, J.A.G., Williams, I.D., Ruzafa, A.P., Milazzo, M., Chemello, R., Marcos, C., Kitsos, M.S., Koukouras, A., Riggio, S., 2000. Evaluating the ecological effects of Mediterranean marine protected areas: Habitat, scale and the natural variability of ecosystems. *Environ. Conserv.* 27 (2), 159–178. https://doi.org/10.1017/S0376892900000199.
Clarke, K.R., Gorley, R.N., 2015. *PRIMER v7: User Manual/Tutorial.* PRIMER-E, Plymouth, UK. https://www.researchgate.net/publication/307447596 via the internet.
Clarke, K.R., Gorley, R.N., Somerfield, P.J., Warwick, R.M., 2014a. *Change in Marine Communities: An Approach to Statistical Analysis and Interpretation*, 3rd edition. Primer-E, Plymouth. https://www.researchgate.net/publication/264543450via the internet.
Clarke, K.R., Tweedley, J.R., Valesini, F.J., 2014b. Simple shade plots aid better long-term choices of data pre-treatment in multivariate assemblage studies. *J. Mar. Biol. Assoc. U. K.* 94 (1), 1–16. https://doi.org/10.1017/S0025315413001227.
Clarke, T.M., Whitmarsh, S.K., Fairweather, P.G., Huveneers, C., 2019. Overlap in fish assemblages observed using pelagic and benthic baited remote underwater video stations. *Mar. Freshw. Res.* 70 (6), 870–880. https://doi.org/10.1071/MF18224.
Connolly, R., Jinks, K., Herrera, C., Lopez-Marcano, S., 2022. Fish surveys on the move: Adapting automated fish detection and classification frameworks for videos on a remotely operated vehicle in shallow marine waters. *Front. Mar. Sci.* 9, 918504. https://doi.org/10.3389/fmars.2022.918504.
Cronin-O'Reilly, S., Cottingham, A., Kalnejais, L.H., Lynch, K., Tweedley, J.R., 2025. Tidal exclusion barriers fragment an invertebrate community into taxonomically and functionally distinct estuarine and wetland assemblages. *J. Mar. Sci. Eng.* 13, 635. https://doi.org/10.3390/jmse13040635.
Davis, T.R., Smith, S.D.A., 2017. Proximity effects of natural and artificial reef walls on fish assemblages. *Reg. Stud. Mar. Sci.* 9, 17–23. https://doi.org/10.1016/j.rsma.2016.10.007.
Dirhamsyah, D., Sutisna, D., Sani, S.Y., Sulha, S., 2024. Marine ranching in Indonesia: Towards sustainable fisheries and aquaculture. *Adv. Biol. Res.* 8, 45-65. https://doi.org/10.9734/bpi/arbs/v8/7292B.
Ditria, E., Buelow, C., Gonzalez-Rivero, M., Connolly, R., 2022. Artificial intelligence and automated monitoring for assisting conservation of marine ecosystems: A perspective. *Front. Mar. Sci.* 9, 918104. https://doi.org/10.3389/fmars.2022.918104.
Ditria, E.M., Lopez-Marcano, S., Sievers, M., Jinks, E.L., Brown, C.J., Connolly, R.M., 2020. Automating the analysis of fish abundance using object detection: Optimizing animal ecology with deep learning. *Front. Mar. Sci.* 7. https://doi.org/10.3389/fmars.2020.00429.
Duarte, A., Weldy, M.J., Lesmeister, D.B., Ruff, Z.J., Jenkins, J.M., Valente, J.J., Betts, M.G., 2024. Passive acoustic monitoring and convolutional neural networks facilitate high-resolution and broadscale monitoring of a threatened species. *Ecol. Indic.* 162, 112016. https://doi.org/10.1016/j.ecolind.2024.112016.

Finucci, B., Hurst, R.J., Bagley, N.W., Al Dhaheri, S.S., Grandcourt, E.M., 2019. Diversity, abundance, behaviour, and catchability of fishes from trap catch and underwater video in the Arabian gulf. *Fish. Res.* 220, 105342. https://doi.org/10.1016/j.fishres.2019.105342.

Florisson, J.H., Tweedley, J.R., Walker, T.H.E., Chaplin, J.A., 2018. Reef vision: A citizen science program for monitoring the fish faunas of artificial reefs. *Fish. Res.* 206, 296–308. https://doi.org/10.1016/j.fishres.2018.05.006.

Folpp, H.R., Schilling, H.T., Clark, G.F., Lowry, M.B., Maslen, B., Gregson, M., Suthers, I. M., Trenkel, V., 2020. Artificial reefs increase fish abundance in habitat-limited estuaries. *J. Appl. Ecol.* 57 (9), 1752–1761. https://doi.org/10.1111/1365-2664.13666.

French, B., Wilson, S., Holmes, T., Kendrick, A., Rule, M., Ryan, N., 2021. Comparing five methods for quantifying abundance and diversity of fish assemblages in seagrass habitat. *Ecol. Indic.* 124, 107415. https://doi.org/10.1016/j.ecolind.2021.107415.

Froese, R., Pauly, D., 2022. *FishBase*. World Wide Web Electronic Publication. www.fishbase.org. version (accessed August 2022).

Gao, Y., Zhang, Y., Zhang, H., Liu, Z., Li, S., Cheng, J., 2023. Spatial distribution of *Larimichthys polyactis* juveniles in the southern Yellow Sea and northern East China Sea in summer. *Mar. Fish.* 45(1), 86. https://doi.org/10.13233/j.cnki.mar.fish.2023.01.002.

Ghazilou, A., Shokri, M.R., Gladstone, W., 2019. Comparison of baited remote underwater video (BRUV) and underwater visual census (UVC) for assessment of reef fish in a marginal reef in the northern Persian gulf. *Iran. Soc. Ichthyol.* 6 (3), 197–207. https://doi.org/10.1016/j.jembe.2012.01.013.

Goetze, J.S., Bond, T., McLean, D.L., Saunders, B.J., Langlois, T.J., Lindfield, S., Fullwood, L.A.F., Driessen, D., Shedrawi, G., Harvey, E.S., 2019. A field and video analysis guide for diver operated stereo-video. *Methods Ecol. Evol.* 10 (7), 1083–1090. https://doi.org/10.1111/2041-210X.13189.

Hammond, M., Bond, T., Prince, J., Hovey, R.K., McLean, D.L., 2020. An assessment of change to fish and benthic communities following installation of an artificial reef. *Reg. Stud. Mar. Sci.* 39, 101408. https://doi.org/10.1016/j.rsma.2020.101408.

Harvey, E.S., Newman, S.J., McLean, D.L., Cappo, M., Meeuwig, J.J., Skepper, C.L., 2012. Comparison of the relative efficiencies of stereo-BRUVs and traps for sampling tropical continental shelf demersal fishes. *Fish. Res.* 125–126, 108–120. https://doi.org/10.1016/j.fishres.2012.01.026.

Hewitt, J., Stobart, B., Díaz, D., Álvarez, F., Alonso, C., Mallol, S., Goñi, R., 2015. Performance of baited underwater video: Does it underestimate abundance at high population densities? *PLoS One* 10 (5), e0127559. https://doi.org/10.1371/journal.pone.0127559.

High, W.L., Ellis, I.E., 1973. Underwater observations of fish behavior in traps. *Helgoländer Meeresun.* 24 (1), 341–347. https://doi.org/10.1007/BF01609524.

Jiao, H., Zheng, D., Zhao, M., You, Z., Huang, C., 2016. Survey and resource evaluation of *Mytilus coruscus* for underwater visual censuses of Yushan Islands. *Chin. J. Oceanol. Limnol.* 32 (2), 84–92. https://doi.org/10.3969/j.issn.0253-4193.2016.02.008.

Jones, R.E., Griffin, R.A., Rees, S.C., Unsworth, R.K.F., 2019. Improving visual biodiversity assessments of motile fauna in turbid aquatic environments. *Limnol. Oceanogr.* 17 (10), 544–554. https://doi.org/10.1002/lom3.10331.

Kitada, S., 2020. Lessons from Japan marine stock enhancement and sea ranching programmes over 100 years. *Rev. Aquac.* 12 (3), 1944–1969. https://doi.org/10.1111/raq.12418.

Krispyn, K.N., Florisson, J.H., Tweedley, J.R., 2025. Artificial reefs act as cleaning stations in temperate waters. *Pac. Conserv. Biol.* 31 (5). https://doi.org/10.1071/PC25049.

La Manna, G., Guala, I., Grech, D., Perretti, F., Ronchetti, F., Manghi, M., Ruiu, A., Ceccherelli, G., 2021. Performance of a baited underwater video system vs. the underwater visual census technique in assessing the structure of fish assemblages in a Mediterranean marine protected area. *Mediterr. Mar. Sci.* 22 (3), 480–495. https://orcid.org/0000-0003-4712-1435.

Laliberté, E., Legendre, P., Shipley, B., 2014. FD: Measuring functional diversity from multiple traits, and other tools for functional ecology. *In*: *R Package Version 1.0-12*. https://cran.r-project.org/web/packages/FD/FD.pdf.

Lecrivain, N., Clement, B., Dabrin, A., Seigle-Ferrand, J., Bouffard, D., Naffrechoux, E., Frossard, V., 2021. Water-level fluctuation enhances sediment and trace metal mobility in lake littoral. *Chemosphere* 264, 128451. https://doi.org/10.1016/j.chemosphere.2020.128451.

Lee, S.-G., Rahman, M.A., 2018. Ecological stock enhancement programs (ESEPs) based fisheries rebuilding plan (FRP) in Korea. *J. Environ. Biol.* 39 (5), 936–942. https://doi.org/10.22438/jeb/39/5(SI)/31.

Lek, E., Fairclough, D.V., Platell, M.E., Clarke, K.R., Tweedley, J.R., Potter, I.C., 2011. To what extent are the dietary compositions of three abundant, co-occurring labrid species different and related to

latitude, habitat, body size and season? *J. Fish Biol.* 78 (7), 1913–1943. https://doi.org/10.1111/j.1095-8649.2011.02961.x.
Lima, J.S., Zalmon, I.R., Love, M., 2019. Overview and trends of ecological and socioeconomic research on artificial reefs. *Mar. Environ. Res.* 145, 81–96. https://doi.org/10.1016/j.marenvres.2019.01.010.
Liu, X., Dong, J., Yu, X., Sun, M., Wang, B., Wang, X., 2014. Fishery resource structure in coastal waters of Liaoning Province. *Mar. Fish.* 36 (4), 289–299. https://doi.org/10.13233/j.cnki.mar.fish.2014.04.003.
Liu, Y., Yu, H., Chen, Y., Xing, B., Tian, T., Wang, Y., Yang, J., 2016. Selectivity of fish traps with different mesh sizes for masked greenling *Hexagrammos otakii*. *J. Dalian Ocean Univ.* 31 (1), 109–112. https://doi.org/10.16535/j.cnki.dlhyxb.2016.01.018.
Lorenzen, K., Leber, K.M., Loneragan, N.R., Schloesser, R.W., Taylor, M.D., 2021. Developing and integrating enhancement strategies to improve and restore fisheries. *Bull. Mar. Sci.* 97 (4), 475–488. https://doi.org/10.5343/bms.2021.0036.
Lowry, M., Folpp, H., Gregson, M., Suthers, I., 2012. Comparison of baited remote underwater video (BRUV) and underwater visual census (UVC) for assessment of artificial reefs in estuaries. *J. Exp. Mar. Biol. Ecol.* 416–417, 243–253. https://doi.org/10.1016/j.jembe.2012.01.013.
Madgett, A.S., Elsdon, T.S., Marnane, M.J., Schramm, K.D., Harvey, E.S., 2023. The functional diversity of fish assemblages in the vicinity of oil and gas pipelines compared to nearby natural reef and soft sediment habitats. *Mar. Environ. Res.* 187, 105931. https://doi.org/10.1016/j.marenvres.2023.105931.
Mallet, D., Pelletier, D., 2014. Underwater video techniques for observing coastal marine biodiversity: A review of sixty years of publications (1952–2012). *Fish. Res.* 154, 44–62. https://doi.org/10.1016/j.fishres.2014.01.019.
Mao, Z., Xue, X., Tian, H., Michael, A.U., 2019. How will China realize SDG 14 by 2030?—A case study of an institutional approach to achieve proper control of coastal water pollution. *J. Environ. Manag.* 230, 53–62. https://doi.org/10.1016/j.jenvman.2018.09.028.
Mason, N.W.H., Mouillot, D., Lee, W.G., Wilson, J.B., 2005. Functional richness, functional evenness and functional divergence: The primary components of functional diversity. *Oikos* 111 (1), 112–118. https://doi.org/10.1111/j.0030-1299.2005.13886.x.
Matsuoka, T., Nakashima, T., Nagasawa, N., 2005. A review of ghost fishing: Scientific approaches to evaluation and solutions. *Fish. Sci.* 71(4), 691. https://doi.org/10.1111/j.1444-2906.2005.01019.x.
Maus, C., Cottingham, A., Bossie, A., Tweedley, J.R., 2024. Assessing the efficacy of a Bouchot-style shellfish reef as a restoration option in a temperate estuary. *J. Mar. Sci. Eng.* 12(1), 87. https://doi.org/10.3390/jmse12010087.
MOA, 2006. *Outline of Action for Aquatic Biological Resources Conservation in China*. http://www.moa.gov.cn/gk/zcfg/__deleted_2018.10.23_09.42.28__flfg/200605/t20060516_610826.htm (In Chinese) (accessed 2006).
Murphy, H.M., Jenkins, G.P., 2010. Observational methods used in marine spatial monitoring of fishes and associated habitats: A review. *Mar. Freshw. Res.* 61 (2), 236–252. https://doi.org/10.1071/MF09068.
Nakamura, M., 1985. Evolution of artificial fishing reef concepts in Japan. *Bull. Mar. Sci.* 37 (1), 271–278. https://www.ingentaconnect.com/content/umrsmas/bullmar/1985/00000037/00000001/art00020#Refs.
Noh, J., Ryu, J., Lee, D., Khim, J.S., 2017. Distribution characteristics of the fish assemblages to varying environmental conditions in artificial reefs of the Jeju Island, Korea. *Mar. Pollut. Bull.* 118 (1), 388–396. https://doi.org/10.1016/j.marpolbul.2017.02.066.
Ou, Z., Wu, Z., Zhang, Z., Tweedley, J., Dong, S., Wang, Z., Gao, D., Liu, M., Xing, B., Tian, T., 2021. A review of baited remote underwater video (BRUV) technology. *J. Fish. Sci. China* 28 (10), 1359–1372. https://doi.org/10.12264/JFSC2020-0333.
Pan, Y., Tong, H., Wei, D., Xiao, W., Xue, D., 2022. Review of structure types and new development prospects of artificial reefs in China. *Front. Mar. Sci.* 9, 853452. https://doi.org/10.3389/fmars.2022.853452.
Patterson, W.F., Dance, M.A., Addis, D.T., 2009. Development of a remotely operated vehicle based methodology to estimate fish community structure at artificial reef sites in the northern Gulf of Mexico. *Proc. Gulf Carribb. Fish. Inst.* 61, 263–270. https://generic.eprints.org:15730/core331.
Pickering, H., Whitmarsh, D., 1997. Artificial reefs and fisheries exploitation: A review of the 'attraction versus production' debate, the influence of design and its significance for policy. *Fish. Res.* 31 (1–2), 39–59. https://doi.org/10.1016/S0165-7836(97)00019-2.
Ramm, L.A., Florisson, J.H., Watts, S.L., Becker, A., Tweedley, J.R., 2021. Artificial reefs in the Anthropocene: A review of geographical and historical trends in their design, purpose, and monitoring. *Bull. Mar. Sci.* 97 (4), 699–728. https://doi.org/10.5343/bms.2020.0046.

Rieucau, G., Boswell, K.M., Kimball, M.E., Diaz, G., Allen, D.M., 2015. Tidal and diel variations in abundance and schooling behavior of estuarine fish within an intertidal salt marsh pool. *Hydrobiologia* 753 (1), 149–162. https://doi.org/10.1007/s10750-015-2202-8.

Rigby, C.L., Dulvy, N.K., Derrick, D., Dyldin, Y.V., Herman, K., Ishihara, H., Jeong, C.-H., Semba, Y., Tanaka, S., Volvenko, I.v., Yamaguchi, A., 2021. *Okamejei kenojei. The IUCN Red List of Threatened Species* e.T161645A124520681. https://doi.org/10.2305/IUCN.UK.2021-1.RLTS.T161645A124520681.en.

Schobernd, Z.H., Bacheler, N.M., Conn, P.B., 2013. Examining the utility of alternative video monitoring metrics for indexing reef fish abundance. *Can. J. Fish. Aquat. Sci.* 71 (3), 464–471. https://doi.org/10.1139/cjfas-2013-0086.

Schroeter, S.C., Reed, D.C., Raimondi, P.T., 2015. Effects of reef physical structure on development of benthic reef community: A large-scale artificial reef experiment. *Mar. Ecol. Prog. Ser.* 540, 43–55. https://doi.org/10.3354/meps.

Scott, M.E., Smith, J.A., Lowry, M.B., Taylor, M.D., Suthers, I.M., 2015. The influence of an offshore artificial reef on the abundance of fish in the surrounding pelagic environment. *Mar. Freshw. Res.* 66(5), 429. https://doi.org/10.1071/MF14064.

Seaman, W., 2000. *Artificial Reef Evaluation: With Application to Natural Marine Habitats*. CRC Press. https://doi.org/10.1201/9781420036633.

Sheehan, E.V., Bridger, D., Nancollas, S.J., Pittman, S.J., 2019. PelagiCam: A novel underwater imaging system with computer vision for semi-automated monitoring of mobile marine fauna at offshore structures. *Environ. Monit. Assess.* 192(1), 11. https://doi.org/10.1007/s10661-019-7980-4.

Song, L., Wang, N., Dong, J., Zhang, Y., Wen, Q., 2010. Impact of fishing on fishery ecosystem: A case study in Changshan archipelago area. *Chin. J. Ecol.* 29 (8), 1578–1584. https://doi.org/10.13292/j.1000-4890.2010.0261.

Streich, M.K., Ajemian, M.J., Wetz, J.J., Stunz, G.W., 2017. A comparison of fish community structure at mesophotic artificial reefs and natural banks in the Western Gulf of Mexico. *Mar. Coast. Fish.* 9 (1), 170–189. https://doi.org/10.1080/19425120. 2017.1282897.

Tang, Y., Wu, C., 1988. On biologic habits and resources distribution of common Japanese conger *Astroconger myriaster* (Brevoort). *J. Zhejiang Coll. Fish. (in Chinese)* 7 (1), 19–26.

Taylor, M.D., Baker, J., Suthers, I.M., 2013. Tidal currents, sampling effort and baited remote underwater video (BRUV) surveys: Are we drawing the right conclusions? *Fish. Res.* 140, 96–104. https://doi.org/10.1016/j.fishres.2012.12.013.

UN, 2015. *Resolution Adopted by the General Assembly on 25 September 2015. Transforming Our World: The 2030 Agenda for Sustainable Development*. General Assembly. Seventieth Session Agenda Items 15 and 116. October 21. https://www.un.org/ga/search/view_doc.asp?symbol=A/RES/70/1&Lang=E (accessed May 2021).

Vivier, B., Dauvin, J.-C., Navon, M., Rusig, A.-M., Mussio, I., Orvain, F., Boutouil, M., Claquin, P., 2021. Marine artificial reefs, a meta-analysis of their design, objectives and effectiveness. *Glob. Ecol. Conserv.* 27, e01538. https://doi.org/10.1016/j.gecco.2021.e01538.

Wakefield, C.B., Lewis, P.D., Coutts, T.B., Fairclough, D.V., Langlois, T.J., 2013. Fish assemblages associated with natural and anthropogenically-modified habitats in a marine embayment: Comparison of baited videos and opera-house traps. *PLoS One* 8 (3), e59959. https://doi.org/10.1371/journal.pone.0059959.

Wan, R., Zhang, T., Li, Z., Ren, Y., Song, P., 2020. Spatial distribution and inter-annual variability of spawning grounds of *Cynoglossus joyneri* in the Yellow Sea coastal waters in summer. *Chin. J. Appl. Ecol.* 31 (3), 1023–1032. https://doi.org/10.13287/j.1001-9332.202003.034.

Wang, Q., Zhuang, Z., Deng, J., Ye, Y., 2006. Stock enhancement and translocation of the shrimp Penaeus chinensis in China. *Fish. Res.* 80 (1), 67–79. https://doi.org/10.1016/j.fishres. 2006.03.015.

Wang, Z., Chen, Y., Zhang, S., Wang, K., Zhao, J., Xu, Q., 2015. A comparative study of fish assemblages near aquaculture, artificial and natural habitats. *J. Ocean Univ. China* 14 (1), 149–160. https://doi.org/10.1007/s11802-015-2455-x.

Whitmarsh, S.K., Fairweather, P.G., Huveneers, C., 2017. What is big BRUVver up to? Methods and uses of baited underwater video. *Rev. Fish Biol. Fish.* 27 (1), 53–73. https://doi.org/10.1007/s11160-016-9450-1.

Wu, Z., Tweedley, J.R., Loneragan, N.R., Zhang, X., 2019. Artificial reefs can mimic natural habitats for fish and macroinvertebrates in temperate coastal waters of the Yellow Sea. *Ecol. Eng.* 139, 105579. https://doi.org/10.1016/j.ecoleng.2019.08.009.

Wu, Z., Zhang, X., Lozano-Montes, H.M., Loneragan, N.R., 2016. Trophic flows, kelp culture and fisheries in the marine ecosystem of an artificial reef zone in the Yellow Sea. *Estuar. Coast. Shelf Sci.* 182, 86–97. https://doi.org/10.1016/j.ecss.2016.08.021.

Xu, S., Qiao, Y., Xu, S., Yue, S., Zhang, Y., Liu, M., Zhang, X., Zhou, Y., 2021. Diversity, distribution and conservation of seagrass in coastal waters of the Liaodong peninsula, North Yellow Sea, northern China: Implications for seagrass conservation. *Mar. Pollut. Bull.* 167, 112261. https://doi.org/10.1016/j.marpolbul.2021.112261.

Ye, Y., Gutierrez, N.L., 2017. Ending fishery overexploitation by expanding from local successes to globalized solutions. *Nat. Ecol. Evol.* 1(7), 0179. https://doi.org/10.1038/s41559-017-0179.

Yu, J., Zhang, L., 2020. Evolution of marine ranching policies in China: Review, performance and prospects. *Sci. Total Environ.* 737, 139782. https://doi.org/10.1016/j.scitotenv.2020.139782.

Yuan, H., Zhang, S., Chen, P., 2024. Research progress and prospects on benefit assessment of marine ranching. *South China Fish. Sci.* 20, 1–13. https://doi.org/10.12131/20240171.

Zhang, F., Li, M., Chen, S., Luo, G., 2022. National marine ranch demonstration zone achieves remarkable results. *China Fish.* (09), 27–29. https://doi.org/CNKI:SINICA.0.2022-09-010.

Zhang, L., Tian, T., Ma, Z., Liu, Y., Yang, J., Liu, H., 2018. Water quality assessment of Dachangshan artificial habitat development demonstration area. *Anim. Husb. Feed Sci.* 10 (4), 226–230. https://doi.org/10.16535/j.cnki.dlhyxb.2018.03.017.

Zhang, W., Ye, Z., Tian, Y., Yu, H., Ma, S., Ju, P., Watanabe, Y., 2022. Spawning overlap of Japanese anchovy *Engraulis japonicus* and Japanese Spanish mackerel *Scomberomorus niphonius* in the coastal Yellow Sea: A prey–predator interaction. *Fish. Oceangr.* 31 (4), 456–469. https://doi.org/10.1111/fog.12595.

Zhang, X.M., Ji, Q.Y., Hu, C.Y., Xu, H.Z., Wang, Y.H., Yang, X.L., Guo, H.Y., 2023. Ecosystem stability of marine ranching and its response to disturbance: Research status, issues, and suggestions. *J. Fish. China* 47 (11), 119509. https://doi.org/10.11964/jfc.20221213862.

Zhang, Y., Ou, Z., Tweedley, J.R., Loneragan, N.R., Zhang, X., Tian, T., Wu, Z., 2024. Evaluating the effectiveness of baited video and traps for quantifying the mobile fauna on artificial reefs in northern China. *Journal of Experimental Marine Biology and Ecology* 573, 152001. https://doi.org/10.1016/j.jembe.2024.152001.

Zhang, Y., Xu, Q., Xu, Q., Liu, H., Zhang, H., Hao, J., Yang, H., 2016. Demersal fish community structure around the Qiansan islets in Haizhou Bay. *J. Fish. Sci. China* 23 (1), 156–168. https://doi.org/10.3724/SP.J.1118.2016.15055.

Zhao, J., Jiang, D., Hao, Z., Guo, D., Wang, L., Li, D., Zhang, Y., 2019. Attaching effect of benthic macroalgae on a new artificial shell reef. *Fish. Sci.* 38(3), 7. https://doi.org/10.16378/j.cnki.1003-1111.2019.03.013.

# 8 The Ecological Importance of Offshore Pipelines as Artificial Reefs and Their Potential in Fishery Management

## *A Case Study in the Northwest Gulf of Mexico*

*Ann Scarborough-Bull*, Milton S. Love, Mary M. Nishimoto, Umarfarooq A. Abdulwahab, and Christopher M. Prosser*
* Corresponding author: ann.bull@ucsb.edu

## 8.1 INTRODUCTION

Global warming, primarily caused by the emission of gases through human activities, has the potential to impact ecosystems worldwide (summarized in The National Research Council 2020). Although a variety of gases (e.g., methane, ozone, and nitrous oxide) play a role in global warming, carbon dioxide ($CO_2$) is a major contributor, and methods to reduce its effects have been the goal of substantial research. As noted by the National Academies of Sciences, Engineering, and Medicine (2019): "Fossil fuel consumption, agriculture, land-use change, and cement production are the dominant anthropogenic sources of $CO_2$ to the atmosphere." While reducing carbon emissions remains a key approach to combatting global warming, a variety of management strategies for removing $CO_2$ from the atmosphere are either being used or have been proposed. These management options include reforestation, soil carbon sequestration, and peatland restoration, as well as geologic storage (Pett-Ridge et al. 2023). As to the latter, "Geologic storage is essential to many types of $CO_2$ removal, including DACS [Direct Air Capture with Storage] and BiCRS [Biomass Carbon Removal and Storage]". $CO_2$ removed directly from the atmosphere or captured in the process of creating energy or fuels must be durably stored to realize the long-term climate benefits of carbon removal. Geologic storage (i.e., where pressurized $CO_2$ is injected into deep porous rocks) is one of the most durable forms of $CO_2$ storage; properly sited and operated sites can be expected to retain more than 99% of injected $CO_2$ over at least 100 years. Storage in sedimentary rocks is the best studied and documented form of geologic $CO_2$ storage (Pett-Ridge et al. 2023).

There are extensive sedimentary areas in the northwest Gulf of Mexico (DeAngelo et al. 2019; Ringrose and Meckel 2019), and there is interest in developing marine infrastructure to transport $CO_2$ to these offshore fields. This is particularly the case in the northwest Gulf of Mexico, based on the "Gulf's unique geology and close proximity to heavy industries that emit significant amounts of carbon pollution" (Lowenthal 2022). More specifically, off the Texas coast where "The large number of industrial sources of carbon dioxide ($CO_2$) in coastal counties and the high density of onshore urbanization and environmentally sensitive areas make this offshore region extremely attractive for long-term storage of carbon dioxide emissions from industrial sources" (Meckel and Trevino 2014).

DOI: 10.1201/9781003676775-8 

Importantly, in all these scenarios, pipelines would carry the $CO_2$ to subsea geological formations. Sea-floor pipelines, carrying petroleum products and other commodities between offshore platforms and the shore, are a major fixture in the Gulf of Mexico. In the Gulf there are at least 21,138 km (13,135 miles) of active and 29,000 km (18,000 mi) of inactive pipelines (BSEE 2025a; Walsh 2023). The percentage of these pipelines that are either lying on the seafloor or buried in sediment is unknown. To our knowledge, there have been no published studies of how pipelines in the Gulf of Mexico impact marine organisms. In this study, we 1) present the northwest Gulf of Mexico area of interest, 2) discuss the artificial infrastructure within the area, 3) expand our discussion of pipelines as artificial reefs to include Australia, 4) present likely temporal and ontogenetic changes in species associations to pipelines, 5) compare pipeline assemblages to associated natural habitats, 6) discuss pipelines and ecological connectivity, 7) discuss pipelines and fisheries, 8) summarize the potential ecological importance for pipelines within the area of interest, 9) present the outstanding questions to be addressed for pipelines, 10) list possible future research efforts, and 11) present potential risks of these pipelines (e.g., biotic homogenization and the facilitation of non-native species) as artificial reef structures.

## 8.2 THE SETTING: NORTHWEST GULF OF MEXICO COASTAL CONTINENTAL SHELF

The Gulf of Mexico is a small ocean basin that originated approximately 160 million years ago (late Jurassic time) due to the rifting of the North American, African, and South American plates. The shallow, primordial Gulf has gradually evolved into its present configuration, was enlarged by continued spreading, deepened due to subsidence, and slowly filled through sedimentation. The spreading phase ended during the Cretaceous Period (approximately 80 million to 100 million years ago). Sedimentation and subsidence will continue as long as rivers flow into the Gulf from the continent and lime-secreting organisms continue to thrive in areas of low stream outflow (Darnell 2015).

The Mississippi River, because of its large drainage area, has been the major contributor of sediments to the Gulf throughout time and, as a consequence, has built an extremely large coastal plain/continental shelf complex in the northwest Gulf of Mexico. The continental shelf of Mexico, Texas, Louisiana, Mississippi, and Alabama is floored with terrigenous sediment containing varying amounts of silt and clay (Rezak et al. 1983). The coastal shelf is also subjected to freshwater input that alters the nearshore salinity gradients.

## 8.3 THE ROLE THAT HARD SUBSTRATE PLAYS IN MARINE WATERS

As noted by Michaelis et al. (2019), "Marine subtidal hard substrates . . . act as settling grounds, shelter, feeding, and nursery grounds for diverse sessile and mobile marine animal communities." Arguably, bottom depth and habitat complexity are two of the most important influences structuring marine benthic communities (e.g., Menge and Sutherland 1976; Martin-Smith 1993; Kovalenko et al. 2012; Sheehan et al. 2015; Hall and Kingsford 2021). Depth is important as it is related to such other environmental factors as temperature, dissolved oxygen content, and plankton availability (Baldwin et al. 2018; Stefanoudis et al. 2019; Ierodiaconou et al. 2023).

Summarizing the importance of habitat complexity, Lavender et al. (2017) state, "Enhancement of habitat complexity is often cited as a key mechanism by which ecosystem engineers [those organisms, such as corals and sponges, that cause changes in the physical state of the environment] increase diversity. Alteration of habitat complexity may modify a number of ecological processes, such as resource availability, recruitment, and predation leading to the development of unique assemblages. Complex habitats may modify predator-prey interactions by providing refuges from predators and decreasing predatory foraging efficiency. However, where abundances of predators are benefited by complexity, predation may increase. Greater niche space in high complexity may promote co-existence by reducing competition. Complex habitats may also act to alter resource availability or colonization dynamics."

Specifically, structurally complex habitats tend to have higher densities and numbers of both invertebrate and fish species than do less complex habitats (Lenihan and Peterson 1998; Sheehan et al. 2015). It is important to note that, even within a reef, some organisms are limited to specific microhabitats. As an example, on the rock reefs off California, a suite of fish species characterized as members of the "sheltering habitat guild" are only found associated with crevices and caves (Love et al. 2006). A similar pattern of species associated with specialized microhabitats was also noted in tropical artificial reefs (Hixon and Beets 1989; Scarborough-Bull and Kendall 1994). As stated by Love et al. (2006), when determining or predicting the species assemblages of a reef, "[i]t is not sufficient to distinguish only between soft and hard bottom types when using habitat to guide fisheries management strategies." Thus, subtle differences within a reef, such as the number and size of sheltering sites, can profoundly affect the associated fish assemblages.

## 8.4 NATURAL INFRASTRUCTURE ON THE NORTHWEST GULF OF MEXICO COASTAL CONTINENTAL SHELF

In general, within the central and western Gulf of Mexico, sediment discharged from the Mississippi River creates extensive mud-bottom habitat (Rezak et al. 1990). Drowned banks are typically submerged freshwater river features, whereas drowned barrier islands are former coastal islands that were submerged by rising sea levels. The inner continental shelf of the northwest Gulf has limited hard structure (e.g., natural reefs made of coral and limestone) and is composed primarily of mud and sand, with interspersed natural shell rubble banks, and a few drowned bank areas (Rezak et al. 1990). Drowned barrier islands and natural banks become prominent features on the inner shelf off Texas and represent relic depositional environments comprised of Eastern Oyster beds (*Crassostrea virginica*) and other shell fragments as dominant features (Rezak et al. 1990; Rodriguez et al. 2004). Oyster beds, combined with a gradation of depths and a variety of substrate types (i.e., mud, sand, and shell) over these drowned barrier islands, provide important habitat to fishes and invertebrates on the inner continental shelf. Drowned barrier islands and oyster and shell deposits (e.g., Sabine and Heald Banks, the Freeport Rocks Bathymetric High (Figure 8.1), and adjacent mud-bottom substrates) located on the inner continental shelf of the northwest Gulf, are important natural habitats. A bathymetric high (ridge) extends northeast to southwest for 20 km, with the ridge crest located between 15 m and 20 m depths. Geologic investigations indicate that substrates along the crest are composed of shell hash and sand, with patches of relic oyster beds, while adjacent mud bottoms are present on both the inshore and offshore areas surrounding the ridge crest. Freeport Rocks Bathymetric High is a relic barrier island on this ridge, with radiocarbon dating suggesting the bank is 37,000–45,000 years old (Rodriguez et al. 2004; Mikulas and Rooker 2008). There are several other submerged banks in its vicinity, presumably formed due to rapid sea-level rise during the late Holocene as remnants of beach ridges (Rodriguez et al. 1999).

A study by Wells et al. (2009) over a natural drowned barrier island in the northwest Gulf of Mexico indicated that the most abundant species collected were predominantly estuarine-associated fishes (e.g., members of the drum and croaker family, Sciaenidae). Fish assemblages on naturally occurring unconsolidated sand or mud bottom in close proximity to artificial reefs (within 60–120 m) indicate that there is persistent and substantial connectivity with the surrounding habitat, acting as a source of biodiversity for artificial reefs (Bohnsack 1989; Rosemond et al. 2018). In addition, low-relief banks (i.e., sand-shell ridges) are prominent features on the inner continental shelf and, along with associated non-structured (i.e., mud bottom) habitats, serve as post-settlement

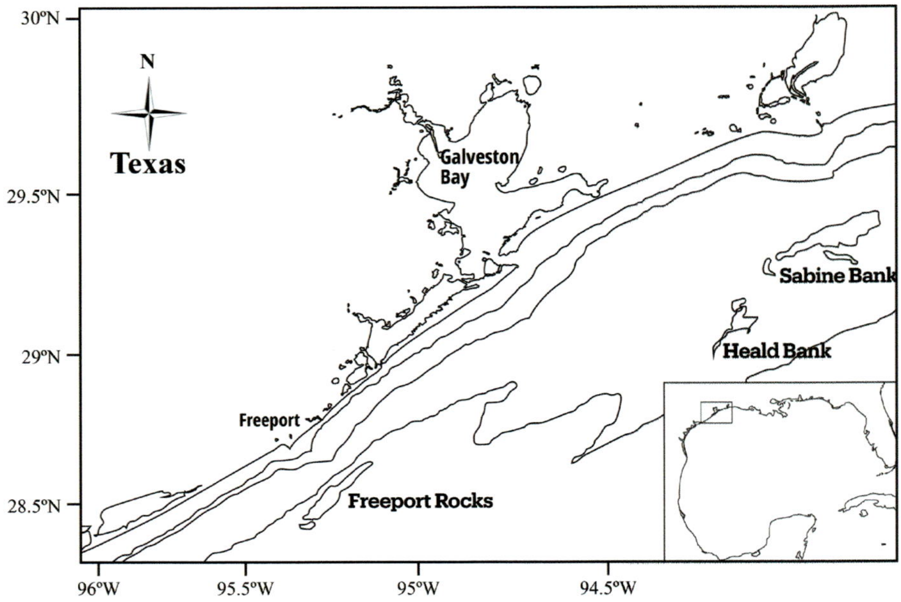


FIGURE 8.1 Inner shelf position of relic banks in the northwest Gulf of Mexico. Mikulas and Rooker (2008).

habitat for some fish species (i.e., snapper, family Lutjanidae; Rooker et al. 2004; Wells et al. 2009). The habitat value of natural sand and relic banks in the northwest Gulf of Mexico may not be equal in all areas due to seascape differences in freshwater/nutrient input, depth, currents, and water chemistry. Sedimentary attributes may also differ due to variation among areas in hydrology and particle source (Brooks et al. 2004). Thus, sandy areas on the continental shelf provide a heterogeneous landscape characterized by differences in sediments, energy regime, and water chemistry (Bortone 1971; Bergen et al. 2001; Sisson et al. 2002). While habitat complexity (i.e., refuge) typically reduces predation-mediated mortality (Hixon and Beets 1989), the relative importance of habitats on these natural banks is still undetermined for post-settlement, economically important species of reef fishes.

## 8.5 ARTIFICIAL REEF INFRASTRUCTURE ON THE NORTHWEST GULF OF MEXICO CONTINENTAL SHELF

The physical footprint (i.e., area covered on the seafloor) of artificial reefs in the ocean off the United States (U.S.) is approximately 19.23 sq km (7.5 sq mi) (Paxton et al. 2023). This area was calculated by compiling spatial data from all 17 U.S. states with artificial reef programs. While this footprint is small compared to the total seafloor, it has increased significantly over the past 50 years, though the rate of expansion has slowed in the last decade (Paxton et al. 2023). It is important to note that artificial reefs may be purpose-built, with the aim of, for instance, fisheries enhancement, habitat repair, or providing recreational opportunities (e.g., Becker et al. 2018; Paxton et al. 2020), or maybe the by-products of such activities as fossil fuel extraction, mariculture, breakwater creation, or shipwrecks.

There is a rich and robust literature regarding the roles that artificial structures in the marine environment may play as habitat for organisms. This research, some of which is summarized in Lima et al. (2019), strongly demonstrates that, as with natural reefs, the biotic communities of these structures and thus their effects on the ecosystem vary greatly and likely reflect such parameters as reef structural complexity, reef siting, and even environmental stochastic processes. The extensive network of oil and gas structures in the Gulf of Mexico, connected by pipelines, may act as an exceptionally large artificial reef ecosystem. This network influences species distribution across the Gulf based on such environmental factors as distance from shore and water conditions. As an example, while some platforms support coral species typical of natural hard bottoms, they also harbor invasive species like the Orange Cup Coral (*Tubastraea coccinea*), demonstrating both the ecological benefits and challenges of these structures.

In the 1980s, Texas was the second Gulf state to initiate an artificial reef program, and currently almost all of these reefs off Texas are based on oil and gas platforms (Stephan et al. 1990; TPWD 2024). In 1989, the Texas Legislature directed the Texas Parks and Wildlife Department (TPWD) to promote, develop, maintain, monitor, and enhance the artificial reef potential in state and federal waters adjacent to Texas to enhance fishery resources and commercial and recreational fishing opportunities. Thirteen ships have been intentionally sunk as part of the Texas Artificial Reef Program, the largest being the USTS Texas Clipper, a 145 m (473 ft) long vessel. There are 82 artificial reef sites off Texas, including 18 separate large areas designated for Rigs-to-Reefs, with several along the inner shelf within the area of interest for this study (Scarborough-Bull and Love 2019; TPWD 2024). Rigs-to-Reefs is the TPWD program that uses petroleum platforms as artificial reefs after they are decommissioned and cleaned. Some of the decommissioned platforms in the Texas Rigs-to-Reefs program remain in the same location where they were once actively producing petroleum, and some are moved to new locations (Scarborough-Bull and Love 2019).

The Texas Artificial Reef Program, created in 1990 as a program within the Coastal Fisheries Division of the Texas Parks and Wildlife Department, promotes, develops, maintains, monitors, and enhances artificial reefs in Texas offshore waters (Scarborough-Bull and Love 2019). Multiple state-designated artificial reefs, managed by TPWD, are present in the northwest Gulf of Mexico (Figure 8.2), which provide habitat for aquatic life. To enhance marine life in the Gulf, the Texas Artificial Reef Program has placed 45 decommissioned petroleum platforms as Rigs-to-Reefs, fly-ash blocks, pipe, concrete blocks, and obsolete ships in 34 different sites along the Texas coast. Liberty ships were a class of cargo ship built in the United States during World War II under the Emergency Shipbuilding Program. Twelve Liberty ships continue as artificial reefs off the coast of Texas with two, the George Vancouver Liberty Ship and Freeport Liberty Ship reefs (Figure 8.2), within sites in the northwest Gulf.

## 8.6 OIL AND GAS INFRASTRUCTURE IN THE NORTHWEST GULF OF MEXICO COASTAL SHELF SETTING

Offshore oil and gas platforms and pipelines occupy many areas on continental shelves throughout the world's oceans. Over the course of their decades-long presence as artificial reefs, these structures become ecologically and economically important habitats that support a wide variety of marine life (van Elden et al. 2022). From a biological viewpoint, increasing evidence suggests that offshore oil and gas infrastructure provides significant ecosystem services while present (Stanley and Wilson 2003). The installation of these platforms and pipelines results in the formation of artificial reefs that create a hard substrate in open waters, which is colonized by a variety of sessile organisms (Scarborough-Bull 1989; Ajemian et al. 2015). Because the physical presence of hard seafloor obstructions, such as pipelines, may exclude some gear types used in commercial and recreational fishing, particularly trawling, pipelines may also act as important de facto marine preserves and refuges for a variety of taxa (Frumkes 2002; Claisse et al. 2014).

The largest artificial reef complex in the world exists in the northwest and north-central Gulf of Mexico, where more than 2,500 petroleum platforms, with pipeline connections to each other and/

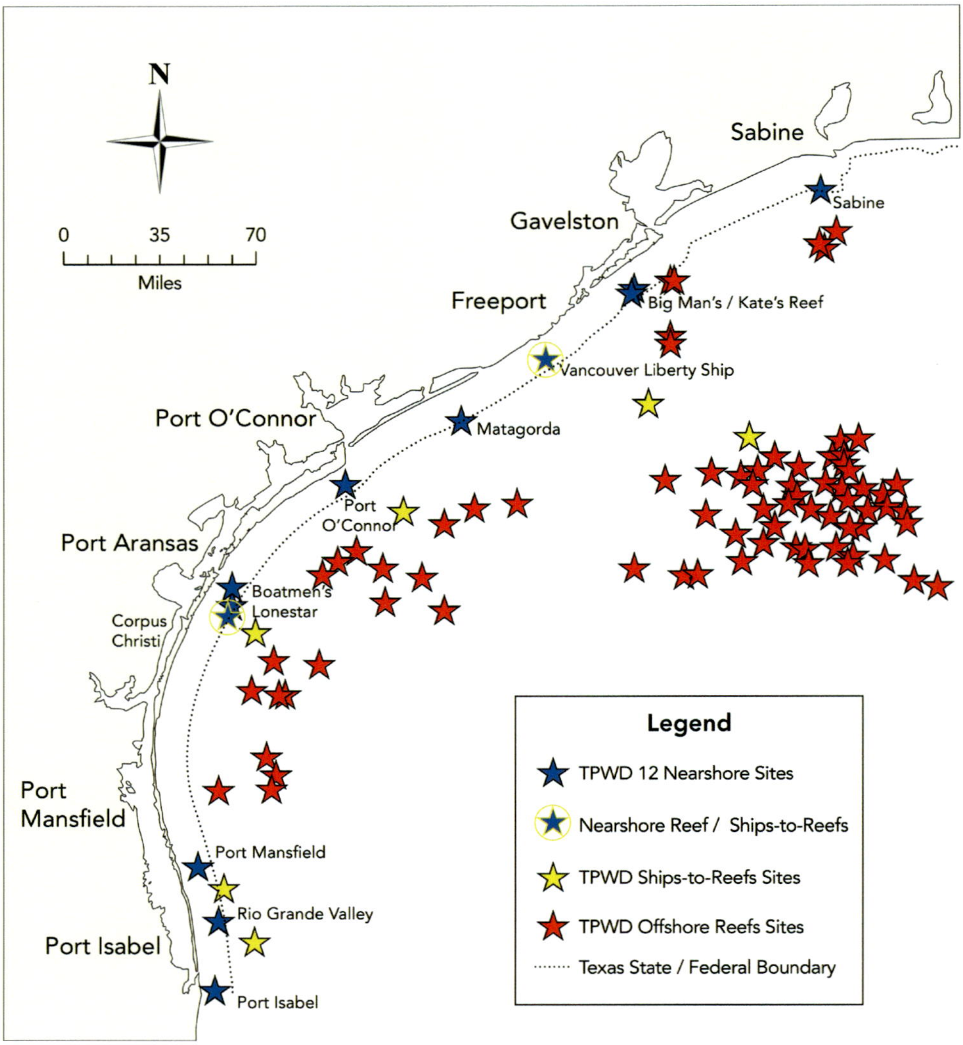


**FIGURE 8.2** The distribution of artificial reef sites managed by the Texas Parks and Wildlife Department within the northwest Gulf of Mexico. The majority of sites lie seaward of the Texas state/federal waters delineation along the inner shelf as denoted by dashed line (TPWD 2024).

or directly to shore, are scattered across the region (Scarborough-Bull and Love 2019) (Figure 8.3) (NOAA NCEI, 2025). Additionally, about 600 platforms have been designated as artificial reefs and are now owned and managed by the Gulf states (i.e., Texas, Louisiana, Alabama, Mississippi, and Florida; Scarborough-Bull and Love 2019; BSEE 2025b). Scientists have hypothesized that artificial reefs, including oil and gas platforms, improve and/or diversify habitat. These structures increase fishery resources, modify the assemblages of organisms in the region, or concentrate existing resources (Davis et al. 1982; Bohnsack et al. 1991). Recent studies suggest that although fish community structure may differ between natural habitat and offshore platforms in the northwest Gulf of Mexico, platforms and pipelines may serve as important habitat for snapper species by potentially diverting fishing pressure from natural habitats; however, future studies that address species-specific life history traits will be needed to better understand the function and performance of artificial infrastructure in supporting fisheries productivity (Streich et al. 2017).

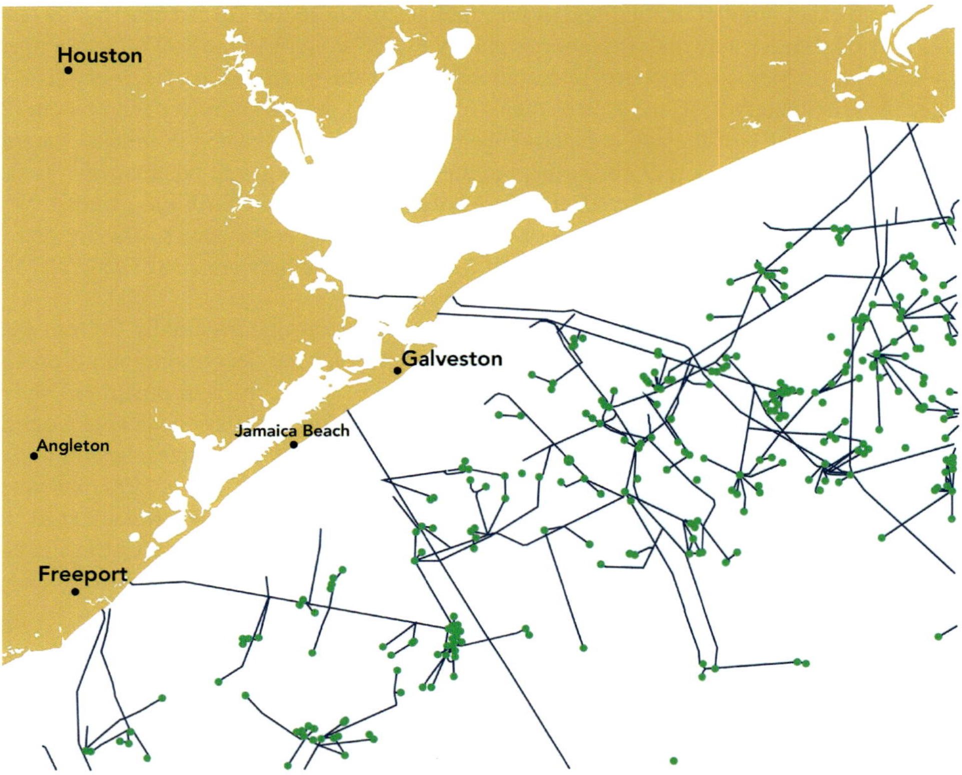


**FIGURE 8.3** Oil and gas platforms (dots) with pipeline connections (lines) to each other and/or directly to shore within the northwest Gulf of Mexico. Source: NOAA National Center for Environmental Information. Depicted sites are located within federal waters beyond the Texas inner shelf (NOAA NCEI 2025).

In the northwest and north-central Gulf, where most of the natural habitat is sedimentary, few representatives of local and coastal fauna are associated with offshore platforms and pipelines (Sonnier et al. 1976; Gallaway and Lewbel 1982; Scarborough-Bull 1989). It is assumed that originally many of the epibiotic organisms and fish species that now inhabit Gulf offshore pipeline infrastructure were transported to these structures as pelagic larvae for extended periods over relatively long distances; however, these organisms may have also been transported via shipping or as attached epifauna on large semisubmersible drilling rigs transported from South America. Once settled on a Gulf platform or pipeline and reached maturity, larvae would subsequently be able to populate nearby structures from already populated man-made habitat.

## 8.7 PIPELINES AS ARTIFICIAL REEF INFRASTRUCTURE

McLean et al. (2022b) state, "Subsea pipelines and wells installed to support the oil and gas industry represent some of the most extensive and numerous anthropogenic structures throughout global marine ecosystems." However, they also state that "globally, there is a paucity of knowledge on the influence of subsea wells and pipelines on marine ecosystems." To that point, there have been no studies of the marine life associated with subsea pipelines in the Gulf of Mexico. However, we believe that the studies that have been conducted around subsea pipelines in other parts of the world, particularly in the tropical waters of Western Australia, can give insights into how pipelines in the Gulf of Mexico may function as fish and invertebrate habitats (Bond et al. 2018a, 2018b, 2018c, 2018d; McLean et al. 2017, 2020).

There is no exact number or total length for pipelines offshore the Texas coast, but Texas is an integral part of a larger network with approximately 21,138 km (13,135 miles) of active pipelines within the Gulf of Mexico. Offshore Texas, about 40% of the pipelines are 5–20 cm (2–8 in) in diameter. Approximately 25% are greater than 46 cm (18 in), with 9% greater than 68 cm (27 in) in diameter. About 91% of the active pipeline length in the Gulf is located in federal waters, and 9% is in state waters. At 15 km (9 nmi) offshore, the state/federal seaward boundary for Texas and Florida is greater than other Gulf coastal states' boundary delineation at 5 km (3 nmi). While a specific number for the northwestern area of Texas is not available, the state's offshore petroleum industry is a significant part of the total Gulf of Mexico pipeline network (BSEE 2025a; INGAA 1996).

Within the northwest Gulf, pipelines extend to/from platforms and eventually to/from shore in major gathering lines (Figure 8.4). It is important to note that the decommissioning process of removing obsolete oil and gas infrastructure considers the fate of obsolete pipelines. McGinnis et al. (2001) note that "[b]oth federal and state regulations allow decommissioned Outer Continental Shelf (OCS) pipelines to be abandoned in place so long as they do not constitute a hazard to navigation, commercial fishing, or unduly interfere with other uses of the OCS." As of 1997, few pipelines had been removed in the course of decommissioning in the Gulf of Mexico (Breaux et al. 1997). That trend continued with most offshore pipelines abandoned in place offshore California as well as within the Gulf of Mexico (Cranswick 2001; McGinnis et al. 2001; Scarborough-Bull and Love 2019).

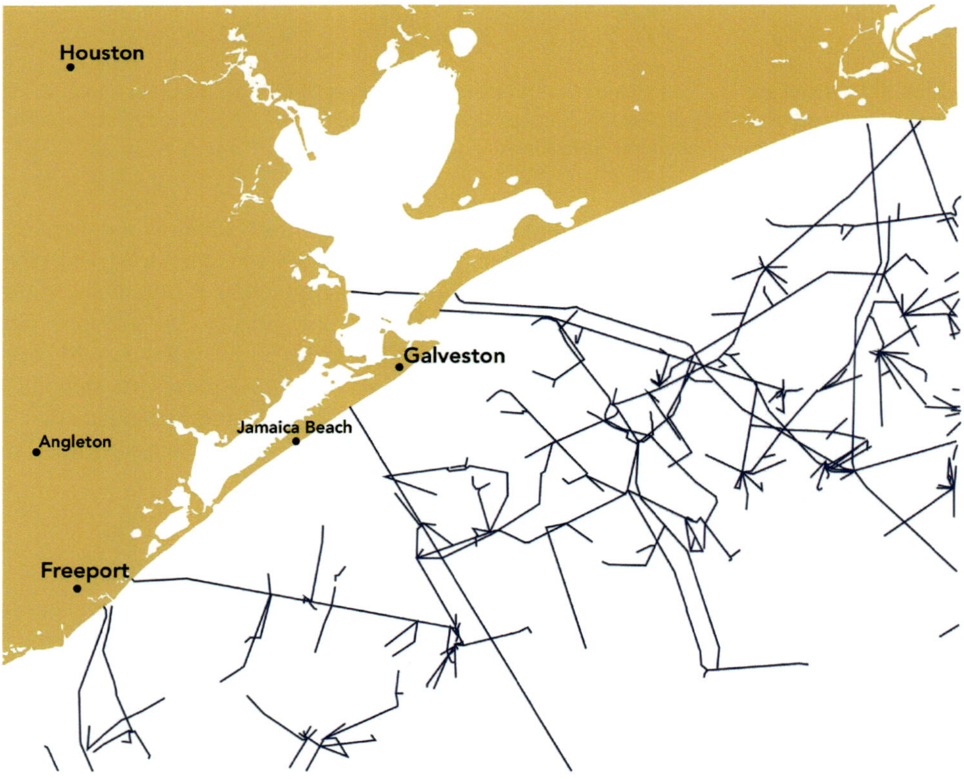


**FIGURE 8.4** Location of active pipelines within the northwest Gulf of Mexico. Several pipelines are sometimes bundled together but are denoted and mapped here as a single line. (NOAA NCEI 2025, BSEE 2025).

While most pipelines have relatively small diameters (0.05 to 1.83 m) (about 2 to 72 inches) and smooth surfaces, when compared to the generally featureless seafloor on which they are situated, the pipelines are often the greatest physical relief and the most complex structure in the area. As noted by Lacey and Hayes (2020),

> The degree to which pipelines and their covering provide a locally novel habitat will depend on the nature of the receiving environment. In soft sediment areas, surface laid pipelines would be expected to create a high degree of local-scale biodiversity through the addition of novel hard substrate, as found for other artificial structures.

Importantly, the complexity of pipelines is directly related to both increased diversity and abundances of a wide range of both economically and non-economically important invertebrates and fishes (e.g., Love and York 2005; Love et al. 2003; Redford et al. 2021; Schramm et al. 2021). Some of the fish that associate with caves or overhangs and pipeline spans and platform support junctions may be places where these species occur (McLean et al. 2022b). Such relationships may help explain why higher abundances of demersal fishery species have been observed on subsea pipelines compared to surrounding sand-dominated communities (Bond et al. 2018a, 2018c; McLean et al. 2022b). Beyond creating novel hard substrates, a number of factors, and in particular the following attributes presented below, can add complexity to a pipeline and thus, even further, contribute to increasing local habitat complexity.

*Spans*: Perhaps the aspect of pipeline complexity that has been most extensively documented to increase species biodiversity and abundance of reef-oriented species, particularly of fishes, is the presence of open pit areas directly below bridging lengths of a pipeline called a span (e.g., McLean et al. 2017, McLean 2022b; Bond et al. 2018a; Redford et al. 2021; Schramm et al. 2021). Spans are created through hydrodynamic current flow that creates scouring in the sediments below the ridge pipeline; the degree of scouring is dependent on a pipeline's orientation to current, the strength of that current, and the composition of the seafloor on which the pipeline is situated (Sumer et al. 2001). It is important to note that even if pipelines are trenched (i.e., partially submerged) and backfilled, "The proportion of pipeline buried . . . and how exposed above the seabed unburied sections are[,] is likely to change over time (especially in high current areas) due to sediment transport and scouring processes" (Schramm et al. 2021). In addition, there is evidence that some "ecosystems engineer" taxa (Wright and Jones 2006), such as lobsters and some fish species (Leckie et al. 2016), excavate the soft sediments under pipelines, creating or accelerating span production. Spans likely increase the diversity and abundance of reef species by providing refuge from predators and (from a predator's perspective) a concomitant increase in prey abundance (Bond et al. 2018a; McLean et al. 2022b). Note that current patterns can create sediment buildup on one side of a pipeline and parallel depletion on the other, leading to differences in invertebrate or fish abundances or assemblages between pipeline sides. This was noted for fish assemblages around submarine power cables off southern California (Love et al. 2017).

*Protective Structures*: Subsea pipelines are subject to damage from such factors as hydrodynamic erosion, snagging by fishing gear or ship anchors, and objects dropped from vessels (Jiang et al. 2020). A number of protective materials, including quarry rock and concrete blocks linked together with steel cable, called concrete pads or mattresses, have been deployed to protect pipelines (Redford et al. 2021). Several studies have demonstrated that both quarry rock and protective concrete pads, which increase surface area and create crevices, have a positive relationship with fish abundances and epifaunal densities (Rouse et al. 2019; Lacey and Hayes 2020; Redford et al. 2021).

*Structure-Forming Invertebrates*: Invertebrates such as sponges and corals are often termed "structure-forming" because other animals utilize them for shelter or as orientation sites. Structure-forming invertebrates are common components of pipeline organismal assemblages and tend to provide a crucial relationship between the reef surface and reef fishes (McLean et al. 2017, 2020; Bond et al. 2018a, 2018b). As an example, Sih et al. (2022), in a study conducted in southern Australia, noted that "[a] combination of epibenthic height and pipeline position were important variables for explaining fish total abundance and richness. Total fish abundance was higher when more complex epibenthic communities were present and whether the pipeline was more exposed . . . percent cover of sponges were the main drivers of species richness."

Lastly, regarding both fish and invertebrates: Where pipelines traverse soft sediments, the species assemblages are quite different (McLean et al. 2022b). However, where natural, complex habitats are situated close to pipelines, the two species assemblages can be similar (Bond et al. 2018b, 2018d).

## 8.8 PIPELINE ASSEMBLAGES COMPARED TO ASSOCIATED NATURAL HABITATS

Three studies have compared a pipeline-associated fish assemblage with those at nearby natural reefs and on the soft substrate sea floor (Love and York 2005; Schramm et al. 2021; Madgett et al. 2023). Two of these studies (Schramm et al. 2021; Madgett et al. 2023; based on data from the same surveys) analyzed these assemblages at sites in the shallow waters (10.6–20 m depths) off Western Australia. These analyses demonstrated that the pipeline assemblages had some, but not complete, species overlap with the natural reefs and little faunal similarity with the soft sediment. These studies also found that the biomass associated with the pipeline and natural reefs was similar and, on average, about 20 times greater than those associated with soft sediment. Similarly, from the third study, overall fish densities were six to seven times greater at a pipeline in comparison to adjacent soft-sediment sea floor off southern California (Love and York 2005). While it is plausible that similar relationships would occur in the northwest Gulf of Mexico, the structural complexity of both the pipeline and associated natural reefs would be a major habitat attribute and would have to be taken into consideration in any study.

## 8.9 TEMPORAL AND ONTOGENETIC CHANGES TO PIPELINE SPECIES ASSOCIATIONS

The species associated with both natural and man-made marine habitats can change in response to a variety of factors that influence the organisms in that biome on both a short-term and long-term basis. Factors that may influence the movements of these organisms include ontogeny, water conditions (seasonal or short-term), food availability, and reproduction. Thus, associations between marine organisms and subsea pipelines may not necessarily be permanent. For instance, the habitat requirements of a taxon may change as its members grow and mature. As an example, Love and York (2005) reported that most of the individuals of larger rockfish species (*Sebastes* spp.) associated with a subsea pipeline off California (USA) were juveniles.

There has been no research on the intra-annual or seasonal variability of pipeline invertebrate and fish assemblages. In addition, there have been few longer-term (i.e., more than one year) studies on pipeline species' associations; thus, the stability of these associations over time is mostly unknown. For instance, we know of only two studies (Bond et al. 2018a; Redford et al. 2021) that have examined differences in day–night assemblages on pipelines. Redford et al. (2021) conducted surveys along pipelines in the North Sea and found differences in the abundance of some invertebrates and fishes between day, night, and crepuscular periods. Bond et al. (2018a), in a study of Western Australia, found that "[t]he fish assemblage sampled along the pipeline during the day differed markedly to that sampled at night. Several ubiquitous predatory species . . . were present during the day but not at night, likely moving off the pipeline to feed at nearby habitats." Parenthetically, this latter study demonstrates that the fish assemblages associated with the pipeline might have a direct influence on the nearby natural habitats.

Similarly, only two studies, both in Australia, conducted repeated, interannual surveys over the same sections of pipeline. Ierodiaconou et al. (2023) characterized both invertebrate and fish assemblages and reported that

> even considering large differences in video quality, fish, and mobile invertebrate assemblages along two flow lines [pipelines] (Flowline 5 and Umbilical 8) changed among years. Assemblages observed in 2020 had higher abundance and species richness than those in previous years for Umbilical 8, while Flowline 5 only had higher abundance for fish in 2020, but not for species richness or for mobile invertebrate abundance or richness, with these remaining relatively constant over time.

McLean et al. (2017), who surveyed only fish assemblages, found that assemblage composition, species richness, and the total number of individuals observed differed among years.

## 8.10 PIPELINES AND ECOLOGICAL CONNECTIVITY

The placement of any artificial structure in marine waters will have at least some effect on the local physical and biological landscape. We note at the outset that whether these effects are "positive" or "negative" depends on one's frame of reference. And, arguably, effects will be larger the greater the structural disparity between the natural sea floor and the artificial reef. This is likely particularly true when this placement creates novel, hard habitat on otherwise soft-bottom seafloors. Among what some might consider negative impacts are "biotic homogenization" [the reduction of biodiversity] (Dafforn et al. 2015), the provision of pathways for the invasion of non-native species (Airoldi et al. 2015), habitat fragmentation, or the creation of long-distance migration corridors for marine species (Bond et al. 2018c).

To that effect, McLean et al. (2022a, 2022b) reviewed the roles that oil and gas structures, including submarine pipelines, might play in "maintaining, altering or enhancing ecological connectivity with natural marine habitats" (McLean et al. 2022a). And, specifically referring to marine fossil fuel infrastructure (e.g., platforms, pipelines, and wellheads), McLean et al. (2022b) reinforced and summarized this hypothesis by noting that "negative impacts can include facilitating the establishment and spread of non-native species . . . disturbing habitats . . . and interfering with hydrodynamic processes and sedimentation patterns." They further state that in circumstances where there is the creation of novel, hard habitat on otherwise soft-bottom seafloors, perceived benefits of pipeline deployment may include that the "structures can facilitate larval settlement and recruitment . . . increase the biomass of fish . . . and promote biodiversity." McLean et al. (2022b) noted that there are few research papers dealing with this subject. Only Russell et al. (2014), Arnould et al. (2015), and Bond et al. (2018d) deal directly with pipelines. In these studies, pinnipeds foraged along pipelines, helping to shape and direct species movements. Overall, however, the degree to which pipelines create any of the impacts listed by McLean et al. (2022b) is unknown.

## 8.11 PIPELINES AND FISHERIES

Many reef fishes are of economic importance, and subsea pipelines often exhibit higher abundances of these species compared to these same species associated with soft seafloors (McLean et al., 2022b). This observed species/habitat association feature leads some fishermen to preferentially target pipelines (Rouse et al. 2017, 2018, 2020; McLean et al. 2022b). As noted by McLean et al. (2022b),

> [t]he concentration of fishery target species on subsea pipelines in the northwest (NW) Australia has influenced commercial trap fishing in the region with fishing effort on pipelines about 11.4 times greater per $km^2$ than effort in the rest of the area. . . . We predict positive relationships between fishery target

species abundance and the complexity of colonizing benthic [invertebrate] communities and the availability of shelter (i.e., pipeline spans, well height).

One of the more important recreational and commercial fisheries in the Gulf of Mexico is the Red Snapper (*Lutjanus campechanus*). It is recognized by many scientists and most fishermen that the presence of offshore infrastructure in the northern Gulf of Mexico has positively affected the distribution and increased the population of Red Snapper by the addition of hard substrate habitat (Peabody and Wilson 2006; Shipp and Bortone 2009; McLaughlin 2013). The major Red Snapper fishing grounds from the commercial industry's inception in the mid-19th century until the mid-20th century were off the nearshore west coast of Florida, the Florida Panhandle with some edging toward Alabama, and the Campeche grounds of southeast Mexico, with few landings off Mississippi, Louisiana, or southward to the Texas–Mexico border (Shipp and Bortone 2009). However, the installation of petroleum platforms and pipelines across the northwest and north-central Gulf of Mexico, and the expansion of Red Snapper into these habitats during the last 60+ years, corresponds with major shifts in harvest locations and areas of Red Snapper concentration (Nieland and Wilson 2003; McLaughlin, 2013). This suggests that habitat was a limiting factor for snapper population abundance during the first 100 years of the fishery and that platforms unintentionally increased reef-like habitat across the Gulf that affected the carrying capacity of Red Snapper in the region (Linton 1994; Nieland and Wilson 2003; Shipp and Bortone 2009).

Offshore infrastructure and fisheries of the Gulf of Mexico continue to coexist. Nearly 90% of commercial Red Snapper landings, approximately 1.8 million kg (4 million pounds) per year, originate in Louisiana waters (McLaughlin, 2013). While the exact amount harvested at petroleum platforms is unknown, it does represent a significant portion of the harvest. In addition to commercial fishers frequenting offshore structures, recreational fishers and SCUBA divers are common platform visitors. Recreational fishing surveys found that 70% of all fishers in coastal Louisiana utilized petroleum platforms as fishing destinations and that the angler catches at these structures were the highest (in comparison to any other Gulf fishing sites) in the published scientific literature (Dimitroff 1982; Witzig 1986; Reggio 1994; Dauterive 2000; McKay et al. 2000; Hiett and Milon 2002).

Streich et al. (2017) found that, although fish assemblages may differ between artificial structures (i.e., platforms, pipelines) and natural reefs or hard-bottom areas in the northwest Gulf, artificial structures attract and retain Red Snapper. Gallaway et al. (2009) found that offshore platforms play a crucial role in the increase of Red Snapper within the western Gulf of Mexico. Red Snapper are attracted to structure or reef habitats at all ages. Pipelines may be especially attractive and useful for shelter and protection to young Red Snapper, whose dominant surrounding habitat is a sandy, featureless bottom. Bond et al. (2018c) found this to be an important role for local reef fishes associated with pipelines off Australia. The present-day Gulf of Mexico Red Snapper fishery is heavily dependent on catches at artificial reefs and offshore infrastructure. Both Streich et al. (2017) and Gallaway et al. (2009) found that Red Snapper production in the northern Gulf likely has been increased by the establishment of significant numbers of artificial structures, including pipelines. Pipelines traverse great distances, depths, geographic locations, and their associated fishes and invertebrates may also enhance local biodiversity.

Whether an increased abundance of commercially valuable species is positive or negative requires more research. First, it has not been possible to ascertain whether those sections of pipelines that harbor relatively high densities of economically important fishes or invertebrates are creating ecosystems that produce or aggregate these organisms. It is arguable that if a species is only being aggregated, and is the target of fishing, the pipeline makes it easier to harvest and thus more susceptible to overharvesting by making it more vulnerable to capture (Bond et al. 2018c; McLean et al. 2022b). Thus, as noted by Bond et al. (2018a), "[d]etermining if a pipeline has enhanced fish abundance and diversity or if the structure is simply acting as a fish aggregating (or accumulation) device (FAD) redistributing existing stock, is an important discussion and one that has been subject to much debate (e.g., Bohnsack 1989; McLean et al. 2017)." In addition, if fishermen are intentionally fishing near pipelines, they may suffer disproportionate gear loss. As noted by Rouse et al.

(2020), "[o]perators and regulators must incorporate gear damage and loss of fishery time into risk calculations if comprehensive assessments are to accompany installation or decommissioning [of pipelines]." For risk assessment purposes, while platforms are in known locations, pipelines extend over many kilometers, and over that expanse, some parts are trenched while others are exposed (Rouse et al. 2020).

## 8.12 CONCLUSION

As discussed previously, there have been some studies on the overall population assemblages of invertebrates and fishes inhabiting subsea pipelines. However, most of these have been limited both temporally (mostly single surveys) and spatially (limited sites). Preliminary research into the types of fish assemblages present at artificial structures such as pipelines is lacking for the northwest Gulf. Subsequent investigation into primary and secondary production or invertebrate assemblages associated with offshore infrastructure in the northwest has not been done in the Gulf of Mexico. Research off California on fish assemblages at platforms (Claisse et al. 2014) found that the oil and gas structures off the coast of California have the highest fish production (as measured by standing biomass) per square meter (about 10 sq. ft) area of seafloor of any marine habitat that has been studied, about 10 times higher than fish communities from other marine ecosystems. The long-term presence of artificial structures in the Gulf of Mexico has likely influenced fishery resources and broader ecosystem dynamics. However, further research is needed to better assess their effect on the greater regional marine community and biodiversity. Such efforts will be essential for informed infrastructure development and sustainable resource management.

Summary Points of Potential Ecological Importance for Pipelines in the Northwest Gulf of Mexico Coastal Shelf Setting:

1) Pipelines can provide significant novel habitat for reef-oriented invertebrates and fishes.
2) It would be expected that the more complexity a length of pipeline exhibits (in the form of spans, pipeline protective constructions, and other complex-forming structures), the greater the diversity and the higher the abundances of reef-oriented species.
3) In areas where exposed subsea pipelines are the only local hard structures, pipelines will harbor higher densities and greater diversity of reef-associated invertebrates and fishes than the surrounding sea floor. Based on limited data, in situations where natural reefs impinge upon, or are close to, pipelines, their animal assemblages may have somewhat similar attributes due to potential connectivity (McLean et al. 2022a).
4) As many reef-associated taxa are often of higher economic value than those living on soft seafloor, the economic value of pipeline-associated taxa is often higher than that of animals living nearby.
5) One of the possible ecological benefits of subsea pipelines, which have large populations of structure-forming invertebrates, is to help recolonize nearby natural hard-bottom habitats whose invertebrate populations have been damaged by fishing gear, for example, bottom trawls.
6) Although more investigation should be undertaken for longer periods, two studies have demonstrated that invertebrate and fish assemblages on subsea pipelines may not be stable over time.
7) From a fishery's standpoint, subsea pipelines may have both positive and negative attributes. Because some taxa of economic importance are found at higher densities in association with pipelines, some fishermen concentrate their efforts there. This can make harvesting more efficient but might also lead to increased depletion of the fishery. In addition, there can be inadvertent gear loss even when the pipeline locations are well known.
8) Importantly for any future subsea pipeline surveys, studies demonstrate that using the common techniques for industry structural surveys as a basis for biological research, conducted with ROVs equipped with older video systems, do not produce optimal survey results. New surveys at slower speeds using HD cameras are optimal (Love et al. 2020; Scarborough-Bull et al. 2023).

Outstanding Questions for Pipelines in the Northwest Gulf of Mexico Coastal Shelf Setting:

1) What constitutes fish and invertebrate assemblages? What factors, such as sea floor depth and surrounding natural habitat, play a role in driving these assemblages?
2) What are the potential effects of pipeline installation and maintenance on the sea floor (e.g., sediment displacement and habitat destruction)?
3) What is the role these structures play in the life history (ontogeny) of species, including residency rates and movement patterns? For instance, do individuals move along pipelines or across depth gradients with ontogenetic shifts? Do juveniles remain associated with a pipeline throughout their life history or move off that pipeline into adjacent habitats? (Bond et al. 2018b).
4) What effects, if any, do pipelines have on predator–prey dynamics, local biodiversity (Bond et al. 2018d), and the introduction of invasive species? Do pipelines cause habitat fragmentation or create long-distance migration corridors for marine species?
5) Are the ecosystems associated with pipelines producing or aggregating animals?
6) What is the relative importance of pipeline assemblages to nearby natural habitats (i.e., the "halo effect", Reeds et al. 2018)? How are communities on subsea pipelines connected to surrounding ecosystems (i.e., connectivity, diel behavior patterns, and food webs)?
7) What is the temporal stability of pipeline assemblages?
8) What, if any, is the contribution of fish populations at pipelines to surrounding fisheries?
9) How might the importance of subsea pipelines as habitats be increased?

Possible Future Research Efforts:

1) Conduct surveys at current pipelines and surrounding habitats in the Gulf of Mexico study area.

   This would not only give more information on what fish and invertebrates live there but also how they relate to surrounding habitats. If surveys along lengths of pipe were repeated over time (both seasonal and interannual surveys), this would help address the issue of assemblage stability. In addition, focus on the question of whether pipelines in the study area are creating ecosystems that support taxa. If, for instance, it is determined that pipelines are nursery grounds for some taxa and that there are no other nursery grounds for those taxa nearby, it may be possible to hypothesize that the pipeline is helping to produce those taxa.
2) Conduct research regarding how, or if, recreational and commercial fishermen utilize pipelines in the study area. In addition, conduct research to understand if trawling occurs in the vicinity of pipelines and if pipelines interfere with trawling activities.
3) Increase surveys of complex habitats in the study area, both natural and artificial, to predict what species would form associations with the new habitat.
4) Consider augmenting both the pipeline and platform to increase the diversity and abundance of reef-oriented species. This has the potential of increasing the value of a pipeline to both commercial and recreational fishermen. This would have to be done with local authorities, such as Texas Parks and Wildlife.
5) Carefully map potential pipeline corridors, paying particularly close attention to the presence of shipwrecks or other cultural resources.

## 8.13 ACKNOWLEDGMENTS

Funding: This study was funded by an ExxonMobil Biomedical Sciences contract to the University of California, Santa Barbara (UCSB), under agreement LAW-2022–0466 to Milton S. Love. Authors Umarfarooq A. Abdulwahab and Christopher M. Prosser are employed by ExxonMobil Biomedical

Sciences, Inc. ExxonMobil Biomedical Sciences, Inc. provided support in the form of salaries for authors U.A.A. and C.M.P., but did not have any additional role in the study design, data collection and analysis, decision to publish, or preparation of the manuscript. The specific roles of these authors, which included reviewing and recommending manuscript revisions, are articulated in the "author contributions" section.

### Author Contributions:

Conceptualization: Ann Scarborough-Bull, Milton S. Love, Mary Nishimoto, Umarfarooq A. Abdulwahab.

Information curation: Ann Scarborough-Bull, Milton S. Love, Mary M. Nishimoto.

Funding acquisition: Milton S. Love, Umarfarooq A. Abdulwahab.

Investigation: Ann Scarborough-Bull, Milton S. Love, Mary M. Nishimoto.

Methodology: Ann Scarborough-Bull, Milton S. Love, Mary M. Nishimoto.

Project administration: Ann Scarborough-Bull, Milton S. Love.

Resources: Ann Scarborough-Bull, Milton S. Love, Mary M. Nishimoto, Umarfarooq A. Abdulwahab, Christopher M. Prosser.

Writing—original draft: Ann Scarborough-Bull, Milton S. Love, Mary M. Nishimoto.

Writing—review: Ann Scarborough-Bull, Milton S. Love, Mary M. Nishimoto, Umarfarooq A. Abdulwahab, Christopher M. Prosser.

Writing—editing and revision: Ann Scarborough-Bull, Milton S. Love.

Competing interests: This research was funded by an ExxonMobil Upstream Research Company contract to the University of California, Santa Barbara (UCSB), contract number LAW-2022–0466. U. A. Abdulwahab and C. M. Prosser are employed by ExxonMobil Biomedical Sciences, Inc. A. Scarborough-Bull, M. S. Love, and M. M. Nishimoto were employed by UCSB under that contract. There are no patents, products in development, or marketed products to declare.

## LITERATURE CITED

Airoldi L, Turon X, Perkol-Finkel S, Rius M. 2015. Corridors for aliens but not for natives: Effects of marine urban sprawl at a regional scale. *Divers Distrib* 21:755–768.

Ajemian MJ, Wetz JJ, Shipley-Lozano B, Shively JD, Stunz GW. 2015. An analysis of artificial reef fish community structure along the northwest Gulf of Mexico Shelf: Potential impacts of "rigs-to-reefs" programs. *PLOS One* 10:e0126354. https://doi.org/10.1371/journal.pone.0126354

Arnould JPY, Monk J, Ierodiaconou D, Hindell MA, Semmens J, Hoskins AJ, Costa DP, Abernathy K, Marshall GJ. 2015. Use of anthropogenic sea floor structures by Australian fur seals: Potential positive ecological impacts of marine industrial development? *PLOS One* 10:e0130581. https://doi.org/10.1371/journal.pone.0130581

Baldwin CC, Tornabene T, Robertson DR. 2018. Below the mesophotic. *Sci Rep* 8:4920b https://doi.org/10.1038/s41598-018-23067-1

Becker, A, Taylor MD, Folpp H, Lowry MB. 2018. Managing the development of artificial reef systems: The need for quantitative goals. *Fish Fisheries* 19:740–752. https://doi/10.1111/faf/12288

Bergen MS, Weisberg SB, Smith RW, Cadien DB, Dalkey A, Montagne DE, Stull JK, Velarde RG, Ranasinghe JA. 2001. Relationship between depth, sediment, latitude, and the structure of benthic infaunal assemblages on mainland shelf of southern California. *Mar Biol* 138:637–647. https://doi.org/10.1007/s002270000469

Bohnsack JA. 1989. Are high densities of fishes at artificial reefs the result of habitat limitation or behavioral preference? *Bull Mar Sci* 44:631–645.

Bohnsack JA, Johnson DL, Ambrose RF. 1991. Ecology of artificial reef habitats. In: Seaman W Jr, Sprague LM, editors. *Artificial Habitats for Marine and Freshwater Fisheries*. New York: Academic Press. p. 61–108.

Bond T, Langlois TJ, Partridge JC, Birt MJ, Malseed BE, Smith L, McLean DL. 2018a. Diel shifts and habitat associations on a subsea pipeline. *Fish Res* 206:220–234.

Bond, T, Partridge JC, Taylor MD, Cooper TF, McLean DL. 2018b. The influence of depth and a subsea pipeline of fish assemblages and commercially fished species. *PLOS One* 13:e0207703. https://doi.org/10.1371/journal.pone.0207703

Bond T, Partridge JC, Taylor MD, Langois TJ, Malseed BE, Smith LD, McLean DL. 2018c. Fish associated with a subsea pipeline and adjacent seafloor of the North West Shelf of Western Australia. *Mar Environ Res* 141:53–65. https://doi.org/10.1016/j.marenvres.2018.08.003

Bond T, Prince J, Partridge JC. 2018d. *The Value of Subsea Pipelines to Marine Biodiversity*. Offshore Technology Conference Asia, Kuala Lumpur, Malaysia, 20–23 2018, OTC-28240-MS. Available from: https://www.researchgate.net/publication/323765132_The_Value_of_Subsea_Pipelines_to_Marine_Biodiversity via the Internet. Accessed 2 July 2025.

Bortone SA. 1971. *Studies on the Biology of the Sand Perch, Diplectrum formosum (Perciformes: Serranidae)*. State of Florida Department of Resources Technical Series No. 65. Available from: https://www.researchgate.net/publication/36002191_Studies_on_the_biology_of_the_sand_perch_Diplectrum_formosum_Linnaeus_Perciformes_Serranidae via the Internet. Accessed 02 July 2025.

Breaux K, Cheramie K, Macklin J, Mars R, Davis D. 1997. Abandoning pipelines working group regulatory issues. In: Pulsipher A, editor. *Proceedings of an International Workshop on Offshore Lease Abandonment and Platform Disposal: Technology, Regulation, and Environmental Effects*. Center for Energy Studies. Baton Rouge, LA: Louisiana State University, p. 65–71.

Brooks RA, Keitzer SC, Sulak KJ. 2004. *Taxonomic Composition and Relative Frequency of the Benthic Fish Community Found on Natural Sand Banks and Shoals in the Northwest Gulf of Mexico. (A Synthesis of the Southeast Area Monitoring and Assessment Program's Groundfish Survey Database, 1982–2000)*. USGS Outer Continental Shelf Studies Ecosystem Program Report USGS-SIR-2004-XXX (CEC NEGOM Program Investigation Report No. 2004–0X, October 2004); Minerals Management Service, OCS Study MMS-2004-XXXX. Available from: https://www.boem.gov/sites/default/files/mm-research/2022-03/2005-USGS-Taxonomic.pdf via the Internet. Accessed 02 July 2025.

Bureau of Safety and Environmental Enforcement (BSEE). 2025a. *Gulf of America Region*. Available from: https://www.bsee.gov/who-we-are/our-organization/regional-offices/gulf-of-america via the Internet. Accessed 14 November 2025.

Bureau of Safety and Environmental Enforcement (BSEE). 2025b. *Rigs-to-Reefs*. Available from: https://www.bsee.gov/what-we-do/environmental-compliance/environmental-programs/rigs-to-reefs via the Internet. Accessed 02 July 2025.

Claisse JT, Pondella DJ II, Love MS, Zahn LA, Williams CM, Williams JP, Bull AS. 2014. Oil platforms off California are among the most productive marine fish habitats globally. *Proceedings of the National Academy of Sciences* 111 (43):15462–15467. https://doi.org/10.1073/pnas.1411477111

Cranswick D. 2001. *Brief Overview of Gulf of Mexico OCS Oil and Gas Pipelines: Installation, Potential Impacts, and Mitigation Measures*. OCS Report MMS 2001–067. Available from: https://www.bsee.gov/sites/bsee.gov/files/federal-register-notice/deepwater/pipeline-pdf.pdf via the Internet. Accessed 14 November 2025.

Dafforn KA, Glasby TM, Airoldi L, Rivero NK, Mayer-Pinto, M Johnston EL. 2015 Marine urbanization: An ecological framework for designing multifunctional artificial structures. *Front Ecol Environ* 13:82–90.

Darnell R. 2015. *The American Sea. A Natural History of the Gulf of Mexico*. Harte Research Institute and Texas A&M University Corpus Christi. p. 554.

Dauterive L. 2000. *Rigs-to-Reefs Policy, Progress, and Perspective*. U.S. Department of the Interior, Minerals Management Service, New Orleans, Louisiana OCS Report. MMS 2000–073. Available from: https://tpwd.texas.gov/publications/pwdpubs/media/rigs_to_reefs_policy_2000_073.pdf via the Internet. Accessed 02 July 2025.

Davis N, VanBlaricom GR, Dayton PK. 1982. Man-made structures on marine sediments: Effects on adjacent benthic communities. *Mar Biol* 70:295–303. https://doi.org/10.1007/BF00396848

DeAngelo MV, Fifariz R, Meckel T, Treviño RH. 2019. A seismic-based $CO_2$-sequestration regional assessment of the Miocene section, northern Gulf of Mexico, Texas, and Louisiana. *Int J Greenhouse Gas* 81:29–37. https://doi.org/10.1016/j.ijggc.2018.12.009

Dimitroff F. 1982. Survey of snapper/grouper fishermen of northwest Florida coast. In: *Proceedings Third Annual Gulf of Mexico Information Transfer Meeting*. Minerals Management Service, New Orleans, Louisiana, p. 56–58. Available from: https://espis.boem.gov/final%20reports/3980.pdf via the Internet. Accessed 02 July 2025.

Frumkes DR. 2002. The status of the California rigs-to-reefs program and the need to limit consumptive fishing activities. *ICES J Mar Sci* 59:S272–S276. https://doi.org/10.1006/jmsc.2002.1289.

Gallaway BJ, Lewbel GS. 1982. *The Ecology of Petroleum Platforms in the Northwestern Gulf of Mexico: A Community Profile*. U.S. Fish and Wildlife Service, Office of Biological Services, Washington, DC. FWS/OBS-82/27. Bureau of Land Management, Gulf of Mexico OCS Regional Office, Open-File Report 82–03. xiv + 92 p. Available from: https://www.arlis.org/docs/vol1/AK_Heritage/FWS/OBS/82/FWS-OBS-82-27.pdf via the Internet. Accessed 02 July 2025.

Gallaway BJ, Szedlmayer ST, Gazey WJ. 2009. A life history review for red snapper in the Gulf of Mexico with an evaluation of the importance of offshore petroleum platforms and other artificial reefs. *Rev Fish Sci* 17:148–167. https://doi.org/10.1080/10641260802160717

Hall AE, Kingsford MJ. 2021. Habitat type and complexity drive fish assemblages in a tropical seascape. *J Fish Biol* 99 (4):1364–1379.

Hiett RL, Milon JW. 2002. *Economic Impact of Recreational Fishing and Diving Associated with Offshore Oil and Gas Structures in the Gulf of Mexico: Final Report*. U.S. Department of the Interior, Minerals Management Service, Gulf of Mexico OCS Region. OCS Study MMS 2002–010. Available from: https://espis.boem.gov/final%20reports/3058.pdf via the Internet. Accessed 02 July 2025.

Hixon MA, Beets JP. 1989. Shelter characteristics and Caribbean fish assemblages: Experiments with artificial reefs. *Bull Mar Sci* 44:666–680.

Ierodiaconou D, McLean DL, Birt MJ, Bond T, Wines S, Glade-Wright O, Morris J, Higgs D, Whitmarsh SK. 2023. Industry remotely operated vehicle imagery for assessing marine communities associated with subsea oil and gas infrastructure on the continental shelf of South-East Australia. *Front Mar Sci* 10.3389. https://doi.org/10.3389/fmars.2023.1095906

Interstate Natural Gas Association of America 1996. *Gulf of Mexico Natural Gas Resources and Pipeline Infrastructure*. 82 p.+ appendices. https://ingaa.org/wpcontent/uploads/2007/06/1372.pdf via the Internet. Accessed 14 November 2025.

Jiang P, Wang H, Zhang X. 2020. Study on the hydrodynamic erosion of subsea pipelines in deepwater environments. *J Offshore Tech* 92 (6):78–90.

Kovalenko KE, Thomaz SM, Warfe DM. 2012. Habitat complexity: Approaches and future directions. *Hydrobiologia* 685:1–17. https://doi.org/10.1007/s10750-011-0974-z

Lacey NC, Hayes P. 2020. Epifauna associated with subsea pipelines in the North Sea. *ICES J Mar Sci* 77:1137–1147. https://doi.org/10.1093/icesjms/fsy196

Lavender JT, Dafforn KA, Bishop MJ, Johnston EL. 2017. Small-scale habitat complexity of artificial turf influences the development of associated invertebrate assemblages. *J Exp Mar Biol Ecol* 492:105–112. https://doi.org/10.1016/j.jembe.2017.01.025

Leckie SHF. 2016. *Scour and Sedimentation around Subsea Pipelines*. [Doctoral Thesis, The University of Western Australia]. Available from: https://api.research-repository.uwa.edu.au/ws/portalfiles/portal/11692687/THESIS_DOCTOR_OF_PHILOSOPHY_LECKIE_Simon_Hamish_Farquharson_2016.pdf via the Internet. Accessed 02 July 2025.

Lenihan HS, Peterson CH. 1998. How habitat degradation through fishery disturbance enhances impacts of hypoxia on oyster reefs. *Ecol Appl* 8:128–140. https://doi.org/10.1890/1051-0761(1998)008[0128:HHDTFD]2.0.CO;2

Lima JS, Zalmon IR, Love M. 2019. Overview and trends of ecological and socioeconomic research on artificial reefs. *Mar Env Res* 145:81–96. https://doi.org/10.1016/j.marenvres.2019.01.010

Linton TL. 1994. A comparison between the fish species harvested in the U.S. Gulf of Mexico and the number of production platforms present. *In*: *Proceedings of the Fifth International Conference for Aquatic Habitat Enhancement*. Long Beach, CA. Bull Mar Sci 55. p.1344.

Love MS, Nishimoto MM, Clark S, Kui L, Aziz A, Palandro D (2020) A comparison of two remotely operated vehicle (ROV) survey methods used to estimate fish assemblages and densities around a California oil platform. *PLoS ONE* 15 (11): e0242017. https://doi.org/10.1371/journal.pone.0242017.

Love MS, Nishimoto MM, Snook L, Schroeder DM, Bull AS. 2017. A comparison of fishes and invertebrates living in the vicinity of energized and unenergized submarine power cables and natural sea floor off southern California, USA. *J Renew Ener*, article ID 8727164. https://doi.org/10.1155/2017/8727164

Love MS, Schroeder DM, Lenarz B, Cochrane GR. 2006. Gimme shelter: The importance of crevices to some fish species inhabiting a deep-water rocky outcrop in southern California. *CalCOOFI* 47:119–126. Available from: https://calcofi.org/downloads/publications/calcofireports/v47/Vol_47_Love_Gimme_Shelter.pdf via the Internet. Accessed 02 July 2025.

Love MS, Schroeder DM, Nishimoto MM. 2003. *Ecological Role of Oil and Gas Production Platforms and Natural Outcrops on Fishes in Southern and Central California: A Synthesis of Information*. Seattle, WA: U. S. Department of the Interior. Available from: https://www.govinfo.gov/content/pkg/

GOVPUB-I-51cf6a08a3bcabd863d066fb8f807ebc/pdf/GOVPUB-I-51cf6a08a3bcabd863d066fb-8f807ebc.pdf via the Internet. Accessed 02 July 2025.

Love MS, York A. 2005. A comparison of the fish assemblages associated with an oil/gas pipeline and adjacent seafloor in the Santa Barbara Channel, Southern California Bight. *Bull Mar Sci* 77:101–117.

Lowenthal AS. 2022. In: *The Opportunities and Risks of Offshore Carbon Storage in the Gulf of Mexico.* Subcommittee on Energy and Mineral Resources, Committee on Natural Resources U.S. House of Representatives, Second Session, Serial Number 117–19. Available from: https://www.congress.gov/event/117th-congress/house-event/114674/text via the Internet. Accessed 02 July 2025.

Madgett AS, Elsdon TS, Marnane MJ, Schramm KD. 2023. The functional diversity of fish assemblages in the vicinity of oil and gas pipelines compared to nearby natural reef and soft sediment habitats. *Mar Env Res* 187:105931. https://doi.org/10.1016/j.marenvres.2023.105931

Martin-Smith KM. 1993. Abundance of mobile epifauna: The role of habitat complexity and predation by fishes. *J Exp Mar Biol Ecol* 174:243–260. https://doi.org/10.1016/0022-0981(93)90020-O

McGinnis MV, Fernandez L, Pomeroy C. 2001. *The Politics, Economics, and Ecology of Decommissioning Offshore Oil and Gas Structures.* University of California, Marine Science Institute, Coastal Research Center, MMS OCS Study 2001–006, Santa Barbara. Available from: https://www.researchgate.net/publication/255992326_The_Politics_Economics_and_Ecology_of_Decommissioning_Offshore_Oil_and_Gas_Structures_Final_Technical_Summary_and_Final_Technical_Report via the Internet. Accessed 02 July 2025.

McKay M, Nides J, Vigil D, editors. 2000. *Proceedings: Gulf of Mexico Fish and Fisheries: Bringing together New and Recent Research.* U.S. Department of the Interior, Minerals Management Service, Gulf of Mexico OCS Region, New Orleans, La OCS Study MMS 2002–004. 674 p. Available from: https://digital.library.unt.edu/ark:/67531/metadc955425/m2/1/high_res_d/3055.pdf via the Internet. Accessed 02 July 2025.

McLaughlin R. 2013. *'Idle Iron' versus 'Rigs to Reefs': Surviving Conflicting Policy Mandates in the Gulf of Mexico.* Presented at a Marine Science, Engineering, Law & Policy Workshop. University of Singapore. November 12–13. Available from: https://cil.nus.edu.sg/wp-content/uploads/2013/06/11.-Prof.-Richard-MCLAUGHLIN-_Idle-Iron_-versus-_Rigs-to-Reefs_-Surviving-Conflicting-Policy-Mandates-in-the-Gulf-of-Mexico.pdf via the Internet. Accessed 02 July 2025.

McLean DL, Ferreira LC, Benthuysen JA, Miller KJ, Schläppy ML, Ajemian MJ, Berry O, Birchenough SNR, Bond T, Boschetti F, et al. 2022a. Influence of offshore oil and gas structures on seascape ecological connectivity. *Glob Change Biol* 28:3515–3536. https://onlinelibrary.wiley.com/doi/10.1111/gcb.16134

McLean DL, Partridge JC, Bond T, Birt MJ, Bornt KR, Langlois TJ. 2017. Using industry ROV videos to assess fish associations with subsea pipelines. *Cont Shelf Res* 141:76–97. https://dx.doi.org/10.1016/j.esr.2017.05.006

McLean DL, Speed CW, Birt MJ, Colquhoun J, Case M, Stowar M, Bond T, Ierodiaconou, D, Whitmarch SK, Taylor MD, et al. 2022b. Habitat value of subsea wells and pipelines for fishery target species in Australia. *Front Mar Sci.* https://doi.org/10.3389/fmars.2022.960496

McLean DL, Vaughan BI, Malseed BE, and M.D. Taylor. 2020. Fish-habitat associations on a subsea pipeline within an Australian Marine Park. *Mar Environ Res* 153. https://doi.org/10.1016/j.marenvres.2019.104813

Meckel T, Trevino R. 2014. *Gulf of Mexico Miocene CO2 Site Characterization Mega Transect.* Final Scientific/Technical Report (Revised). Reporting Period December 8, 2009—September 30, 2014. DOE Award Number DE-FE0001941. Available from: https://netl.doe.gov/sites/default/files/event-proceedings/2014/carbon_storage/9-Tip-Meckel.pdf via the Internet. Accessed 02 July 2025.

Menge BA, Sutherland JP. 1976. Species diversity gradients: Synthesis of the roles of predation, competition, and temporal heterogeneity. *Am Nat* 110:351–369.

Michaelis R, Hass HC, Mielck F, Papenmeier S, Sander L, Ebbe B, Gutow L, Wiltshire KH. 2019. Hard-substrate habitats in the German Bight (South-Eastern North Sea) observed using drift videos. *J Sea Res* 144:78–84. https://doi.org/10.1016/j.seares.2018.11.009

Mikulas JJ, Rooker JR. 2008. Habitat use, growth, and mortality of post-settlement lane snapper (*Lutjanus synagris*) on natural banks in the northwestern Gulf of Mexico. *Fish Res* 93:77–84. https://doi.org/10.1016/j.fishres.2008.02.012

National Academies of Sciences, Engineering, and Medicine. 2019. *Negative Emissions Technologies and Reliable Sequestration: A Research Agenda.* Washington, DC: The National Academic Press. https://doi.org/10.17226/25259 Available from: https://nap.nationalacademies.org/catalog/25259/negative-emissions-technologies-and-reliable-sequestration-a-research-agenda via the Internet. Accessed 02 July 2025.

National Oceanic and Atmospheric Administration, National Center for Environmental Information (NOAA, NCEI). 2025. Available from: https://www.ncei.noaa.gov/maps/gulf-data-atlas/atlas.htm?plate=Offshore%20Structures via the Internet. Accessed 02 July 2025.

National Research Council. 2020. *Climate Change: Evidence and Causes: Update 2020*. Washington, DC: National Academic Press. Available from: https://nap.nationalacademies.org/catalog/25733/climate-change-evidence-and-causes-update-2020 via the Internet. Accessed 02 July 2025.

Nieland DL, Wilson CA. 2003. Red Snapper recruitment to and disappearance from oil and gas platforms in the northern Gulf of Mexico. In: Stanley D, Scarborough-Bull A, editors. *Fisheries, Reefs, and Offshore Development*. Am Fish Soc Vol. 36, p. 73–82. https://fisheries.org/bookstore/all-titles/afs-symposia/x54036xm/

Paxton AB, Shertzer KW, Kyle W, Bacheler NM, Kellison GT, Riley KL, Taylor JC. 2020. Meta-analysis reveals artificial reefs can be effective tools for fish community enhancement but are not one-size-fits-all. *Front Mar Sci* 7:282. https://doi.org/10.3389/fmars.2020.00282

Paxton AB, Stewart DN, Mille KJ, Renchen J, Harrison ZH, Byrum JS, Brinton C, Nelson A, Simpson E, Clarke PJ, LaPorta C, Barrett PD, Rousseau M, Newton DC, Rigby RB, Williams DT, Shipley JB, Murakawa P, Runde BJ, Riley KL, Bacheler NM, Kellison GT, Taylor JC. 2023. Artificial reef footprint in the United States ocean. *Nature Sustainability* 7:140–147.

Peabody MB, Wilson CA. 2006. *Fidelity of Red Snapper (Lutjanus campechanus) to Petroleum Platforms and Artificial Reef Infrastructure in the Northern Gulf of Mexico*. OCS Study MMS 2006–05, New Orleans, LA. 64 p. Available from: https://www.govinfo.gov/content/pkg/GOVPUB-I-440faa3fb-76f06e85b7b62c8e556e1c2/pdf/GOVPUB-I-440faa3fb76f06e85b7b62c8e556e1c2.pdf via the Internet. Accessed 02 July 2025.

Pett-Ridge J, Kuebbing S, Mayer A, Hovorka S, Pilorgé H, Baker S, Pang S, Scown C, Mayfield K, Wong A. 2023. *Roads to Removal: Options for Carbon Dioxide Removal in the United States, Lawrence Livermore National Laboratory*. LLNL-TR–852901. https://doi.org/10.2172/2301853 Available from: https://www.osti.gov/servlets/purl/2301853 via the Internet. Accessed 02 July 2025.

Redford M, Rouse S, Hayes P, Wilding TA. 2021. Benthic and fish interactions with pipeline protective structures in the North Sea. *Front Mar Sci* 8:652630. https://doi.org/10.3389/fmars.2021.652630

Reeds KA, Smith JA, Suthers IM, Johnston EL. 2018. An ecological halo surrounding a large offshore artificial reef: Sediments, infauna, and fish foraging. *Mar Environ Res* 141:30–38. https://doi.org/10.1016/j.marenvres.2018.07.011

Reggio VC Jr. 1994. *Artificial Reefs: Oases for Marine Life in the Gulf*. MMS PR-03/18/94 Rigs to Reefs/Artificial Reefs #40018. BOEM-Newsroom/Press-Releases/1994/40018.aspx. Available from: https://www.boem.gov/sites/default/files/boem-newsroom/Press-Releases/1994/40018.pdf via the Internet. Accessed 02 July 2025.

Rezak R, Bright TJ, McGrail DW. 1983. *Reefs and Banks of the Northwestern Gulf of Mexico: Their Geological, Biological, and Physical Dynamics*. Executive Summary: Northern Gulf of Mexico Topographic Features Monitoring and Data Synthesis. Contract No. AA851-CT1–55 Submitted to the U.S. Department of the Interior Minerals Management Service Outer Continental Shelf Office New Orleans, LA. Available from: https://www.govinfo.gov/content/pkg/GOVPUB-I-e244822ba0a30f3a5155ae29e66441d5/pdf/GOVPUB-I-e244822ba0a30f3a5155ae29e66441d5.pdf via the Internet. Accessed 02 July 2025.

Rezak R, Gittings SR, Bright TJ. 1990. Biotic assemblages and ecological controls on reefs and banks of the northwest Gulf of Mexico. *Amer Zool* 30 (1) (February):23–35. https://doi.org/10.1093/icb/30.1.23

Ringrose PS, Meckel TA. 2019. Maturing global $CO_2$ storage resources on offshore continental margins to achieve 2DS emissions reduction. *Sci Rep* 9:17944. https://doi.org/10.1038/s41598-019-54363-z

Rodriguez AB, Anderson JB, Siringan FP, Taviani M. 1999. Sedimentary facies and genesis of Holocene sand banks on the east Texas inner continental shelf. *Soc Econ Pa No. 64* 165–178. https://doi.org/10.2110/pec.99.64.0165

Rodriguez AB, Anderson JB, Siringan FP, Taviani M. 2004. Holocene evolution of the east Texas coast and inner continental shelf: Along-strike variability in coastal retreat rates. *J Sed Res* 74:405–421. https://doi.org/10.1306/092403740405

Rooker JR, Landry AM, Geary BW, Harper JA. 2004. Assessment of a shell bank and associated substrates as nursery habitat of post settlement Red Snapper. *Estuar Coast Shelf S* 59:653–661. https://doi.org/10.1016/j.ecss.2003.11.009

Rosemond RC, Paxton AB, Lemoine HR, Fegley SR, Peterson CH. 2018. Fish use of reef structures and adjacent sand flats: Implications for selecting minimum buffer zones between new artificial reefs and existing reefs. *Marine Ecology Progress Series* 587:187–199.

Rouse S, Hayes P, Wilding TA. 2020. Commercial fisheries losses arising from interactions with offshore pipelines and other oil and gas infrastructure and activities. *ICES J Mar Sci* 77:1148–1156. https://doi.org/10.1093/icesjms/fsy116

Rouse S, Kafas A, Catarino R, Peter H. 2018. Commercial fisheries interactions with oil and gas pipelines in the North Sea: Considerations for decommissioning. *ICES J Mar Sci* 75:279–286. https://doi.org/10.1093/icesjms/fsx121

Rouse S, Kafas A, Hayes P, Wilding TA. 2017. Development of data layers to show the fishing intensity associated with individual pipeline sections as an aid for decommissioning decision-making. *Underwater Technol* 34 (4):171–178. https://doi.org/10.3723/ut.34.171

Rouse S, Lacey NC, Hayes P, Wilding TA. 2019. Benthic conservation features and species associated with subsea pipelines: Considerations for decommissioning. *Front Mar Sci* 6. https://doi.org/10.3389/fmars.2019.00200

Russell DJF, Brasseur SMJM, Thompson D, Hastie GD, Janik VM, Aarts G, McClintock BT, Matthiopoulos J, Moss SEW, McConnell B. 2014. Marine mammals trace anthropogenic structures at sea. *Curr Biol* 24:638–639.https://dx.doi.org/10.1016/j.cub.2014.06.033

Scarborough-Bull A. 1989. Some comparisons between communities beneath the petroleum platforms off California and in the Gulf of Mexico. In: Reggio VCJ, editor. *Petroleum Structures as Artificial Reefs: A Compendium.* Fourth International Conference on Artificial Habitats for Fisheries, Rigs-to-Reefs Special Session. (Miami, FL: OCS Study MMS). Available from: https://babel.hathitrust.org/cgi/pt?id=uc1.31822009421603&seq=5 via the Internet. Accessed 02 July 2025.

Scarborough-Bull A, Kendall JJ. 1994. An indication of the process: Offshore platforms as artificial reefs in the Gulf of Mexico. *Bull Mar Sci* 55:1086–1098. Available from: https://www.researchgate.net/publication/233670206_An_Indication_of_the_Process_Offshore_Platforms_as_Artificial_Reefs_in_the_Gulf_of_Mexico via the Internet. Accessed 02 July 2025.

Scarborough-Bull A, Love MS. 2019. Worldwide oil and gas platform decommissioning: A review of practices and reefing options. *Ocean Coast Manage* 168:274–306. https://doi.org/10.1016/j.ocecoaman.2018.10.024

Scarborough-Bull A, Nishimoto MM, Love MS, Clark S, Seeto K, McCrea M, Park C, Kui L, Jainese C, Aziz A, Angle J. 2023. Comparison of methods (ROV, Diver) used to estimate invertebrate assemblages and densities on an offshore California oil platform. *Continental Shelf Research* 252 (1):22. https://doi.org/10.1016/j.csr.2022.104856.

Schramm KD, Marnane MJ, Elsdon TS. 2021. Fish associations with shallow water subsea pipelines compared to surrounding reef and soft sediment habitats. *Sci Rep* 11:6238. https://doi.org/10.1038/s41598-021-85396-y

Sheehan EV, Bridger D, Attrill MJ. 2015. The ecosystem service value of living versus dead biogenic reef. *Estuar Coast Shelf* S 154:248–254. https://doi.org/10.1016/j.ecss.2014.12.042

Shipp RL, Bortone SA. 2009. A perspective of the importance of artificial habitat on the management of Red Snapper in the Gulf of Mexico. *Rev Fish Sci* 17 (1):41–47. https://doi.org/10.1080/10641260802104244

Sih TL, Cure K, Yilmaz IN, McLean D, Macreadie PI. 2022. Marine life and fisheries around offshore oil and gas structures in southeastern Australia and possible consequences for decommissioning. *Front Mar Sci* 9. https://doi.org/10.3389/fmars.2022.979212

Sisson JD, Shimeta J, Zimmer CA, Traykovski P. 2002. Mapping epibenthic assemblages and their relations to sedimentary features in shallow-water, high-energy environments. *Cont Shelf Res* 22:565–583. https://doi.org/10.1016/S0278-4343(01)00074-7

Sonnier F, Teerling J, Hoese HD. 1976. Observations on the offshore reef and platform fish fauna of Louisiana. *Copeia* 1:105–111. Available from https://spo.nmfs.noaa.gov/sites/default/files/pdf-content/1976/742/hastings.pdf from the Internet. Accessed 02 July 2025.

Stanley DR, Wilson CA. 2003. Seasonal and spatial variation in the biomass and size frequency distribution of fish associated with oil and gas platforms in the Northern Gulf of Mexico. In: Stanley DR, Scarborough-Bull A, editors, *Fisheries, Reefs, and Offshore Development.* Amer Fish SS 36, Bethesda, MD. p. 123–154.

Stefanoudis PV, Gress E, Pitt JM, Smith SR, Kincaid T, Rivers M, Andradi-Brown DA, Rowlands G, Woodall LC, Rogers AD. 2019. Depth-dependent structuring of reef fish assemblages from the shallows to the rariphotic zone. *Front Mar Sci* 6:00307. https://doi.org/10.3389/fmars.2019.00307

Stephan CD, Dansby BG, Osburn HR, Matlock GC, Riechers RK, Rayburn R. 1990. *Texas Artificial Reef Fishery Management Plan. Fisheries Management Plan Series Number 3.* Texas Parks and Wildlife Department Coastal Fisheries Branch, Austin. Available from: https://tpwd.texas.gov/publications/pwd-pubs/media/pwd_pl_v3400_0332.pdf via the Internet. Accessed 02 July 2025.

Streich MK, Ajemian MJ, Wetz JJ, Stunz GW. 2017. A comparison of fish community structure at mesophotic artificial reefs and natural banks in the Western Gulf of Mexico. *Mar Coastal Fish* 9:170–189. https://doi.org/10.1080/19425120.2017.1282897

Sumer BM, Truelsen C, Sichmann T, Fredsøe J. 2001. Onset of scour below pipelines and self-burial. *Coast Eng* 42:313–335. https://doi.org/10.1016/S0378-3839(00)00066-1

Texas Parks and Wildlife Department (TPWD). 2024. *Artificial Reefs*. Rigs-to-reefs Texas. Available from: https://tpwd.texas.gov/landwater/water/habitats/artificial_reef/rigs-to-reefs.phtml via the Internet. Accessed 02 July 2025.

van Elden S, Meeuwig JJ, Hobbs RJ. 2022. Offshore platforms as novel ecosystems: A case study from Australia's Northwest Shelf. *Ecol Evol* 12 (2):e8496. https://doi.org/10.1002/ece3.8496

Walsh D. 2023. Seafloor cables and pipelines: Are they secure? *Proceedings of the United States Naval Institute* 149:1. Available from: https://www.usni.org/magazines/proceedings/2023/march/seafloor-cables-and-pipelines-are-they-secure via the Internet. Accessed 02 July 2025.

Wells RJD, Harper JO, Rooker JR, Landry AM, Dellapenna TM. 2009. Fish assemblage structure on a drowned barrier island in the northwestern Gulf of Mexico. *Hydrobiologia* 625:207–221. Available from: https://www.tamug.edu/rooker/pdf/Wells_Harper_2009.pdf via the Internet. Accessed 02 July 2025.

Witzig J. 1986. Rigs fishing in the Gulf of Mexico—1984 marine recreation fishing survey results. In: *Proceedings Sixth Annual Gulf of Mexico Information Transfer Meeting*. Minerals Management Service, p. 103–105 OCS Study/MMS 86–0073, New Orleans, LA. Available from: https://espis.boem.gov/final%20reports/3792.pdf via the Internet. Accessed 02 July 2025.

Wright JP, Jones CG. 2006. The concept of organisms as ecosystem engineers ten years on: Progress, limitations, and challenges. *Bioscience* 56 (3):203–209. Available from: https://www.researchgate.net/publication/202001805_The_Concept_of_Organisms_as_Ecosystem_Engineers_Ten_Years_On_Progress_Limitations_and_Challenges#:~:text=subject%20to%20copyright.-,The%20Concept%20of%20Organisms%20as%20Ecosystem%20Engineers%20Ten%20Years%20on,203%2D209 via the Internet. Accessed 02 July 2025.

# 9 Changes in Florida's (USA) Artificial Reef Program

## *Managing Expectations and Challenges to Benefit Diverse Stakeholders and Ecosystems*

*Carolyn Kalinowski* and Keith Mille*
* Corresponding author: carolyn.kalinowski@myfwc.com

## 9.1 INTRODUCTION

As of 1 November 2025, over 4,476 artificial reefs have been deployed in state and federal coastal waters off the State of Florida (USA), utilizing both public and private funds. The Florida Fish and Wildlife Conservation Commission's Artificial Reef Program (hereafter referred to as the Florida Artificial Reef Program) has provided direct funding of over US$26,575,000 for projects to construct, deploy, and monitor many of these reefs (Kalinowski, 2025). In the United States, states other than Florida retain direct ownership of artificial reef permits and directly oversee the construction of artificial reef sites. In Florida, permits to construct artificial reefs are the responsibility of local coastal governments, such as cities and counties. Additionally in Florida, the Florida Artificial Reef Program provides technical guidance and funding to facilitate these activities. This approach has historically enabled the deployment of a wide variety and large volume of artificial reef materials, while allowing local resource coordinators to tailor the goals of these reefs to the needs of their local communities and stakeholders.

In recent years, the Florida Artificial Reef Program has documented drastic changes in artificial reef activities throughout Florida. Changes in material availability and increased construction costs have resulted in the selection of prefabricated concrete modules as the dominant material type. Additionally, the emergence of highly engineered structures has led to expanding and, occasionally, conflicting expectations for the performance of artificial reefs. These changes have also caused the Florida Artificial Reef Program managers to reconsider the role of the State in guiding artificial reef activities to ensure successful artificial reefs that meet their intended objectives while benefiting the marine ecosystem.

In this chapter, the different eras of Florida's artificial reef activities are reviewed by outlining the materials, goals, methods, and research that define each era. Additionally, we focus on the recent trends and challenges that have arisen regarding artificial reef activities and pose questions about how to best address these challenges as artificial reef managers. Finally, we discuss the themes that emerged from discussions at the 12th CARAH (Conference on Artificial Reefs and Related Aquatic Habitats)—a forum where our intent was to gain global perspectives on Florida's approach to artificial reef management and incorporate input offered by colleagues during the conference.

In the context of this paper, the following definitions apply:

"Artificial reef" means one or more objects of natural or human origin that are intentionally placed on the seafloor to enhance marine life for human use.

"Permit" means the legal authority required to deploy artificial reefs.

 DOI: 10.1201/9781003676775-9

"Permitted area" means the location in which artificial reefs are legally allowed to be placed.
"Deployment" means the act of placing an artificial reef on the seafloor.
"Construction" means the fabrication of an artificial reef, including both the material and the placement on the seafloor.
"Patch reef" means the reef complex created by one or more artificial reefs in close proximity to one another.
"Module" means a structure that is created through the fabrication of materials such as metal or concrete specifically for the purpose of being deployed as an artificial reef.
"Project" means the entire process of creating an artificial reef, including planning, permitting, fabrication, and deployment.
"Secondary-use concrete" means concrete that has had another purpose prior to being used for artificial reefs, such as construction or bridge materials.
"Hybrid reefs" means artificial reef materials that are intended to serve multiple, sometimes conflicting, purposes. These are typically prefabricated concrete and require a high level of engineering to design.

## 9.2 BACKGROUND

The state of Florida (USA) is home to many unique geophysical, biological, and socio-economic attributes. This diversity of attributes presents a variety of conditions for artificial reef managers to consider. The Florida peninsula experiences both tropical and subtropical climates, with 2,170 km of coastline extending along both the Atlantic Ocean and the Gulf of Mexico (Figure 9.1). The habitats of these coastal waters include coral reefs, mangroves, salt marshes, tidal flats, and surf zones. Additionally, the offshore habitats include substrates with sandy shoals, hardbottom ridges, muddy zones, carbonate reefs, and quartz sand. The bathymetry also displays extreme variation in Florida's continental shelf profile; the shelf remains shallow (less than 40 m depth) as far as 150 km from shore in the Gulf of Mexico (i.e., the Gulf of America), while quickly dropping off in the Atlantic Ocean to 40 m depth as close as 2 km to shore (Figure 9.1). These shelf profile differences influence the habits and preferences of recreational anglers and divers alike. Additionally, Florida's geographic position makes it particularly susceptible to intense hurricane activity, which can have extreme impacts on its artificial reefs. For example, in 2018, Hurricane Michael displaced artificial reef materials by almost 0.5 km along the seafloor (Renchen et al., 2020).

The warm temperate and subtropical climates, and the over 500 species of saltwater fish managed by the Florida Fish and Wildlife Conservation Commission, make Florida a popular destination for engaging in marine-related activities. In 2024, over 143 million visitors joined Florida's 23 million residents, including four million licensed anglers and one million registered vessels (Florida Fish and Wildlife Conservation Commission, 2021). Anglers partake in inshore and offshore fishing, and spearfishing, maintaining Florida's status as the self-proclaimed "Fishing Capital of the World" and generating an estimated economic impact of US$9.2 billion (Wallmo et al., 2021b). In addition to fishing, Florida's robust recreational diving industry reports 4.56 million visitor-days per year, with a substantial number of dive trips off Northwest and Southeast Florida (Figure 9.2) to artificial reefs (Huth et al., 2014; Wallmo et al., 2021a).

Florida's Artificial Reef Program is also unique. While other states in the United States directly manage and oversee artificial reef permits, local coastal Florida governments, such as counties and municipalities, apply for the artificial reef permits and oversee their planning and deployment. The Florida Artificial Reef Program provides technical and financial assistance to these governmental entities by coordinating with a variety of artificial reef associations, non-profit organizations, environmental consultants, volunteer dive teams, saltwater fishing clubs, and academic institutions.

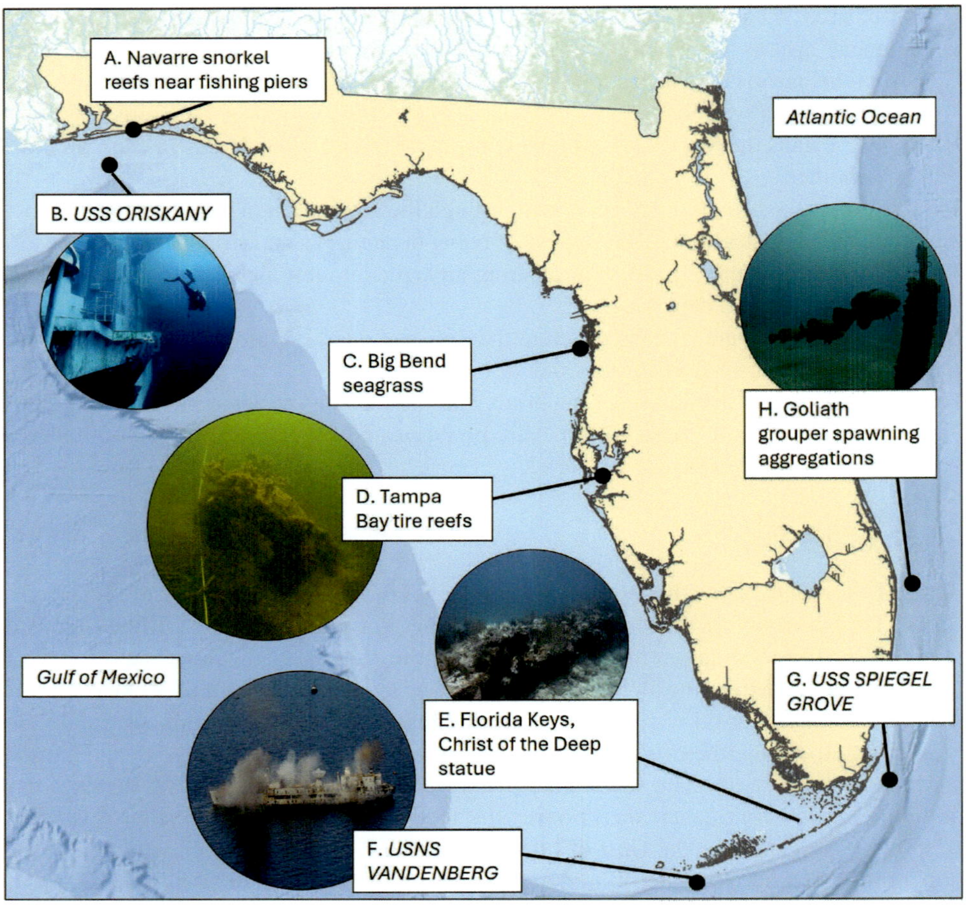


**FIGURE 9.1** Map of referenced artificial reef locations. A. Navarre snorkel reefs are located approximately 350 m from a popular fishing pier and may be contributing to regular reports of turtle bycatch. B. The *USS ORISKANY* is located off Pensacola in the Northwest Panhandle, and is a popular dive site. It is also the largest vessel to be deployed as an artificial reef, measuring 303 m. C. The Big Bend area hosts seagrass beds—some of the sensitive habitats that are important to consider for artificial reef planning. D. Tampa Bay is one of the locations of historical tire reefs, where tires are now being assessed for removal. E. Florida's Reef Tract is the only natural coral reef system in the continental United States, and another sensitive habitat area that is important to consider when planning artificial reefs. This is also the location of the Christ of the Deep statue, a popular dive site near natural reefs. F. The *USNS VANDENBERG*, a 160m United States Navy ship deployed in 2009, is one of the most popular fishing and diving sites in the Florida Keys. G. The *USS SPIEGEL GROVE* is another United States Navy vessel. Deployed in 2022, the 155m Spiegel Grove now serves as a popular fishing and diving site located in the Florida Keys.

In 2025, the Florida Artificial Reef Program recorded a total of 4,476 artificial reefs. These reefs include 1,951 prefabricated module reefs, 1,437 secondary-use concrete reefs, 566 vessels and barges, 281 metal reefs, 161 rock or limestone boulder reefs, and 80 "other" reefs that utilize materials that are no longer deployed (e.g., tires, fiberglass boats, and car bodies; Figure 9.2, Table 9.1). The average artificial reef has 3 m of vertical relief, is located at a depth of 20 m, weighs 270 metric tons, and covers a 200-square-meter area. While the depth of Florida's artificial reefs ranges from 1 meter to 156 meters, the majority are located between 5 and 33 meters deep, which is ideal for fishing and diving (Kalinowski, 2025).

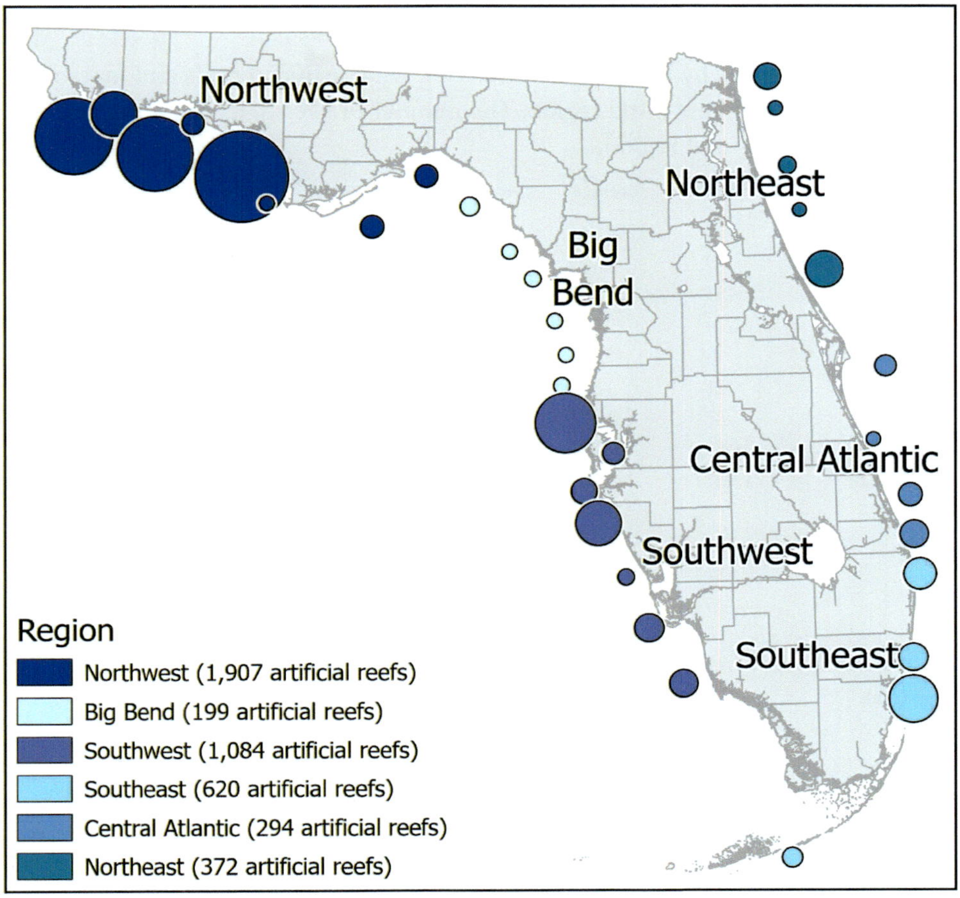


**FIGURE 9.2** Map of artificial reefs in Florida by region. Circles indicate the relative number of artificial reefs for each county, with larger circles representing more artificial reefs. The color of each circle is associated with the region that is represented.

## 9.3 EARLY DAYS: PRE-1980

While artificial reefs in Florida were likely deployed or utilized in some manner long before humans began recording them, the "Early Days" of Florida's artificial reef deployments can be generally described as those reefs deployed as the era before the 1980s. Early examples include vessels that were sunk during World War II or were accidentally sunk during transit, both of which later became popular fishing sites. The earliest recorded artificial reef permits for intentionally placed materials began in 1960, with dismantled concrete from a bridge replacement project.

During this earlier period, approximately 110 artificial reefs were recorded (Figure 9.3). More than half of these artificial reefs were vessels and barges (n = 68), while the remaining reefs comprised secondary-use concrete (27), metal structures such as car bodies or scrap metal, and tires (Kalinowski 2025; Figure 9.3). These artificial reefs primarily served as waste disposal options, with the secondary benefits of creating fish habitat and new fishing locations (Stone, 1972; Stone et al., 1974). Since then, it has become known that tires are an inappropriate material for artificial reefs, as they often subsided or were displaced while leaching toxins and exhibiting little faunal

**TABLE 9.1**

**Artificial reefs in Florida by location and material. Total artificial reefs and material tonnages are listed by county and region. The number of each artificial reef material type is also listed by county and region.**

| County/Region | Reefs | Tonnage | Module | Concrete | Vessel | Metal | Rock | Other |
|---|---|---|---|---|---|---|---|---|
| Escambia | 496 | 264,383 | 390 | 42 | 31 | 30 | 1 | 2 |
| Santa Rosa | 49 | 2,079 | 45 | | 2 | 1 | | 1 |
| Okaloosa | 484 | 27,734 | 387 | 35 | 38 | 23 | | 1 |
| Walton | 74 | 3,843 | 68 | 3 | 1 | 2 | | |
| Bay | 613 | 43,381 | 456 | 41 | 41 | 64 | | 11 |
| Gulf | 21 | 1,194 | 4 | 9 | 4 | 4 | | |
| Franklin | 87 | 5,028 | 38 | 35 | 13 | 1 | | |
| Wakulla | 83 | 2,262 | 50 | 27 | 1 | 1 | | 4 |
| **Northwest Total:** | **1,907** | **349,903** | **1,438** | **192** | **131** | **126** | **1** | **19** |
| Taylor | 53 | 1,436 | 40 | 6 | | 7 | | |
| Dixie | 27 | 1,525 | 6 | 2 | | 1 | 18 | |
| Levy | 31 | 741 | 12 | 19 | | | | |
| Citrus | 25 | 6,186 | 6 | 18 | | | | 1 |
| Hernando | 26 | 2,162 | 4 | 12 | | 10 | | |
| Pasco | 37 | 5,786 | 1 | 19 | 7 | 10 | | |
| **Big Bend Total:** | **199** | **17,836** | **69** | **76** | **7** | **28** | **18** | **1** |
| Pinellas | 367 | 30,445 | 13 | 277 | 17 | 12 | 1 | 47 |
| Hillsborough | 75 | 17,988 | 6 | 67 | 1 | | | 1 |
| Manatee | 103 | 67,931 | 49 | 39 | 3 | | 8 | 4 |
| Sarasota | 252 | 23,499 | 159 | 70 | 6 | 16 | 1 | |
| Charlotte | 36 | 33,140 | 4 | 25 | 4 | | 3 | |
| Lee | 130 | 30,185 | 19 | 80 | 16 | 4 | 11 | |
| Collier | 121 | 34,639 | 9 | 74 | 5 | 9 | 24 | |
| **Southwest Total:** | **1,084** | **237,827** | **259** | **632** | **52** | **41** | **48** | **52** |
| Monroe | 65 | 52,696 | 9 | 31 | 24 | 1 | | |
| Miami-Dade | 273 | 195,076 | 68 | 48 | 106 | 16 | 35 | |
| Broward | 128 | 50,506 | 26 | 9 | 74 | 11 | 5 | 3 |
| Palm Beach | 154 | 168,046 | 23 | 41 | 39 | | 51 | |
| **Southeast Total:** | **620** | **466,325** | **126** | **129** | **243** | **28** | **91** | **3** |
| Martin | 120 | 43,660 | 17 | 71 | 22 | 7 | 2 | 1 |
| St. Lucie | 86 | 34,532 | 16 | 48 | 20 | 2 | | |
| Indian River | 17 | 48,217 | 5 | 11 | | | 1 | |
| Brevard | 71 | 3,943 | 8 | 13 | 14 | 36 | | |
| **Central Atlantic Total:** | **294** | **130,352** | **46** | **143** | **56** | **45** | **3** | **1** |
| Volusia | 191 | 86,464 | 1 | 174 | 15 | 1 | | |
| Flagler | 14 | 6,802 | 4 | 10 | | | | |
| St. Johns | 42 | 29,195 | 3 | 25 | 4 | 6 | | 4 |
| Duval | 107 | 24,423 | 3 | 44 | 54 | 6 | | |
| Nassau | 18 | 6,620 | 2 | 12 | 4 | | | |
| **Northeast Total:** | **372** | **153,505** | **13** | **265** | **77** | **13** | **0** | **4** |
| **Statewide Total:** | **4,476** | **1,355,748** | **1,951** | **1,437** | **566** | **281** | **161** | **80** |

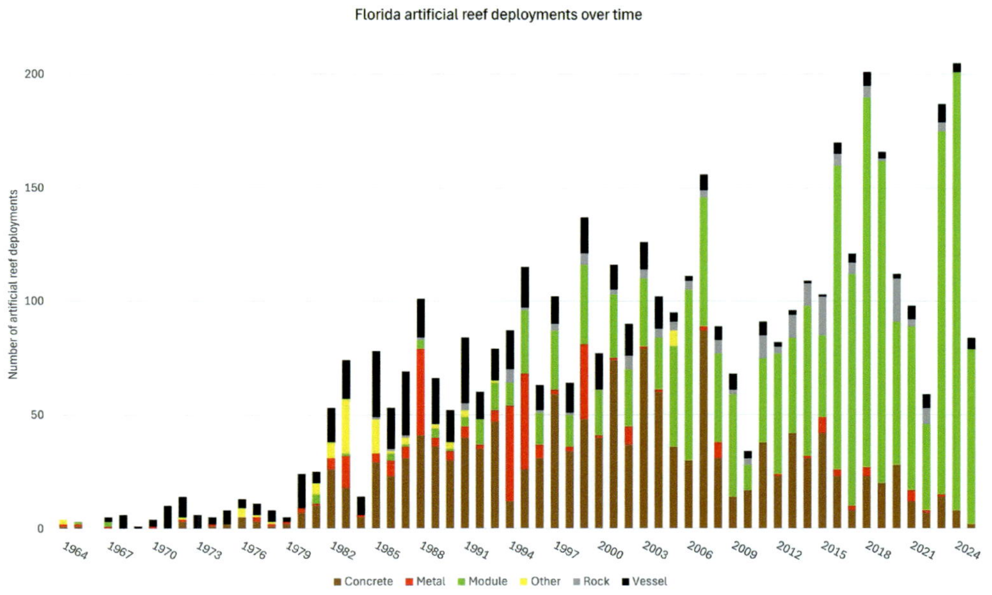


**FIGURE 9.3** Artificial reef deployments over time indicating the number of each type of artificial reef deployed for each year. Concrete reefs are represented by brown bars, metal reefs are represented by red, modules are represented by green, others are represented by yellow, rock are represented by gray, and vessels are represented by black.

colonization or success as artificial reefs (Miller and Chadik, 1993). While relatively few of these tire reefs occurred, they are frequently referenced as examples of "mistakes to avoid" and indicate the need for intentional planning of artificial reef materials (Sherman and Spieler, 2006).

Few regulations governed the placement of materials offshore during this early period of Florida's artificial reef activities, leading to little or no planning and haphazard material deployment practices. Additionally, there were no maps of existing habitats and little information about the seafloor. This occasionally resulted in the placement of materials directly on unspoiled natural hardbottoms. In some cases, materials were intentionally placed in close proximity to natural hardbottom resources, such as the Christ of the Deep statue that was deployed in 1965 and is now a well-known feature of the natural Dry Rocks Reef (in the Florida Reef Tract; Figure 9.1). Early technology for locating materials often used LORAN (i.e., the Long Range Navigation system developed during World War II), which was also unreliable, hampering both planning and monitoring.

These first artificial reefs, however, also prompted the first research on artificial reefs, as scientists sought to understand the fundamental mechanisms of artificial reef function. Work by Bohnsack and Stone is still incredibly valuable and referenced today (Bohnsack, 1979; Bohnsack and Sutherland, 1985; Stone et al., 1979). The first CARAH occurred in 1974, showing foresight regarding the global implications of artificial reefs (Bortone, 2015).

## 9.4 PEAK CONSTRUCTION: 1980s–2010

Deployment of artificial reefs began to increase during the 1980s and steadily grew through the 2000s, as projects oriented around experimental materials gave way to projects aimed at developing more efficient deployment practices. This period is referred to here as the Peak Construction era. Over 2,400 public deployments occurred during this period, focusing on fishing opportunities, habitat creation, waste disposal, and dive site creation. As diving became a more popular activity, derelict vessels were sought as a materials of opportunity to create attractive dive sites.

Commonly deployed materials of this time period mostly involved secondary-use concrete (1,093) and specifically designed modules (569) as researchers, managers, and industry leaders began experimenting with concrete and metal to create prefabricated modules (Figure 9.3). Vessels (424) were also an extremely popular material. One to two dozen vessels were deployed each year until vessel availability, expense, and remediation requirements restrained vessel deployments in 2005 (Environmental Protection Agency and Maritime Administration, 2006). Other metal structures (245) included decommissioned military tanks and vehicles, civilian automobile bodies, storage tanks, and scrap metal. The deployment of less desirable artificial reef materials (e.g., automobiles, which are easily displaced by storms and are short-lived) waned in the 1990s as managers realized the potential environmental harm and lack of ecological benefits from these weak structures. Limestone boulders (60), available from quarries in central Florida, came into consideration as a material because they were relatively easy to deploy with few environmental concerns. Other materials (60), such as tires and fiberglass boats, were deployed during the 1980s and also fell out of favor by the 1990s because of their negative environmental effects and limited benefits (Atlantic and Gulf States Marine Fisheries Commissions, 2020).

Experimentation in the types of materials was also mirrored by experimentation in deployment methods as more marine contractors became involved with artificial reefs. Contractors became proficient in using barges and construction equipment to move large quantities of materials into the water. Advances in technology allowed for higher precision with regard to location, as developments in depth sounders and eventually GPS (i.e., Global Positioning System) allowed for even more accurate placement and improved the ability to locate reefs once deployed. Managers focused artificial reefs in areas of sand away from existing hardbottom to concentrate on creating new habitat with limited impacts to existing hardbottom resources. SCUBA diving also became more popular for both recreation and research. Moreover, the availability of Nitrox (a gas mixture of nitrogen and oxygen) and the development of improved underwater cameras allowed researchers and managers to spend more time underwater evaluating the success of deployments, survey methods, material designs, and fish usage of the reefs.

The technological advances spurred artificial reef researchers such as Bortone, Caddy, Lindberg, Patterson, and Seaman—to look at more complex questions involving the attraction-production debate (Bortone and Van Orman, 1985; Caddy, 2008; Lindberg, 1997; Patterson III et al., 2009; Seaman et al., 1989). Eight CARAHs occurred during this time, gathering artificial reef practitioners from around the world as experimentation continued globally.

## 9.5 PRESENT DAY: 2010–2025

Artificial reefs in recent years have undergone drastic changes when compared with previous eras. Approximately 1,863 deployments have occurred since 2010, almost three-quarters of which consist of prefabricated concrete modules (n = 1,389). While there are large numbers of modular artificial reefs in Florida, they are often much smaller in area and volume than historical, non-modular artificial reefs in terms of actual seafloor coverage and weight. This discrepancy inflates the apparent amount of artificial reef deployments that are occurring (Keenan et al., 2022; Paxton et al., 2024). For example, one barge load of prefabricated modules can create multiple artificial reef sites when compared with concrete or boulders, where only one artificial reef is deployed. During the last decade in particular, secondary-use concrete (316), limestone boulders (99), and vessels (59) were less commonly used for artificial reefs.

The recent trend in the artificial reef materials used reflects a shift in both the expectations stakeholders have for artificial reefs and the conditions under which marine contractors operate.

Recent goals tend to be multipurpose, attempting to focus on a range of outcomes for a single reef. These goals include increasing fishing opportunities and fish habitat, creating dive and snorkel sites, generating tourism, restoring habitat, protecting shorelines, attenuating waves, and creating memorial sites, art exhibits, and ecosystem service replacements. Highly engineered "hybrid" reef modules are designed to address a suite of goals. For example, some mitigation sites are expected to act as shoreline protection and habitat creation. Wave attenuation structures have been proposed for coral restoration sites. Art sculptures have been deployed to garner attention and funding, while simultaneously creating dive sites and serving as new habitat. While it is true that any artificial reef can serve multiple functions, the most successful artificial reefs are designed with a primary goal in mind and with realistic expectations.

The trend of expecting an artificial reef to serve many functions is partly due to recent technological advances that allow for diverse artificial reef designs and precision placement. Concrete molds, dry-forming, and 3D printing are used by contractors to create replicas of almost any design. Side-scan and multibeam sonar create clear pictures of the seafloor, allowing for the accurate localization of natural resources so that artificial reef deployments can avoid impacting them. Increased use of onboard GPS and dynamic positioning allows contractors to achieve extreme accuracy in the placement of individual modules. Artificial reef managers can use these advances to plan more deliberately and consider connectivity to existing artificial reefs and natural resources. Managers also face pressure to consider how to best benefit a variety of user groups who visit these reefs in diverse habitats, each subject to different fisheries management regulations. These considerations range from creating easily accessible snorkeling and kayaking sites to establishing shallow dive sites or deep-water fishing sites offshore in federal waters.

Recent research by Stallings and Paxton focuses on the intentional planning of artificial reefs by considering existing resources, comparative functions with natural reefs, spatial planning, social and economic impacts of reefs, and interactions among reef usage, fisheries, and fisheries management (Chong et al., 2022; Paxton et al., 2022, 2025; Stallings et al., 2025).

## 9.6 CHALLENGES

### 9.6.1 Marine Contracting

As artificial reef managers, one of the most impactful challenges for the State of Florida Artificial Reef Program is the recent shift in marine contracting. The current marine contracting environment looks quite different from that of the early days of artificial reef deployments. As of 2025, there are only a few marine contractors working in Florida who are qualified, interested in artificial reef work, and operating at feasible prices. Many smaller contracting companies do not have the experience required, while larger firms do not see a sufficient profit margin from relatively small artificial reef projects. This problem with contracting for artificial reef deployments is exacerbated by exponential increases in construction costs and regulations coupled with a decrease in material availability across industries since 2020 (Shah, 2025). In 2019, artificial reef managers reported prices ranging from US$34 to US$53 per ton to deploy secondary-use concrete. By contrast, in 2025, artificial reef managers reported deployment prices ranging from US$281 to US$450 per ton (Kalinowski, 2025).

One side effect of this issue is the recent shift toward prefabricated modules (Figure 9.4). These projects are often "turn-key", which means they have clearly stated costs and are logistically easier to secure because a single contractor completes all steps of the process; material acquisition or fabrication, transportation, and deployment. When compared with multistep projects involving concrete, boulders, or vessels, where the material, transportation, preparation, and deployment may each involve separate contractors, modules require much less administrative oversight and can be more affordable.

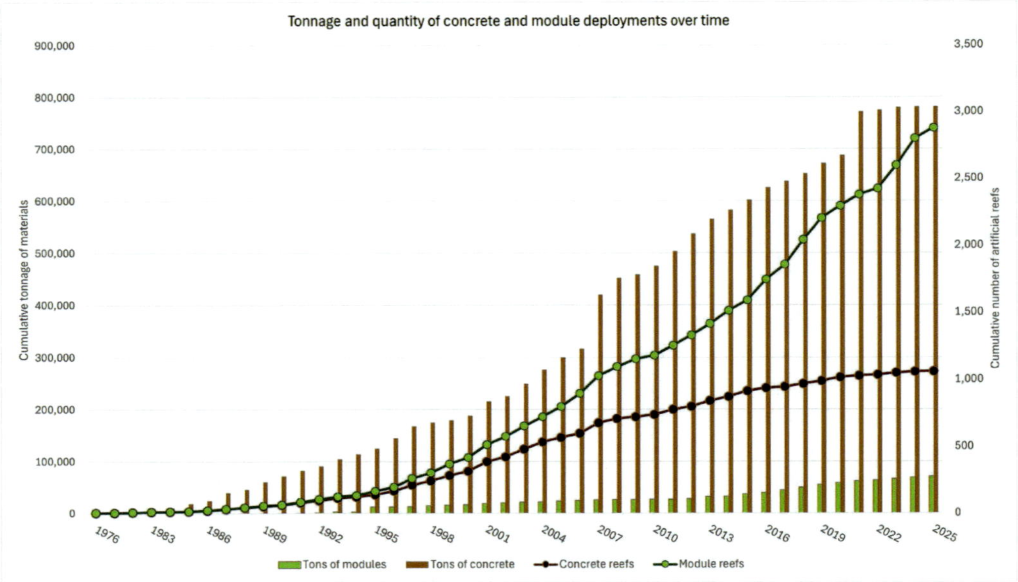


**FIGURE 9.4** The cumulative tonnage of modules and concrete along with the cumulative number of artificial reefs deployed per year. Cumulative tonnage of concrete reefs is represented by brown bars, and that of module reefs is represented by green bars. Cumulative numbers of concrete reefs are represented by the brown line, and cumulative numbers of module reefs are represented by the green line.

## 9.7 SENSITIVE HABITAT

Marine ecosystems in Florida face a suite of issues including decreased water quality, shoreline erosion, habitat loss, coral bleaching, and increased fishing pressure (Gilliam et al., 2023). In recent years, artificial reefs are regularly proposed as solutions for addressing these ecosystem issues. These projects propose to deploy artificial reefs in areas of sensitive habitat, such as seagrass beds in the Big Bend region or coral reefs in the Florida Reef Tract (Figure 9.1). Historically, the preference has been to keep artificial reef activities away from these sensitive areas to prevent damage (Lindberg and Seaman 2011), but current habitat concerns are forcing managers to reconsider the relative risks and rewards.

One example lies in the Florida Keys, home to the majority of the Florida Reef Tract, the only natural barrier coral reef system in the continental United States (Banks et al., 2008). This area is faced with increasing pressures from fishing and diving, acidification, coral bleaching, and natural reef flattening. Here, managers are looking for new methods to restore the functions of degrading reefs and natural hardbottom. Artificial reefs have been proposed as "habitat support structures" to address some of these concerns by increasing the structural complexity and available complex substrate. Artificial reefs have also been proposed as mechanisms to shift user pressure from these sensitive natural reefs towards preferred recreational use areas. Reviews of recreational use activity before and after the 2002 deployment of the U. S. Navy vessel, the Spiegel Grove, found a decrease in use of natural reefs and an increase in local charter business (Leeworthy et al., 2006). Thus, managers should address two central questions: First, can careful placement of artificial structures benefit the existing resources in the Florida Keys? Second, can new artificial reef sites be strategically placed to shift fishing and diving pressure from natural reefs.

In another example, the previous use of tires as an artificial reef material has resulted in tires becoming scattered throughout some seagrass beds, where they are viewed as sub-par habitat (CSA Ocean Sciences Inc, 2019). Removing these tires disturbs the surrounding habitats as well as the habitat that was created; often, juvenile grunts and snappers can be found using the tires for cover. Consequently, managers need to consider this question: Is it better to leave the reefs in place, remove them entirely, or remove and replace them with a preferred structure, such as a concrete module?

## 9.8 REGULATORY BOTTLENECKS

Recently, new regulations and interpretations of existing permitting language have created "bottlenecks" (i.e., restrictions) for local coastal governments. Full archaeological assessments prior to permit issuance and approved engineering plans for artificial reefs are cost-prohibitive for local coastal governments. Currently, State Artificial Reef Program managers are working with the federal entities to relax these regulations, while also considering the potential need to provide these services directly to counties by completing the pre-permitting archaeological surveys for them.

New protected species regulations, such as those for sea turtles, have also created obstacles for artificial reef development. Large areas (specifically, the area extending from the entire inshore Florida coastline out to depths of 20 m) have been proposed as critical habitat for the Green Sea Turtle, *Chelonia mydas* (Fish and Wildlife Service, 2023). Combined with recent (2017) material requirements to prevent turtle bycatch (National Marine Fisheries Service, 2017), the permitting timeline has significantly increased, impeding the progress and momentum upon which artificial reef managers rely.

## 9.9 CONFLICTING USE

As artificial reef managers consider the needs and preferences of local stakeholders and resources, conflicts have arisen at these public sites. One example can be seen with the increased bycatch of turtles at piers located in close proximity to artificial reef sites, particularly at Navarre Pier in Northwest Florida (Figure 9.1). While these nearshore artificial reef sites provide additional habitat and nearshore fishing access, they are also used as resting areas for sea turtles; emerging research speculates that the close proximity of the reefs to fishing piers contributes to this turtle bycatch (Brannum et al., 2025; Reimer et al., 2023).

Additionally, the recent recovery and reopening of the Goliath Grouper (*Epinephelus itajara*) fishery has been contentious for divers and anglers. A few artificial reefs are known to be Goliath Grouper spawning aggregation sites (Collins et al., 2015), creating exciting attractions for divers, but also creating a potential vulnerability for the species if they are targeted by recreational anglers during spawning activities (Figure 9.1). In 2023, the Florida Fish and Wildlife Conservation Commission considered potential management plans that involved creating closed areas on some of these artificial reef spawning sites. When discussed with the public, stakeholders disagreed about the proposed closures, indicating a desire to have their own stakeholder-specific form of recreation prioritized. No closures were enacted, and the question remains: How do managers design and place artificial reefs when the benefits impact stakeholders in different, and sometimes conflicting, ways?

## 9.10 COMPETING EXPECTATIONS

In addition to conflicts regarding the use of artificial reefs, stakeholders frequently have competing expectations and perceptions for each site. For example, these structures are often expected to achieve multiple goals, representing a disconnect between engineering and biology. The most successful artificial reefs have always been those with clear, well-defined goals (Paxton et al., 2022). Presently, the Florida Artificial Reef Program helps guide stakeholders through the trade-offs that occur when reefs are designed with many purposes. One example is inshore reefs intended for wave attenuation that are also proposed to benefit shoreline restoration, coral outplanting (i.e., transplanting corals onto a hard structure), and fish habitat, while serving as snorkel attractions. Critiques of this multi-use approach for the goals of an artificial reef deployment argue that vulnerable coral outplants should not be placed in an area that experiences high wave activity. Another example can be witnessed by the fact that early artificial reefs were most often deployed in offshore waters to serve as fishing sites, but more recently, they are being placed in shallower waters to cater to recreational divers as well. A consequence is that the stability of these shallower reefs can be compromised as they are more easily impacted by tropical storms, particularly when using materials such as vessels, which are more susceptible to movement. Additionally, recent partnerships between local coastal governments and the military have utilized military training exercises involving missile strikes to deploy vessels. The military provides the vessel and the deployment, while the local coastal government provides an active, permitted artificial reef area. While this allows the local coastal government to acquire a vessel as an artificial reef at no cost, using missile strikes sacrifices material stability and habitat integrity. Because artificial reefs can offer both social and ecological benefits, managers must identify a reef project's primary goal and consider feasibility and practicability when planning new artificial reef sites and material designs.

## 9.11 UNCOORDINATED EFFORTS

While the approach taken in Florida toward artificial reefs allows for local coastal governments to have more control over the artificial reef activities in their communities, it can also lead to uncoordinated and, occasionally, duplicated efforts across the state. There have been a few recent successful partnerships between coastal counties and agencies. For example, the sinking of the *DEEP STIM III* involved funding from several different coastal governments and nonprofit organizations. Additionally, local governments often work with local universities to conduct monitoring studies of their reefs. Despite this, it is difficult to focus the efforts of many coastal governments, nonprofit organizations, and academic researchers to achieve any broad, coordinated goals. Recent attempts by county officials in southeast Florida to coordinate their activities by contracting with a marine construction vendor were unsuccessful. This effort failed, despite the potential for a higher-value project, due to the intricacies of differing purchasing requirements among the entities.

Difficulties in coordinating artificial reefs in Florida are also due to regional fluctuations in engagement, with some programs becoming inactive and requiring extensive assistance to become active again. When local artificial reef coordinators eventually retire or change careers, they take with them a wealth of institutional knowledge and momentum for reef projects. Additionally, local coordinators often face similar issues at the same time, such as a lack of vendors or permitting obstacles. While the Florida Artificial Reef Program attempts to regularly host regional and statewide workshops for artificial reef professionals, limited staff and resources make this difficult. Consequently, the question remains: How can a state artificial reef program maintain the independence and specificity of the local programs while also coordinating efforts across the state?

## 9.12 WEAK MONITORING

Florida's active Artificial Reef Program has been quite successful in deploying artificial reef materials, but the projects often lack any monitoring once deployed. While a few sites are highly visited (e.g., *USS ORISKANY, USS SPIEGEL GROVE*, Figure 9.1) or well-studied, basic statewide monitoring has been neglected. The program presently has only three full-time staff dedicated to artificial reefs across the state and minimal resources for monitoring. Verifying the coordinates and physical conditions of the more than 4,000 public artificial reef sites is nearly impossible with these limited resources. Consequently, the Florida Artificial Reef Program relies heavily on the local reef coordinators and scientists (Seaman, 2004). This reliance makes it difficult to provide an accurate evaluation of the current status and success of Florida's many artificial reefs. However, existing statewide habitat and fisheries monitoring efforts are beginning to include artificial reefs (Keenan et al., 2022; Switzer et al., 2020).

## 9.13 FISHERIES MANAGEMENT

Another pressing issue for artificial reef managers to consider is the lack of integration with fisheries management. It is understood that artificial reefs can have fisheries-level impacts, but neither state nor national fisheries management entities prescribe artificial reefs as a fisheries management option (Gallaway et al., 2009; Jaxion-Harm and Szedlmayer, 2015; Shipp and Bortone, 2009; Strelcheck et al., 2007). Moreover, fisheries management decisions generally do not incorporate artificial reefs as part of fishery management plans. The Florida Fish and Wildlife Research Institute and NOAA Fisheries have started to consider habitat classifications in models used for stock assessments, thereby providing a glimmer of hope for incorporating artificial reefs into fisheries management decisions (Karnauskas et al., 2017; Kenworthy et al., 2024; Powers et al., 2023). Nevertheless, this effort is still limited by the lack of comprehensive monitoring of existing reef sites.

## 9.14 FUTURE DIRECTIONS

Florida's Artificial Reef Program intends to address the challenges identified above to improve the success of artificial reefs throughout the state. To do this, the program plans to guide discussions around highly engineered hybrid reefs, focusing on functionality and feasibility. The Florida Artificial Reef Program is also participating in projects taking place in sensitive areas, including coral reefs in the Florida Keys, seagrass beds in Florida's Big Bend region, and a tire removal project in Tampa Bay (Figure 9.1). The Florida Artificial Reef Program will also provide historical context for these projects and identify subject-matter experts to help prevent unintended consequences in these sensitive areas.

Intentional spatial planning to consider sensitive habitats, stakeholder expectations, and resource needs will be critical to guiding artificial reef projects in Florida. Currently, the State's Seafloor Mapping Initiative is mapping the entire Florida seafloor down to 200 meters and making the data publicly available (Florida Department of Environmental Protection, 2023). These large-scale, high-resolution data will benefit artificial reef managers immensely, allowing them to capture the current state of artificial reefs in Florida, including their location, seafloor coverage, relief, and stability. Ideally, this information will lead to improved habitat classification and can be incorporated into artificial reef planning that considers habitat and reef connectivity on a larger scale.

The Florida Artificial Reef Program is also proposing to fund additional statewide socioeconomic studies to further understand reef use, perceptions, expectations, and economic value (Ropicki et al., 2021). As costs have increased, reef project expenses are being further scrutinized. Thus, having recent data that indicate the social and economic benefits of artificial reefs can help secure funding and prioritize resources.

The Florida Artificial Reef Program anticipates the need to revisit previously permitted areas with existing artificial reefs. These areas may have additional space available and may be more feasible, given the regulatory challenges of acquiring new permitted areas. Revisiting these areas also includes learning best-practice lessons from historical sites that have required remediation—such as tire artificial reefs—and adding new materials between existing patch reefs to promote connectivity. Also worth considering are areas that were previously ignored due to sensitive resources; these may serve as future sites when considering artificial reefs to address habitat deficits.

Finally, the ideal outcome for artificial reef activities conducted by the State is to integrate their anticipated benefits and fisheries impacts into fisheries management plans. Potential methods for achieving this include working within existing marine protected areas or closures, enacting new closures around highly utilized artificial reefs, designing reefs with consideration for existing fishing practices and the species that frequent them, and considering the impact of artificial reefs on each fishery.

## 9.15 CARAH DISCUSSION TOPICS

At the 12th CARAH, Carolyn Kalinowski, the Florida Artificial Reef Program Biologist, gave a presentation highlighting the different eras of Florida's Artificial Reef Program. Below are the questions that were posed for consideration at the conference:

- What challenges and successes have others identified while working in areas with sensitive habitats, such as coral or seagrasses?
- What challenges and successes have others identified in implementing comprehensive spatial planning or ecosystem-based management?
- What are some of the pros and cons of hybrid and engineered reefs?
- What is holding us back currently?

The discussion that followed with conference participants in response to these questions was extremely beneficial and informed the Florida Artificial Reef Program managers of some of the current global opinions on artificial reef activities. Conference participants had generally favorable opinions of engineered or hybrid reefs, with minimal concerns regarding conflicting expectations, and more emphasis placed on documented successes. Regarding fisheries management, some successes in integrating artificial reefs into fisheries management were identified. Dr. Jeremy Claisse with California Polytechnic State University, Pomona, discussed local successes with artificial reefs by identifying habitat bottlenecks and designing structures to support specific life history stages at specific locations experiencing habitat deficits (Pondella et al., 2018, 2022). Dr. Matthew Taylor with the New South Wales Department of Primary Industries' Port Stephens Fisheries Institute discussed the incorporation of stock enhancements on artificial reefs as part of increasing inshore recreational angler satisfaction. His research found benefits to using artificial reefs as tools to artificially disperse stocked fish (Becker et al., 2024). Representative researchers from other countries also mentioned dealing with permitting restrictions. For example, new permit areas are not allowed in France; therefore, all activity is confined to existing reef sites. This leads to deploying reefs on top of already existing reefs, or combining a new reef with other planned construction activities.

In contrast to the Florida Artificial Reef Program, noticeable differences in monitoring efforts became apparent. For example, some countries (such as France and Australia), with programs smaller than Florida's, have far greater monitoring efforts on their artificial reefs and are able to incorporate monitoring results into subsequent reef projects.

One topic in particular that was referred to often was the desire for increased coordination and communication between artificial reef colleagues around the world, which reflects some of the desires and trends seen within Florida to coordinate efforts and collaborate on research studies.

The Florida Artificial Reef Program staff attended the 12th CARAH with the goal of gaining an international perspective on artificial reefs and hearing how other countries are approaching some of the same challenges found in Florida. Based on the aforementioned discussions that occurred during the conference, there are multiple areas in which Florida Artificial Reef Program managers can learn from practices in other countries to improve the management of Florida's reefs.

One such topic was the idea of designing artificial reefs for placement in areas with limited space. France is one country where new artificial reefs are not typically permitted, which limits space and requires projects to occur in conjunction with other planned construction—such as mooring buoys or marinas—or to take place on top of existing artificial reefs. Examples included eco-mooring buoys with Red Lionfish (*Pterois volitans*) exclusion features and a "Pegasus" project that complexified existing artificial reef materials by adding a newly designed 3D-printed module to be placed over the existing concrete. It is worth considering these efforts when working in areas with Florida's sensitive habitats or when considering visiting previous sites for additional deployments.

It is also worth examining the artificial reef activities in China when considering reef deployments in especially deep water or when deploying large metal structures. Additional references to floating artificial reefs, or FADs (Fish Aggregation Devices), were made during the conference, which may have implications for projects proposed in Florida.

Discussions with attendees from the United States also provided beneficial insights into how to successfully use artificial reefs to impact specific fisheries. Dr. Jeremy Claisse's work quantifying productivity on a mitigation reef provides a template that can be followed when planning artificial reefs in the Florida Keys, where extensive habitat degradation has occurred. By identifying the life history stages that are habitat-limited and mimicking that habitat with well-designed artificial reefs, it can be possible to increase productivity for a given species. The Florida Artificial Reef Program intends to work with local researchers to identify fish species that are suitable for this type of approach and create a framework that can be followed to achieve these outcomes.

## 9.16 CONCLUSION

Florida's Artificial Reef Program has changed since its inception, transitioning from opportunistic deployments of inexpensive, repurposed materials to a more complex system of intentional planning and multifaceted ecological and socio-economic goals. Each era of artificial reef development reflects both advances in technology and shifting expectations from stakeholders, researchers, and resource managers. These changes have expanded the potential benefits of artificial reefs but have also come with against new challenges related to material availability, construction costs, regulatory constraints, marine contracting, and conflicting uses.

While Florida's Artificial Reef Program has been successful in terms of the quantity of artificial reefs deployed throughout the state, the present era of artificial reef management in Florida now demands a more integrated and deliberate approach. Managers must navigate sensitive habitats, reconcile competing user expectations, incorporate large-scale spatial data, and address the persistent shortcomings in monitoring and fisheries integration. At the same time, increasing interest in hybrid reef structures has created new opportunities to address habitat deficits, enhance recreational opportunities, and improve ecological resilience, provided these designs are matched with clear, realistic objectives.

The discussions at the 12th CARAH highlighted that many of Florida's challenges are shared globally, and international perspectives offer valuable insights. Programs in Europe and Australia demonstrate the benefits of comprehensive monitoring and strategic placement in constrained areas. Florida's path forward will require embracing intentional spatial planning, strengthening partnerships across agencies and counties, investing in socio-economic and ecological monitoring, and incorporating artificial reefs more directly into broader fisheries and habitat management frameworks. By applying lessons learned from both Florida's history and international programs, the State can continue to guide artificial reef practices that are ecologically sound, beneficial to recreational users, and likely to achieve long-term success.

## REFERENCES

Atlantic and Gulf States Marine Fisheries Commissions ARS. 2020. *Guidelines for marine artificial reef materials*. 3 ed.: James R. Ballard. p. 1–187.

Banks KW, Riegl BM, Richards VP, Walker BK, Helmle KP, Jordan LKB, Phipps J, Shivji MS, Spieler RE, Dodge RE. 2008. The reef tract of continental southeast Florida (Miami-Dade, Broward and palm beach counties, USA). In: Riegl BM, Dodge RE, editors. *Coral reefs of the USA*. Dordrecht: Springer Netherlands. p. 175–220.

Becker A, Dworjanyn SA, Mos B, Niella YV, Smith JA, Lowry MB, Taylor MD. 2024. Residency and movement patterns of juvenile yellowfin bream acanthopagrus australis, stocked on estuarine artificial reefs. *Marine & Freshwater Research.* 75(15).

Bohnsack JA. 1979. *The ecology of reef fishes on isolated coral heads: An experimental approach with emphasis on island biogeographic theory.* University of Miami.

Bohnsack JA, Sutherland DL. 1985. Artificial reef research: A review with recommendations for future priorities. *Bulletin of Marine Science.* 37(1):11–39.

Bortone S. 2015. CARAH (international conference on artificial reefs and related aquatic habitats): An historical perspective of accomplishments. *Journal of Applied Ichthyology.* 31:3–14.

Bortone SA, Van Orman D. 1985. Biological survey and analysis of Florida's artificial reefs. Gainesville, Fla., Marine Advisory Program, Florida Cooperative Extension Service.

Brannum RJ, Minch K, Wideroff M, Koperski ME, Shea CP, Foley AM, Trindell RN. 2025. Pier pressure: 23 years of incidental sea turtle captures at recreational fishing piers along Florida's gulf coastline. *Frontiers in Marine Science.* 12.

Caddy JF. 2008. The importance of cover in the life histories of demersal and benthic marine resources: A neglected issue in fisheries assessment and management. Bulletin of Marine Science. 83(1):7–52.

Chong L, Collins AB, Abeels H, Braswell A, Jackson S, Camp E. 2022. Artificial reefs in Florida 101—Why are they built? Part 1 of an artificial reef series. *EDIS.* 2022(2).

Collins A, Barbieri LR, McBride RS, McCoy ED, Motta PJ. 2015. Reef relief and volume are predictors of Atlantic goliath grouper presence and abundance in the eastern gulf of Mexico. *Bulletin of Marine Science.* 91(4):399–418.

CSA Ocean Sciences Inc. 2019. *Tampa bay hard bottom mapping project.* St. Petersburg, FL. No. TBEP Technical Report #08–19.

Environmental Protection Agency US, Maritime Administration US. 2006. National guidance: Best management practices for preparing vessels intended to create artificial reefs. In: Agency USEP, Administration USM, editors.

Fish and Wildlife Service I. 2023. Endangered and threatened wildlife and plants; designation of critical habitat for green sea turtle. In: Service FaW, editor.: Federal Register: Endangered and Threatened Wildlife and Plants; Designation of Critical Habitat for Green Sea Turtle. p. 1–195.

Florida Seafloor Mapping Initiative. 2023. Tallahassee, FL: Florida Department of Environmental Protection; [accessed 2025 December 19]. https://www.floridagio.gov/pages/fsmi.

Gallaway BJ, Szedlmayer ST, Gazey WJ. 2009. A life history review for red snapper in the Gulf of Mexico with an evaluation of the importance of offshore petroleum platforms and other artificial reefs. *Reviews in Fisheries Science.* 17(1):48–67.

Gilliam FS, Murrell MC, Beck MW. 2023. Editorial: Environmental threats to the state of Florida—Climate change and beyond: Volume ii. *Frontiers in Ecology and Evolution.* 11.

Huth WL, Morgan OA, Burkart C. 2014. Measuring Florida artificial reef economic benefits: A synthesis. Final Report to the FWC, Contract Number: FWC-11231, 116 pages.

Jaxion-Harm J, Szedlmayer S. 2015. Depth and artificial reef type effects on size and distribution of red snapper in the northern gulf of Mexico. *North American Journal of Fisheries Management.* 35(1):86–96.

Kalinowski C. 2025. Florida artificial reef database. In: Program FAR, editor. Tallahassee, FL: Florida Fish and Wildlife Conservation Commission. https://myfwc.com/fishing/saltwater/artificial-reefs/locate/.

Karnauskas M, Walter J, Campbell M, Pollack A, Drymon M, Powers S. 2017. Red snapper distribution on natural habitats and artificial structures in the northern gulf of Mexico. *Marine and Coastal Fisheries.* 9:50–67.

Keenan SF, Switzer TS, Knapp A, Weather EJ, Davis J. 2022. Spatial dynamics of the quantity and diversity of natural and artificial hard bottom habitats in the eastern gulf of Mexico. *Continental Shelf Research.* 233:104633.

Kenworthy M, Burgess E, Renchen J, Espittia J, McCawley J. 2024. *Applying a holistic management approach to managing Florida's inshore recreational fisheries.* Paper presented at: 9th World Fisheries Congress. Seattle, USA.

Leeworthy VR, et al. 2006. Can artificial reefs alter user pressure on adjacent natural reefs? *Bulletin of Marine Science.* 78(1):29–37.

Lindberg WJ. 1997. Can science resolve the attraction-production issue? *Fisheries.* 22(4):10–13.

Lindberg WS, Seaman W. 2011. Guidelines and management practices for artificial reef siting, use, construction, and anchoring in southeast Florida. In: Protection FDoE, editor. Miami, FL: Maritime. p. 150.

Miller WL, Chadik PA. 1993. *A study of waste tire leachability in potential disposal and usage environments.* Amended Final Report to Florida Department of Environmental Regulation No. SW67.

National Marine Fisheries Service US. 2017. U.S. Army corps of engineers Jacksonville district's programmatic biological opinion (jaxbo). In: Commerce USDo, editor: National Oceanic and Atmospheric Administration, National Marine Fisheries Service, Southeast Regional Office. p. 334.

Patterson III WF, Dance MA, Addis DT. 2009. *Development of a remotely operated vehicle based methodology to estimate fish community structure at artificial reef sites in the northern Gulf of Mexico*. Paper presented at: Proceedings of the Gulf and Caribbean Fisheries Institute.

Paxton AB, Runde BJ, Smith CS, Lester SE, Vozzo ML, Saunders MI, Steward DaN, Lemoine HR, Valdez SR, Gittman RK. 2025. Leveraging built marine structures to benefit and minimize impacts on natural habitats. *BioScience*. 75(2):172–183.

Paxton AB, Steward DaN, Harrison ZH, Taylor JC. 2022. Fitting ecological principles of artificial reefs into the ocean planning puzzle. *Ecosphere*. 13(2):e3924.

Paxton AB, Steward DaN, Mille KJ, Renchen J, Harrison ZH, Byrum JS, Brinton C, Nelson A, Simpson E, Clarke PJ. 2024. Artificial reef footprint in the United States ocean. *Nature Sustainability*. 7(2):140–147.

Pondella D, Williams JP, Williams CM, Claisse J, Witting D. 2018. *Restoring a nearshore rocky reef ecosystem in the challenge of an urban setting*. Paper presented at: American Fisheries Society Symposium.

Pondella DJ, Claisse JT, Williams CM. 2022. Theory, practice, and design criteria for utilizing artificial reefs to increase production of marine fishes. *Frontiers in Marine Science*. 9:983253.

Powers SP, Drymon JM, Hightower CL, Carleton LM, Hoenig JM. 2023. Direct estimates of reef fish abundance across an artificial reef network. *Canadian Journal of Fisheries and Aquatic Sciences*. 80(4):676–689.

Reimer J, Siegfried T, Roberto E, Piacenza SE. 2023. Influence of nearby environment on recreational bycatch of sea turtles at fishing piers in the eastern gulf of Mexico. *Endangered Species Research*. 50:279–294.

Renchen J, et al. 2020. *Assessment of artificial reefs impacted by Hurricane Michael*. 2020 Florida Artificial Reef Summit, Virtual, Florida Fish & Wildlife Conservation Commission; Division of Marine Fisheries Management.

Ropicki A, Adams C, Lindberg B, Stevely J. 2021. The economic benefits associated with Florida's artificial reefs: Fe649, 7/2021. *EDIS*. 2021(4):8.

Seaman W. 2004. Artificial Reef Monitoring in Florida Coastal Counties. Florida Sea Grant, University of Florida, National Sea Grant College Program, Florida; U.S. Department of Commerce, NOAA: 1–15.

Seaman W, Buckley RM, Polovina JJ. 1989. Advances in knowledge and priorities for research, technology and management related to artificial aquatic habitats. *Bulletin of Marine Science*. 44(2):527–532.

Shah J. 2025. *Trends in the cost of construction materials*. Tallahassee, FL: Florida TaxWatch.

Sherman RL, Spieler RE. 2006. Tires: Unstable materials for artificial reef construction. *Transactions on Ecology and the Environment*. 88:215–223.

Shipp RL, Bortone SA. 2009. A perspective of the importance of artificial habitat on the management of red snapper in the Gulf of Mexico. *Reviews in Fisheries Science*. 17(1):41–47.

Stallings CD, Emory ME, Peake JA, Schram MJ, Wall KR, Williams I. 2025. Ten years (2013–2023) of fish assemblage data collected seasonally with diver surveys on artificial and natural reefs. Ecology. 106(5):e70108.

Stone RB. 1972. Artificial reefs of waste material for habitat improvement. *Marine Pollution Bulletin*. 3(2):27–28.

Stone RB, Buchanan CC, Steimle FW. 1974. *Scrap tires as artificial reefs*. US Environmental Protection Agency.

Stone RB, Pratt H, Parker Jr R, Davis G. 1979. A comparison of fish populations on an artificial and natural reef in the Florida keys. *Mar Fish Rev*. 41(9):1–11.

Strelcheck AJ, Cowan J, Patterson W. 2007. *Site fidelity, movement, and growth of red snapper: Implications for artificial reef management*. Paper presented at: American Fisheries Society Symposium. American Fisheries Society.

Switzer T, Tyler-Jedlund M, Keenan S, Weather E. 2020. Benthic habitats, as derived from classification of side-scan-sonar mapping data, are important determinants of reef-fish assemblage structure in the eastern Gulf of Mexico. *Marine and Coastal Fisheries*. 12:21–32.

*The economic impacts of saltwater fishing in Florida*. 2021. Tallahassee, FL: Florida Fish and Wildlife Conservation Commission [accessed 2025 December 19]. https://myfwc.com/conservation/value/saltwater-fishing/.

Wallmo K, Edwards P, Steinback S, Wusinich-Mendez D, Allen M. 2021a. Economic impact analysis of snorkeling and scuba diving on Florida reefs. NOAA National Ocean Service, National Coral Reef Conservation Program. NOAA Technical Memorandum CRCP 42. Silver Spring, MD. 48 pp. https://doi.org/10.25923/g8ex-r982.

Wallmo K, Lovell S, Gregg K, Allen M. 2021b. Economic impact analysis of recreational fishing on Florida reefs. NOAA National Ocean Service, National Coral Reef Conservation Program. NOAA Technical Memorandum CRCP 41. Silver Spring, MD. 11 pp. https://doi.org/10.25923/rsgj-ta64.

# 10 Citizen Science-Community Engagement in Research and Coastal Stewardship in Florida's (USA) Northern Gulf of America

## *The Taylor County Artificial Reef Monitoring Program*

*Victor Blanco**
* Corresponding author: victorblancomar@ufl.edu

## 10.1 INTRODUCTION

In the United States, the state of Florida supports one of the largest artificial reef programs globally, with thousands of permitted deployments along both its Gulf and Atlantic coasts (Camp et al., 2022). State guidance emphasizes appropriately located, durable structures that add habitat complexity while benefiting fisheries without displacing conservation priorities (Florida Fish and Wildlife Conservation Commission, 2003). County-level projects in Florida often rely on partnerships among agencies, universities, and volunteers to plan, deploy, and monitor reefs (Chong et al., 2022).

Since 1990, Taylor County, Florida, has deployed over 1,360 metric tons (1,500 US tons) of artificial reef material at permitted sites in the Gulf of Mexico (i.e., Gulf of America), notably at Buckeye Reef. These artificial reefs enhance marine biodiversity, provide habitat for sportfish, and support recreational activities such as fishing and diving. Historically, monitoring was conducted by the Taylor County Reef Research Team, composed of local volunteer divers. This early program began in 1995 and ended in 2003. Revitalized efforts through structured training, systematic data collection, and strengthened partnerships with the Florida Fish and Wildlife Conservation Commission (FWC) restarted in 2018 (Taylor County Board of County Commissioners, 2024).

This chapter presents a technical and ecological assessment of the Buckeye Reef system based on multi-year monitoring, emphasizing the role of citizen science, and reef design. More importantly, this chapter also provides information on how a citizen-based approach to monitoring can provide data that are useful in evaluating artificial reef projects. Within this context, Taylor County's Buckeye Reef complex (*sensu* Grove and Sonu, 1983) was developed as a nearshore recreational fishing and diving destination, providing an opportunity to evaluate how reef materials and configuration influence fish assemblages, sportfish occurrence, and material longevity.

The goal of the artificial reef monitoring project was to periodically provide reliable, science-based data to evaluate the ecological and fisheries characteristics of the Artificial Reef Program at Buckeye Reef. Results from this monitoring effort serve to stimulate management actions that

 DOI: 10.1201/9781003676775-10

promote the use and enjoyment of artificial reefs by the public (Bortone et al., 2011). The program's objectives include the following:

1. Ensure that the Buckeye Reef site complied with the conditions defined in all authorizing permits and other applicable laws or regulations.
2. Coordinate and train a group of volunteer divers in data collection procedures that included quality assurance and data management. These procedures also involved developing plans to organize and conduct surveys, maintain data quality, and perform data analyses of selected sites.
3. Descriptively and comparatively quantify the fish assemblages associated with the artificial reefs over time at the Buckeye Reef monitoring site to document changes resulting from the increased habitat provided by the artificial reefs.
4. Document the occurrence of any non-native marine fish, especially the Red Lionfish (*Pterois volitans*), to facilitate their early detection and future eradication efforts.
5. Periodically and transparently disseminate the results of the analyses of the monitoring data analyses to interested parties to facilitate management actions.

A widely cited definition states that citizen science involves "the participation of volunteers in scientific endeavors that contribute to expand knowledge of the natural world". Concomitantly, the unique benefit of citizen science for the field of ecology lies in understanding processes occurring at broad geographic scales and on private lands, which are impossible to sample extensively with traditional field research models (Dickinson et al., 2010). Cigliano et al. (2015) examined the expanding role of citizen science in monitoring marine and coastal environments and highlighted its potential to generate meaningful scientific and conservation outcomes. The authors identified six major impact areas as a result of citizen-science monitoring: policy, education, community capacity-building, site management, species management, and research.

They also posited that these outcomes demonstrate the unique value of public participation in monitoring and stewardship of marine systems. Because marine ecosystems are vast, variable, and difficult to survey, the involvement of trained volunteers such as divers, anglers, and coastal residents enhances both the spatial and temporal reach of data collection. They emphasized that high-quality training, standardized protocols, and strong partnerships between scientists and communities are essential for ensuring data reliability and maximizing conservation benefits. Ultimately, it was concluded that well-designed marine citizen-science initiatives can improve ecological understanding, support management decisions, and foster stronger public engagement with ocean conservation.

Different authors have concluded that citizen-science data can be scientifically useful and, under the right conditions, comparable to data collected by professional researchers. However, data quality depends strongly on project design, protocols, training, and verification. Well-designed programs that use standardized methods, provide participant training, incorporate quality control (expert validation, replication, or automated checks), and match tasks to volunteer skill levels can routinely produce data that support ecological inference, long-term trend detection, and management decisions.

Two well-established examples illustrate these points. First, Forrester et al. (2015) compared long-term coral-reef monitoring data collected by professional teams and Reef Check volunteers (https://www.reefcheck.org/) and found they were able to detect long-term ecological change. This demonstrated that volunteer programs can identify signals of environmental change comparable to professional monitoring when methods and sampling effort are sufficient. Second, Kosmala et al. (2016) reviewed data quality across many ecological citizen-science projects and showed that features such as simple, task-specific protocols, training, redundancy (i.e., multiple independent observations), and expert validation strongly increase the reliability of volunteer data. These examples indicate that projects with such reliable and replicable practices can meet the statistical standards required for research and management.

## 10.2 MATERIALS AND METHODS

### 10.2.1 Study Area

The Buckeye Reef complex is located approximately 35.4 km offshore from Steinhatchee, Florida, in the Gulf of America (Figure 10.1). The artificial reef monitoring focused on six patch reefs (Figure 10.1, Table 10.1) deployed between 1996 and 2013. The reefs are composed of large and medium pieces of scrap metal, concrete cube modules, secondary-use concrete culverts, and prefabricated concrete pyramids (tetrahedrons).

## 10.3 VOLUNTEER RECRUITMENT AND TRAINING

Volunteers were recruited via social media and regional dive networks (dive shops and dive clubs). Participant divers were assigned different activities during the surveys: fish counting and identification, reef structure assessment, photography/videography, and diver safety. Training for volunteers included fish identification sessions, in-person workshops, and practical exercises on fish count

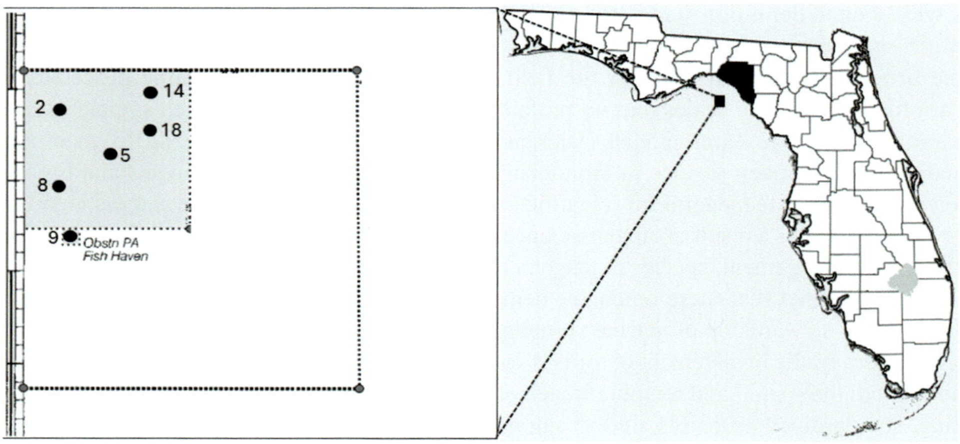


**FIGURE 10.1** Relative location of the Buckeye Reef artificial reef site and the monitored reef patches in the Gulf of America.

**TABLE 10.1**
**Location and description of monitored artificial reef patches (i.e., groups *sensu* Grove and Sonu, 1983) at Buckeye Reef—Taylor County, Florida. LINK Excel.Sheet.12 "D:\\TAYLOR UF IFAS\\PERSONAL TRAINING\\2025\\03 CARAH\\FOR THE BOOK\\Buckeye Reef Diver Surveys Database 2024 FINAL.xlsx" "Resume!R67C1:R73C6" \a \f 4 \h.**

| Reef | North | West | Material | Depth (m) | Year Deployed |
|---|---|---|---|---|---|
| 2 | 29° 38.877′ | 83° 54.766′ | Lindberg Cube | 15.2 | 2013 |
| 5 | 29° 38.660′ | 83° 54.500′ | Scrap Metal | 14.9 | 2002 |
| 8 | 29° 38.442′ | 83° 54.753′ | Culverts | 15.8 | 2013 |
| 9 | 29° 38.169′ | 83° 54.733′ | Tetrahedrons | 14.9 | 2003 |
| 14 | 29° 39.008′ | 83° 54.244′ | Scrap Metal | 14.6 | 1996 |
| 18 | 29° 38.804′ | 83° 54.262′ | Scrap Metal | 14.6 | 2013 |

techniques. An achievement score of 90% on the fish identification examination was required for participants to conduct fish counts. The training sessions included at least six hours of lectures, role-playing, fish identification practice (using cutouts, pre-recorded videos, and pool sessions, when possible), and a final examination.

## 10.4 SURVEY METHODS

Each site was assessed twice a year during late spring (April) and summer (May to September) in 2021, 2023, and 2024. Two survey techniques were adopted and applied: the Modified Stationary Visual Census or SVC (Bohnsack et al., 1986) and the Reef Environmental Education Foundation (REEF) Roving Diver Technique or RDT. The adopted SVC method employed was a modification of the original 7.5m radius protocol, surveying an imaginary cylinder with a 7.5m diameter over the survey area. The main reason for this modification was that the average underwater visibility at Buckeye Reef was at or below 10 m most of the time. These two methods allowed data collection on fish species composition, abundance, and diversity. The SVC is a method used to quantify the abundance and diversity of reef fish assemblages in marine environments. This technique involves visually sampling the reef fish assemblage by conducting censuses within a defined radius cylinder at stationary points, haphazardly selected by a diver. The method is simple, fast, objective, and repeatable, providing data on frequency of occurrence, abundance, fish length, and assemblage composition. The method is particularly useful in environments with highly diverse and abundant reef fish assemblages (Bohnsack and Bannerot, 1986).

The RDT is a non-quantitative, species-richness survey technique designed to document the presence and relative abundance of fish species across a reef site. Unlike fixed transect- or stationary-count methods, the RDT allows divers to move freely (i.e., "rove") throughout a site, which increases the likelihood of observing a greater diversity of species. During a survey, trained volunteers actively search for and identify fish species, recording the data on waterproof sheets of paper affixed to a clipboard or slate. Rather than counting exact numbers, divers record relative abundance using REEF's subjective, standardized "log (logarithmic)-scale" subjective abundance categories (Single, Few, Many, and Abundant). This approach maximizes accessibility for a wide range of diver skill levels while producing scientifically useful data.

This technique emphasizes species detection, making it particularly effective for documenting biodiversity patterns, identifying rare or cryptic species, and building long-term datasets across large spatial and temporal scales. Because the RDT does not require rigid transect placement, it can be used across varied habitats.

A pre-dive briefing was conducted with the volunteer divers to review survey and safety protocols. Given a max depth of 15.8 m, most dives were planned for a 20-minute bottom time (5 min SVC and 10 min RDT), followed by a 3-minute safety stop (for diver decompression) at 5 m, a 30-minute surface interval, and a maximum of three dives per day for each diver. Additionally, reef assessments were conducted that included inspecting the reef for structural condition, dimensions, deployment configuration, and top/bottom elevations. Site visibility and depth were also recorded. Any evidence of sea turtles, marine mammals, or invasive species (e.g., Red Lionfish) presence or activity was also noted.

The Margalef Biodiversity Index, Density Index (Den), Sighting Frequency (i.e., Frequency of Occurrence or %SF), and Abundance Score were calculated for the sportfish species based on the data collected from all surveyed reef sites. To analyze these data, we used the same assumptions outlined in the REEF protocol for the analysis of relative abundance data.

The Margalef Richness Index in Biodiversity was calculated as (S−1)/Log (n), where "S" is the total number of species and "n" is the total number of individuals in the sample. The Density Index (Den) is a measure of how many individuals of a species are observed on the basis of a scale of 1 to 4. It is representative of the abundance category (1–4) that was most frequently recorded for the species when it was observed. Abundance category weights are Single = 1, Few = 2, Many = 3, and

Abundant = 4. This weighted density average is calculated as ((nS×1) + (nF×2) + (nM×3) + (nA×4))/ Total Number of Surveys in which Species were Reported, where "nS" is the number of surveys that reported the species as Single, "nF" is the number of surveys that reported the species as Few, "nM" is the number of surveys that reported the species as Many, and "nA" is the number of surveys that reported the species as Abundant.

The Sighting Frequency Index (%SF) is a measure of how often a species was observed among surveys. It indicates the percentage of times out of all surveys that the species was recorded. The %SF parameter for a given species is calculated as the Number of Surveys in which the species was reported divided by the total number of surveys conducted.

The Abundance Score is calculated by simultaneously examining the Sighting Frequency (%SF) and Density index (Den) so that data summaries can be interpreted for different species. The Den and %SF scores can be multiplied (%SF × Den) to provide a measure of species abundance that accounts for zero observations. More details on these analyses can be found at https://www.reef.org/interpreting-reef-data.

One diver recorded videos and took photographs during each survey. The videos were later reviewed and compared to the fish census data obtained by the divers for quality assurance and control. These videos were used to verify any uncertain fish identifications and to re-train volunteers. As part of this quality assurance process, the videos were reviewed, and all fish were counted and identified. These data were then compared to the field data, and any discrepancy observed were adjusted using the data from the videos. Finally, a post-program evaluation was submitted by all volunteer participants to assess their perceptions of the artificial reef monitoring program.

## 10.5 RESULTS

A total of 120 divers participated as volunteers to collect data in the monitoring program between 2021 and 2024. Each year, these divers contributed approximately 2,500 volunteer hours/year (about 360 diving hours), valued at $67,000/year for the monitoring program. A total of 75% of the participants passed the rigorous training exams and actively contributed to data collection. Partnerships with local stakeholders, including dive shops and community groups, enhanced volunteer recruitment and public awareness of the reef monitoring program. Post-survey evaluations of volunteers indicated that 95% of those participating in the program agreed that the experience was satisfactory to very satisfactory. Of those evaluated (n = 27), 91% reported learning new information, 85% planned to adopt good practices for coastal ecosystem conservation, 66% reported that they would advocate for coastal ecosystem conservation by sharing the knowledge gained, and 90% reported that they would participate in future monitoring events and invite new members to participate.

A total of 48 surveys were performed between 2018 and 2024 as part of the Buckeye Reef citizen science monitoring program. A total of 41 fish species from 26 families were recorded between 2021 and 2024. Sites with large scrap metal structures (Reefs 5, 14, and 18) supported the highest diversity of fish species. Scrap metal at Reef site 14 had the most diverse fish composition, with 23 species with recorded using the RDT method, while Reef site 2 (concrete cubes) had the lowest diversity, with 13 species detected using this method (Table 10.2).

The expectation of higher fish species diversity at artificial reef sites with larger area coverage of the bottom and greater structural complexity is based on several ecological principles and research findings. First, larger artificial reef sites provide more complex habitats, which are more likely to support a wider range of fish species (Spieler et al., 2001; Riera et al., 2024). Second, more complex structures offer different levels of shelter, food sources, and breeding grounds, thereby attracting a variety of fish species (Hixon and Beets, 1993; Johnson et al., 2019). Larger sites may also foster more stable environments, reducing the risk of overfishing and allowing for a more diverse range of species to thrive (Lindberg et al., 2006). Finally, artificial reefs mimic natural reef ecosystems, which are known to support a greater diversity of fish species (Jones et al., 2020; Marshall et al., 2025).

**TABLE 10.2**
**Number of fish species and families by survey method (SVC and RDT) and by reef site at Buckeye reef.**

| | | Number of Species | | Number of Families | |
|---|---|---|---|---|---|
| Reef | Material | SVC | RDT | SVC | RDT |
| 2 | Concrete Cubes | 13 | 15 | 8 | 9 |
| 5 | Scrap metal | 16 | 23 | 12 | 15 |
| 8 | Concrete Culverts | 14 | 20 | 10 | 14 |
| 9 | Pyramids | 13 | 21 | 9 | 14 |
| 14 | Scrap metal | 14 | 23 | 10 | 16 |
| 18 | Scrap metal | 14 | 18 | 10 | 12 |

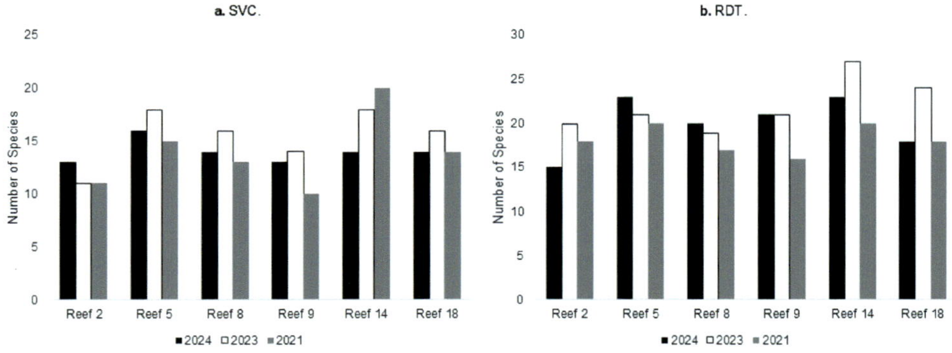


**FIGURE 10.2** Annual variation in the number of species by reef site and survey method (a. SVC; b. RDT) at Buckeye Reef.

These factors contribute to the overall complexity and stability of the reef ecosystem, which are essential for maintaining fish biodiversity.

Below is presented an example of the application and utility of the data obtained by citizen scientists. This section focuses primarily on the sportfish species identified and observed during the visual fish-monitoring surveys, considering their importance to the recreational fishing industry using the reef and the interest of the scuba diving community, as defined by Ropicki et al (2021). Based on the total fish count per sportfish species at Buckeye Reef using the modified SVC (number of fish per 65.5 $m^2$), the Greater Amberjack emerged as the most abundant species, followed by White Grunt, Gray Snapper, and Gag. The most diverse fish family was Serranidae (including groupers of the family Epinephelidae) with five species, followed by the family Sciaenidae (drums) with four species; the families Carangidae (jacks) and Haemulidae (grunts) with three species each; families Lutjanidae (snappers), Labridae (wrasses), and Carcharhinidae (requiem sharks) with two species each; and finally the rest of the families with only one species represented.

Overall, both fish survey methods showed consistent patterns across the years, with species richness generally peaking in 2023, remaining intermediate in 2024, and falling to its lowest in 2021. Under the SVC method (Figure 10.2a), species counts ranged from eight to 20 species per site, with Reef 14 consistently supporting the highest richness across all years. In 2024, species richness at most sites remained comparable to that of 2023, though typically one to three species lower, which suggests slight interannual variability but overall stability in community composition. Reef 5 and Reef 14 showed the most notable increases from 2021 to 2023, indicating potential reef maturation or improved habitat complexity over time.

The RDT method (Figure 10.2b) indicated a higher species richness overall compared to the species richness recorded via SVC, as expected due to the method's broader search area and mobility. Species richness ranged from approximately 15 to 27 species per site. Again, Reef 14 exhibited the highest richness, with close to 27 species documented in 2023. This pattern of 2023 > 2024 > 2021 was consistent across nearly all sites, with the largest differences occurring at Reef 14 and Reef 5. The elevated species counts in 2023 across both methods may reflect improved environmental conditions, increased diver effort, or continued ecological development of the reef modules.

Together, these results demonstrate that species richness at Buckeye Reef is site-dependent, with Reef 14 emerging as the most diverse location across all survey years and methods. The consistent improvement from 2021 to 2023 suggests a trajectory of ecological enhancement, which is likely related to reef maturity, structural complexity, or the improved fish count skills of the volunteer divers. The shifts in abundance might also result from increased fishing pressure on declining species, potentially creating an ecological niche that allows other, less-frequently targeted species to increase in abundance and occupy available habitat. The slight decline in 2024 across most sites may reflect natural year-to-year variability rather than a long-term trend; however, continued monitoring will help clarify whether richness stabilizes or fluctuates with shifting environmental conditions.

Across all sites, biodiversity, as measured using the Margalef index, generally increased from 2021 to 2023, followed by a slight stabilization or modest decline in 2024 (Figure 10.3). The exception to this trend was observed at Reef 14, which exhibited the highest biodiversity values overall, decreasing from 3.3 in 2021 to nearly 2.9 in 2023 and reaching its lowest value of 2.2 in 2024. Other sites, such as Reefs 8 and 9, showed prominent increases from 2021 to 2023, reaching values near or above 2.8 before dipping slightly in 2024. These results indicate a moderate level of diversity within the sampled community, suggesting that the assemblage has a moderate number of species, with a relatively even distribution of species abundances. The Margalef richness index is particularly useful for assessing the diversity of species in ecosystems, as it accounts for both the number of species and the evenness of their distributions. The values of the index at Buckeye Reef suggest that the assemblage was experiencing a balance between species richness and evenness, which is often a sign of a healthy and well-functioning ecosystem. A higher value indicates a greater number of species present, which suggests a more diverse ecosystem. However, the index does not account for the relative abundance of species, meaning that a community with a high proportion of rare species may have a lower index value than a community with a high number of common species.

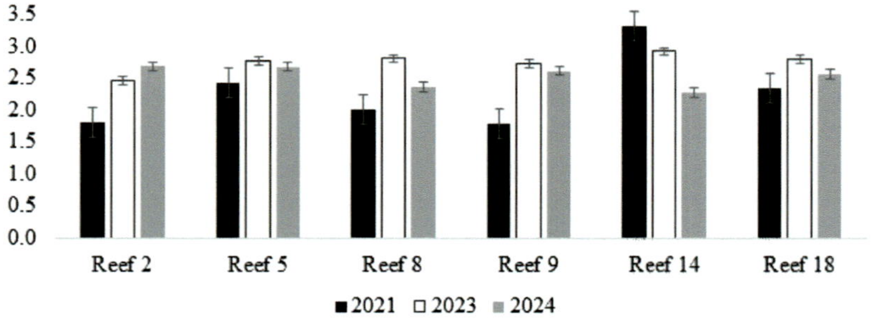


**FIGURE 10.3** Annual variation of the Margalef Biodiversity Index by reef site at Buckeye Reef.

Therefore, while a higher index value suggests greater species richness, it is important to consider the distribution of species within the community to understand its overall diversity and ecological health.

During the study period, the White Grunt (Den = 3.2) and Gray Snapper (Den = 3.0) exhibited the highest average Density Index values, followed by the Greater Amberjack (Den = 2.6) and Gag (Den = 2.4), as estimated from Figure 10.4. REEF (2024) reports that Den indicates which abundance category the species was most often recorded in when it was observed. For example,

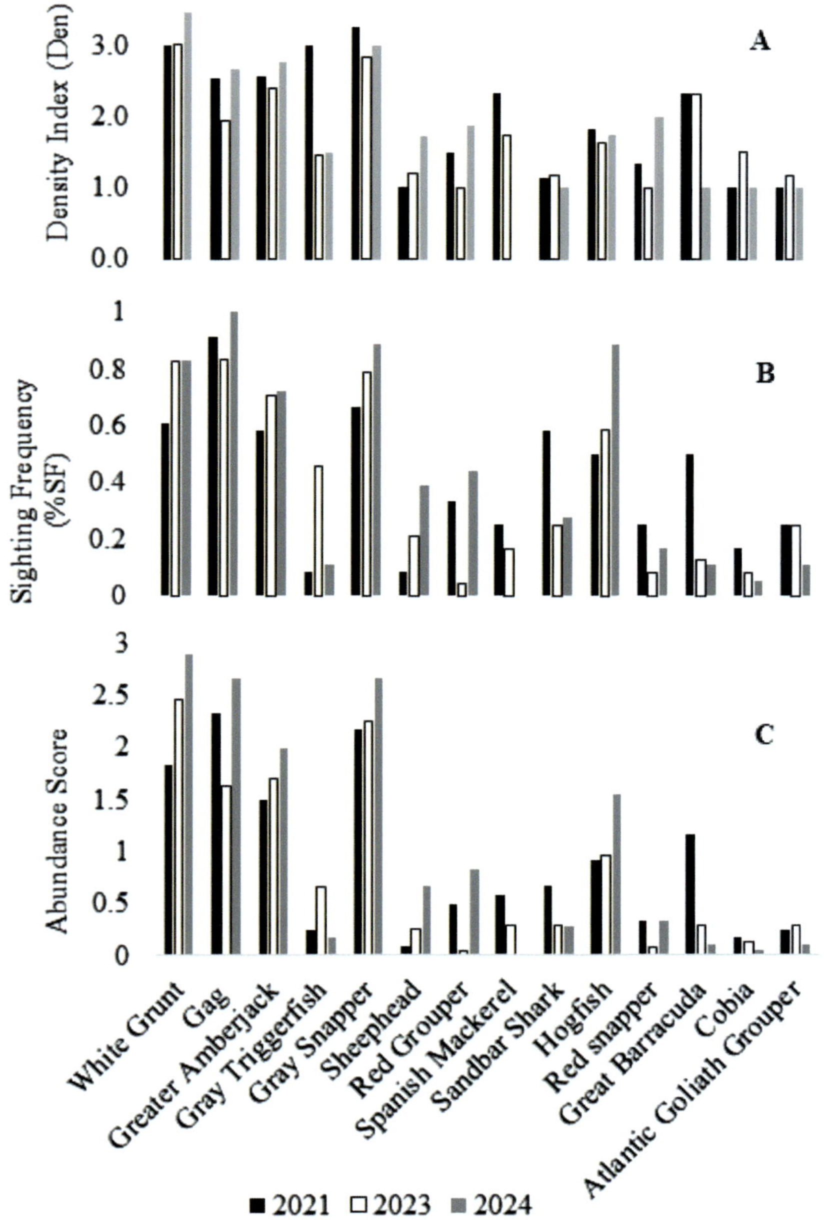


**FIGURE 10.4** Annual variation in (A) Density Index (Den), (B) Sighting Frequency Index (%SF), and (C) Abundance Score for sportfish species recorded using the Roving Diver Technique (RDT) from 2021 to 2024 at Buckeye Reef.

a Den value of 3.2, like that of the White Grunt, would be reflective of a species that was most often recorded in Category 3 (Many), but because the density index is greater than 3, there were some abundances recorded for this species in the larger abundance category (likely Category 4 or Abundant). The density index is used as an abundance guide because area is not rigorously controlled in the Roving Dive Technique method. It should also be kept in mind that the Den attribute is reflective of sighting distributions in the four different abundance categories (Single, Few, Many, and Abundant). Different distributions of sightings across these categories could potentially yield similar values of Den (i.e., it does not account for non-sightings).

When the variability of Den values in the 2024 surveys was compared with results from previous years (Figure 10.4a), most sportfish species showed an increase in Den, indicating higher relative abundance. For the remaining species, Den values were lower, reflecting fewer observations within the upper relative abundance categories. For example, Gray Snapper Den value in 2023 was 2.8, which represents more visualizations in the "Few" relative abundance category, while in 2021, it was 3.3, which represents more observations in the "Many" relative abundance category.

Results for Sighting Frequency Index (%SF) indicate that the Gag had the highest value, followed by the Gray Snapper, Hogfish, White Grunt, and Greater Amberjack (Figure 10.4b). The %SF indicates the percentage of times out of all surveys that the species was recorded. For all three survey periods (2021, 2023, and 2024), the Gag maintained the species %SF, with an average of 92% frequency. This means that Gag were counted in most of the total surveys performed during the monitoring events.

For species with a lower sighting index, as observed for the Gray Triggerfish, Sheepshead, Red Grouper, and Spanish Mackerel, a value equal to or below 0.4 indicates a low sighting frequency of reef fish. This means that the surveyor observed these reef fish at a significantly lower rate compared to other fish species or during other survey periods. The index is used to assess the abundance and health of reef fish populations, and a value of 0.4 suggests that the reef fish populations may be underrepresented in the survey data. This could be due to various factors, such as the timing and location of the survey or the presence of other, more visible species. It is important to consider these factors when interpreting the results and making any management decisions based on the survey data.

The Den and the %SF scores can be multiplied to provide a measure of species abundance. Figure 10.4c shows that at least 54% of the species presented an annual increase in their Abundance Score values, including the Gag, White Grunt, Gray Snapper, Greater Amberjack, Red Grouper, and Red Snapper. In contrast, species like the Gray Triggerfish, Spanish Mackerel, Great Barracuda, Cobia, and Atlantic Goliath Grouper decreased in abundance compared to previous years. An Abundance Score below 1 suggests that the number of fish species observed is significantly lower than the expected range for those categories, as observed for the Sheepshead, Spanish Mackerel, Red Grouper, Red Snapper, Cobia, and Goliath Grouper. This information is crucial for understanding the biodiversity and health of reef ecosystems, as it can indicate potential threats to the marine environment and the need for conservation efforts. Oppositely, a value of 2 or more in the ecological Abundance Score, as observed for White Grunt, Gag, Greater Amberjack, and Gray Snapper, indicates that the surveyor observed a relatively small number of fish species. This score suggests that the surveyor did not see many fish during the survey, which may be a result of the surveying conditions or the specific location of the survey. It is important to consider the context of the survey, such as the time of year, the depth of the reef, and the presence of other factors that could affect fish visibility and abundance. This score is part of a broader system used to assess the health and diversity of marine ecosystems, helping researchers and conservationists understand the impact of human activities on marine life.

Nevertheless, the assessment of artificial reef materials showed that structural integrity was high (>70%), though some degradation was noted at older scrap metal sites. Reef design and structural complexity were key drivers of biodiversity and fish abundance. Reefs composed of scrap metal with a large bottom coverage outperformed smaller, simpler structures. Temporal trends suggest that some sportfish populations, such as Gray Snapper and Greater Amberjack, are increasing, while others fluctuate, possibly due to fishing pressure and environmental variability.

The results of this study were presented by volunteers and the project coordinator to a variety of local stakeholders, including fishing clubs, marinas, dive shops, recreational diving groups, and the Florida Fish and Wildlife Conservation Commission (FWC). To effectively communicate findings and foster broader community engagement, the program developed multiple outreach materials, including fact sheets, educational posters, and a YouTube channel that hosts a repository of survey-dive videos. These resources have been used at academic conferences, public meetings, and community events to increase awareness of the artificial reef monitoring efforts.

The findings of this study have directly informed decision-making for new deployments at the Buckeye Reef site. The documented ecological performance of existing structures, along with the visibility generated through public outreach, has strengthened the county's ability to secure funding and partnerships. As a result, additional support has been obtained from the FWC, the Coastal Conservation Association (CCA), and local chambers of commerce, thereby advancing continued reef enhancement and long-term monitoring initiatives.

## 10.6 CONCLUSION

The Buckeye Reef monitoring program demonstrates the significant ecological value of artificial reefs in enhancing biodiversity and supporting recreational fisheries along the Gulf of America. Sustained monitoring efforts, combined with adaptive management, remain essential for maintaining these benefits and ensuring the long-term resilience of local marine resources. A central component of the program's success was the participation of trained volunteer divers. Through targeted training, volunteers improved their species identification skills and developed a deeper understanding of local marine ecology—which, in turn, elevated community awareness and engagement.

The involvement of citizen scientists greatly expanded data collection capacity and cultivated a sense of stewardship within the community. Effective citizen science initiatives rely on strong planning, consistent training, and clear leadership, all of which were foundational to the success of this program. The monitoring initiative highlights the social, economic, and recreational contributions of artificial reefs while advancing habitat restoration and sustainable resource use. Furthermore, the program directly aligns with Florida Fish and Wildlife Conservation Commission (FWC) priorities by generating essential data to support artificial reef management, site evaluation, and policy decisions.

Educational outreach and community participation have also increased public understanding of the ecological and economic importance of these reef systems, thereby fostering broader support for conservation actions.

Despite its success, the program faced several challenges, including limited sampling frequency, weather-related constraints, and variable access to equipment. Future improvements could include increasing monitoring intervals, investing in more durable or function-specific reef materials, and incorporating more rigorous statistical analyses to better detect long-term ecological trends and management needs.

## 10.7 ACKNOWLEDGMENTS

We express our sincere gratitude to the Florida Fish and Wildlife Conservation Commission (FWC) for their financial support and to the dedicated volunteer divers whose time, expertise, and commitment made this project possible.

## REFERENCES


Bohnsack JA, Bannerot S. 1986. *A Stationary Visual Census Technique for Quantitatively Assessing Community Structure of Coral Reef Fishes*. NOAA Technical Report NMFS 41.

Bortone SA, Brandini FP, Fabi G., Otake S. 2011. *Artificial Reefs in Fisheries Management*. CRC Press, Taylor and Francis Group, Boca Raton, London, New York.

Camp E, Chong L, Collins A, Abeels H, Mille K, Sipos M, Hall-Scharf B, Zangroniz A, Jackson S, Blanco V, et al. 2022. An update on Florida's Artificial Reefs: Recent deployments and trend. FA242. *EDIS* 2022(2). https://doi.org/10.32473/edis-FA242-2022

Chong L, Collins A, Abeels H, Braswell A, Ropicki A, Jackson S, Camp E. 2022. Artificial Reefs in Florida 101—Why are they built? Part 1 of an Artificial Reef series. FOR379. *EDIS* 2022(2). https://doi.org/10.32473/edis-fr450-2022

Cigliano JA, Meyer R, Ballard H, Freitag A, Phillips T, Wasser A. 2015. Making marine and coastal citizen science matter. *Citizen Science: Theory and Practice* 1(2):1–9. https://doi.org/10.5334/cstp.36

Dickinson JL, Zuckerberg B, Bonter D. 2010. Citizen science as an ecological research tool: Challenges and benefits. *Annual Review of Ecology, Evolution, and Systematics* 41:149–172. https://doi.org/10.1146/annurev-ecolsys-102209-144636

Florida Fish and Wildlife Conservation Commission. 2003. *State of Florida Artificial Reef Strategic Plan*. Division of Marine Fisheries Management, Tallahassee, FL.

Forrester G, Baily P, Conetta D, Forrester L, Kintzing E, Jarecki L. 2015. Comparing monitoring data collected by volunteers and professionals shows that citizen scientists can detect long-term change on coral reefs. *Journal for Nature Conservation* 24:1–9. https://doi.org/10.1016/j.jnc.2015.01.002

Grove RS, Sonu CJ. 1983. Review of Japanese fisheries reef technology. Report 83-RD 137. Southern California Edison Company, Rosemead, California. In: Seaman W, Sprague L (eds). 1991, *Artificial Habitats for Marine and Freshwater Fisheries*. Academic Press, San Diego.

Hixon M, Beets J. 1993. Predation, prey refuges, and the structure of coral-reef fish assemblages. *Ecological Monographs* 63(1):77–101. https://doi.org/10.2307/2937124

Johnson D, Christie M, Moye J. 2019. Habitat complexity and recruitment of coral reef fishes. *Marine Ecology Progress Series* 621:1–12. https://doi.org/10.3354/meps12987

Jones S, Asher J, Boland R, Kanenaka B, Weng K. 2020. Fish biodiversity patterns of a mesophotic-to-subphotic artificial reef complex and comparisons with natural substrates. *PLOS ONE* 15(4):e0231668. https://doi.org/10.1371/journal.pone.0231668

Kosmala M, Wiggins A, Swanson A, Simmons B. 2016. Assessing data quality in citizen science. *Frontiers in Ecology and the Environment* 14(10):551–560. https://doi.org/10.1002/fee.1436

Lindberg W, Frazer T, Portier K. 2006. Density-dependent habitat selection and performance by a harvested reef fish. *Ecological Applications* 16(2):473–486. https://doi.org/10.1890/04-1251

Marshall B, Russell S, Florisson J, Saunders B, Newman S, Harvey E. 2025. The influence of artificial reef structural complexity on fish assemblage composition. *Marine Environmental Research* 208. https://doi.org/10.1016/j.marenvres.2025.107103

REEF. 2024. *Interpreting REEF data*. Reef Environmental Education Foundation. https://www.reef.org/interpreting-reef-data

Riera E, Mauroy B, Francour P. 2024. Establishing complexity targets to enhance artificial reef designs. *Scientific Reports* 14, Article 22060. https://doi.org/10.1038/s41598-024-72227-z

Ropicki A, Adams C, Lindberg B, Stevely J. 2021. The economic benefits associated with Florida's artificial reefs. FE649. *EDIS* 2021(4):8–18. https://doi.org/10.32473/edis-fe649-2021

Spieler R, Gilliam D, Sherman R. 2001. Artificial reef design: Void space, complexity, and attractants. *ICES Journal of Marine Science* 59(Supplement):S196–S200. https://doi.org/10.1006/jmsc.2001.1163

Taylor County Board of County Commissioners. 2024. *Final Report of the 2024 Buckeye Reef Monitoring Program, Taylor County, Florida*. Victor Blanco (ed). Grant Agreement FWC–23055, Perry, FL. 85 p.

# 11 Social Sciences in Artificial Reefs

## *Past "Gaps" and Future Directions*

*Jorge Ramos**
* Corresponding author: jhramos@ualg.pt

## 11.1 INTRODUCTION

Since its inception, the deployment of artificial reefs has grown to the point of being fully established as an environmental management alternative (Bohnsack and Sutherland 1985; Baine 2001; Jensen 2002; Ramm et al. 2021). The idea of artificial reefs as an environmental management option has sparked interest in some areas of science, such as biology or even materials and structural engineering. However, artificial reefs generally have not attracted the attention of a large part of society. Lacking this power of attraction, it is up to enthusiasts of the topic to find ways to incorporate artificial reefs as a management alternative, including the social benefits that can be derived from their deployment. A specific, more recent focus toward this effort has been on disseminating information and educating younger generations about the topic of artificial reefs (Rendle and Rodwell 2014; Salaün et al. 2022a).

One of the several premises underlying the motivation for reef development is the capacity of such structures to enhance biological assemblages of the species that initially colonize them. Subsequent to this initial colonization, understanding the maturation of food chains, along with the potential increase in biodiversity and biomass production, is essential. An additional premise is that these structures also benefit people in the local communities where they are located (Holland 1991; Bortone et al. 1994; Kirkbride-Smith et al. 2016). The first known examples of artificial reef structures used for human benefit are found in Japan (Nakamura 1985; Thierry 1988; Otake 2020). Subsequently, the concept of deploying artificial reefs has spread to many other countries, generating significant interest and potential for future development.

However, despite significant interest in the natural sciences, few studies in the social sciences have addressed the deployment of artificial reefs and how they may benefit the social and economic structure of local communities (Tickell et al. 2019). Over the more than 50 years since the first Conference on Artificial Reefs and Related Artificial Habitats (i.e., CARAH) in 1974, the research focus has consistently been on the ability of these artificial structures (regardless of whether they are created for the purpose of giving materials a secondary usage) to diversify natural habitats for aquatic life and, consequently, foster aquatic biodiversity (Bortone 2015; Riera et al. 2024). However, what has been observed in research related to artificial reefs is a much greater occurrence of studies in the natural sciences, particularly in fisheries biology, and, to a lesser extent, research in other areas of knowledge, including the social sciences. It should be noted that structural engineering and the availability of new materials have also undergone significant development in the design and deployment of artificial reefs (Na 2023). More recently, beginning in the 2010s, Industry 4.0 emerged, representing the fourth industrial revolution, characterized by the integration of advanced digital technologies into production processes (Dunn et al. 2022).

DOI: 10.1201/9781003676775-11

**TABLE 11.1**
**Examples of some published studies in reference to socioeconomic objectives on artificial reefs (AR indicates artificial reef).**

| Method | N | Study objective | Reference |
|---|---|---|---|
| Interview | 4 | Social benefits of ARs | (Tessier et al. 2015) |
| | 4 | Environmental perception of ARs | (Kirkbride-Smith et al., 2016) |
| | 8 | Use of ARs for fishing | (Islam et al., 2014) |
| | 4 | Use of ARs as diving area | (Polak and Shashar 2013) |
| | 3 | Local ecological knowledge | (Lima et al., 2018) |
| | 4 | Economic value | (Westerberg et al., 2013) |
| Questionnaire | 4 | Use of ARs as diving area | (Stolk et al., 2005) |
| Likert scale | 3 | Environmental perception of ARs | (Shani et al., 2012) |
| | 1 | Use of ARs for fishing | (Al Ismaili et al., 2024) |

Industry 4.0 encompasses artificial intelligence, the Internet of Things (IoT), big data, and other developments linked to robotics and automation to create highly efficient production processes. The overall goal of these developments is to optimize production, reduce costs, and increase the flexibility and customization of products and services. In particular, Industry 4.0 developments using 3D printers to design modular structures have shown significant improvements in the design and configuration of artificial reefs (Cruz et al. 2025). In contrast, in the social sciences, there has been only modest growth in both scientific and technological development, as evidenced by the number of published articles resulting from these studies. Some studies have focused on the increased economic capacity derived from the maturation of successful reef structures that contribute to surplus production and thus the generation of wealth for surrounding human communities (Al Ismaili et al. 2024).

There are few studies published in the primary literature in other areas of the social sciences (e.g., geography, economics, sociology). However, there have been some published scientific research results (see Table 11.1 for examples) related to tourism, particularly diving-related activities (Stolk et al. 2005; Shani et al. 2012; Oliveira et al. 2015) and sport fishing (Jensen 2002; Fabi et al. 2011; Herbert et al. 2017). There are also some artificial reef studies showing their role as exclusion areas, in addition to their benefit to traditional coastal communities and artisanal fishermen (Lima et al. 2018, 2019b). Another example is related to the social benefits provided by artificial reefs as documented by Tessier et al. (2015). However, despite some specific cases, these types of investigations are still not widely disseminated, given that a considerable portion of them are in the gray or secondary literature and are most often related to specific reports of particular cases—often in languages other than English. Many of these study results fulfilled their mission of reporting on a specific type of reef project. Notably, however, many of these studies have not been submitted for publication in scientific journals and are therefore relegated to obscurity (Bohnsack and Sutherland 1985; Ramm et al. 2021).

The objective of this chapter is to provide a retrospective of artificial reef research in the social sciences, with an added focus on the potential of this research to influence the future of our overall understanding of artificial reefs. The approach here involved consulting the literature in the area of artificial reefs specifically directed toward socioeconomic research. This investigation intends to cover the most relevant topics related to artificial reefs and the social sciences, including the importance of involving students and local communities. Additionally, it explores the future of research employing emerging technologies, such as Industry 4.0 developments, which are sparking interest in all areas of knowledge and from which the social sciences may benefit enormously.

## 11.2 SOCIAL SCIENCES AND ARTIFICIAL REEFS

### 11.2.1 The Need to Monitor Artificial Reef Projects

Artificial reefs have long demonstrated their potential as management options in the aquatic environment, which is likely one of the main reasons they are so prevalent in coastal development projects worldwide (Pickering et al. 1999; Watanuki and Gonzales 2006). The entities promoting and financing artificial reef projects vary from country to country, as do the legal frameworks for reef project implementation. Some projects, due to their large scale, are promoted by government entities, whereas others, on a smaller scale, are developed by local communities and may have varying levels of support. Certain projects are often derived from private initiatives and may even be promoted through initiatives published on websites. Some are large-scale projects while others are much smaller. Interest in these projects is also uneven and can vary in duration and extent (Murray 1994; Aabel et al. 1997; Saharuddin et al. 2012).

Funding for artificial reef project initiatives most often requires monitoring activities in both the pre-deployment and deployment phases and, although less common, also involves some ongoing monitoring in the post-deployment phase. These monitoring efforts require the use of certain predefined indicators to assess the success or failure of reef development (Brochier et al. 2021; Gibson Banks et al. 2021).

There has been a wide range of interests and monitoring possibilities, with some being more traceable than others. In the past, investigating bio-colonization processes was invariably the most popular and interesting way for scientists to justify the usefulness of an artificial reef over time. More recently, it has been emphasized that investigations should involve not only the maintenance of biodiversity in a given location, but also the utility of an artificial reef to people. The results of this type of study would then be a fundamental part of a presentation when explaining the success of such initiatives (Elrick-Barr et al. 2022). The more apparent the benefits of artificial reefs are to a community, the greater the probability of their acceptance by society (Westerberg et al. 2013; Glarou et al. 2020).

The social sciences are broad in scope, encompassing several academic disciplines (Figure 11.1). The requirement to include the social sciences in monitoring efforts before, during, and after reef deployments stems from the need for a broader scope of scientific knowledge and, therefore, a better and more robust justification for the resources invested in this type of structure (Silva et al. 2019; Seaman 2022).

### 11.2.2 Breakdown by Geographic Area

In the United States, artificial reef usage has focused more on the recreational benefits that these structures provide (Creque et al. 2006). The engineering aspects related to the materials and shapes of the structures, along with the fisheries biology associated with the added value the structures create, have been important features of artificial reef studies. Some studies, however, have also been developed by investigating the benefits that the implemented structures provide to people who engage in recreational activities in these areas by conducting various types of temporal comparisons (Pondella et al. 2022).

In Europe, Spain, France, and Italy have recently developed multidisciplinary studies on artificial reef deployments. In the North Sea, the United Kingdom (UK), the Netherlands, and Germany have conducted some multidisciplinary investigations related to rigs-to-reefs. In the particular case of Portugal, in the early 2000s, it was deemed important to create multidisciplinary research teams to justify the funds invested in reef structures (Monteiro and Santos 2000). Much knowledge was produced employing this multidisciplinary approach. However, as funding related to reef development is often limited, it was only later that this type of multidisciplinary team was re-employed, and to a lesser extent. This was the case with the Nazaré artificial reef off the western coast of mainland

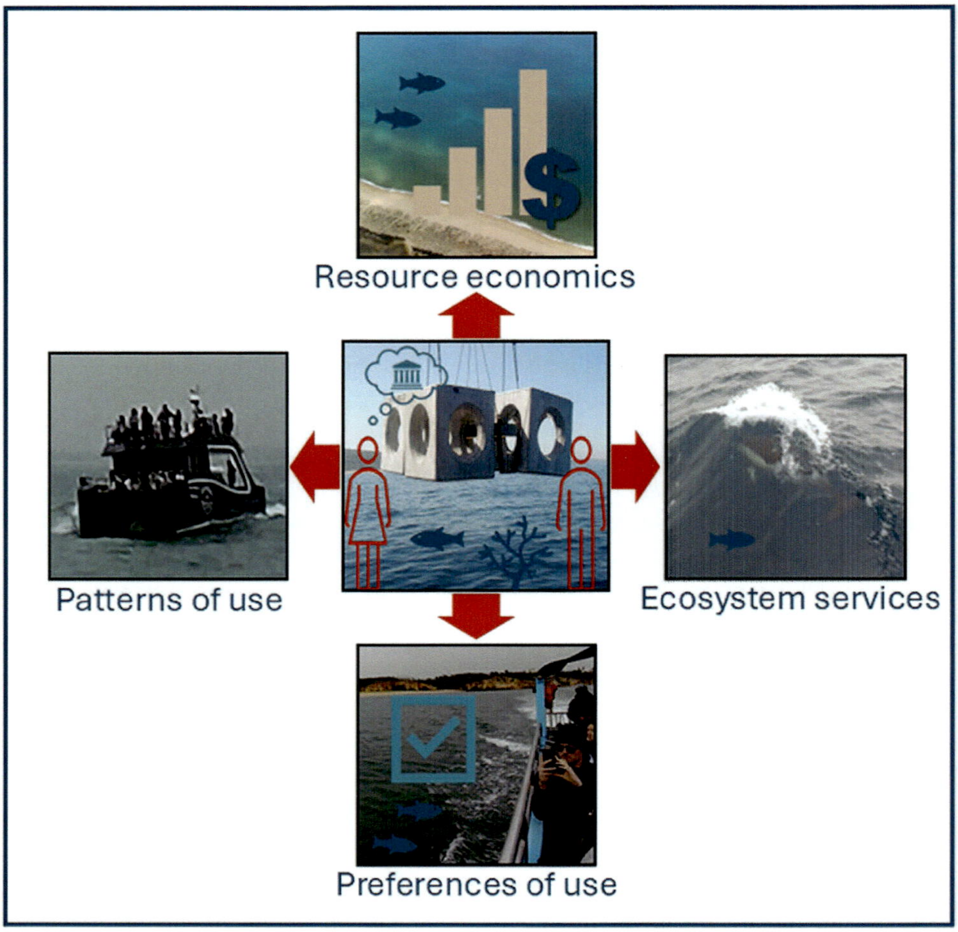


**FIGURE 11.1** Key social science research topics when studying artificial reefs (past and present).

Portugal (Ramos et al. 2021). In France, there were even some situations where socioeconomic research gaps persisted until the mid-2010s (Tessier et al. 2015).

While it has been shown that reefs are not a "panacea" for fisheries problems in coastal areas (Gannon et al. 1985; MacDonald 1994; Anderson 2023), it is still possible to observe the effects of artificial reef deployments on important socio-ecological systems over time in some areas. Importantly, it has been demonstrated that artificial reefs can add value to local coastal communities (Blount et al. 2021; Taylor and Suthers 2021; Salaün et al. 2022b). These local communities can benefit from good economic performance if appropriate co-management is implemented, incorporating local fishermen to avoid user congestion (Chong et al. 2024). In Japan, there are many similar examples of co-management. Several case studies have shown that the active and relevant involvement of fishermen through cooperative associations is crucial to the success of Japanese fisheries management, which is supported by financial subsidies and technical assistance from various government structures. The continuity of the system is due to the fishermen's commitment to conflict prevention, compliant behavior, and cultural values (e.g., Lim et al. 1995; Watanuki and Gonzales 2006).

### 11.2.3 Pioneering Studies

Although biological and reef colonization studies have been conducted since the early 1960s (Randall 1962), socioeconomic studies only began in the late 1980s following recommendations from the seminal work of Bohnsack and Sutherland (1985). These early studies emerged in the United States and were related to sport fishing and diving. The focus of these investigations was to understand the human need to utilize artificial reefs and the research instrument was surveys directed toward potential and actual users of reef structures. The pioneers of these studies were Walter Milon, who developed several studies involving economic analyses of artificial habitats (Milon 1988, 1991), and Karl Samples, who studied the economics of conflicting interests (Samples 1989).

In the 1990s, there was only a modest production of articles in this area of investigation, with the exception of those by a few authors. Among them was Stephen Holland, who explored research related to the uses of artificial habitats by people, primarily in the U.S. state of Florida. In the article, the author reviewed previous studies and identified variables that affect the use of artificial reefs by recreational fishermen. An index for searching for artificial reefs using various components was also created (Holland 1991).

Similar studies were conducted by David Whitmarsh, Helen Pickering, and Anthony Jensen, who were virtually "pioneers" in incorporating the legal and economic aspects of reef development. Their published articles remain among the most cited on the socioeconomic features of artificial reefs to this day and, as such, are considered reference articles (Pickering et al. 1999).

## 11.3 THE GROWTH OF SOCIAL SCIENCES IN ARTIFICIAL REEF STUDIES

### 11.3.1 Methodological Approaches

As discussed in the previous section, considerable scientific research was being conducted on artificial reefs in the 1990s. Although some research appeared sporadically with greater or lesser emphasis on the social sciences, there was little scientific production in this field until the end of the decade. Consequently, methodological approaches lacked input from potential users, with little contribution from the broadening application of new methods and materials. The few studies published until the late 1990s only refer to the use of social science methods such as directly inquiring with or interviewing potential reef users or making direct observations using binoculars. This meant that there was no significant progress made regarding the approach of the social sciences toward our understanding of artificial reefs.

However, this gap in our progress toward expanding and including social understanding was identified. This led to a first attempt to combine efforts and write about methods for evaluating artificial habitats for application in the field of social sciences, which was incorporated into the book edited by William Seaman and published in 2000 (Seaman 2000). This book compiled various techniques for evaluating reefs, including a chapter entirely dedicated to social and economic research methods, written by the individuals cited earlier (Milon et al. 2000).

These methods included traditional survey approaches with questionnaires and semi-structured interviews, the results of which were subsequently published (e.g., Whitmarsh et al. 2008; Polak and Shashar 2013). Exploring these methods, one can also find studies focused on multiple users and their stakeholders, sharing diverse and sometimes conflicting interests (Sutton and Bushnell 2007). Some studies also included direct observation of locations where reefs were known to have been established (Ramos et al. 2006; Calandra et al. 2022). Some other research appeared that combined many of these techniques (Ramos et al. 2007; Islam et al. 2014; Schuett et al. 2016; Vona et al. 2022). Another methodological approach to gather data that began to be explored somewhat later for monitoring users of reef areas was the inclusion of high-resolution, high-range cameras, mounted to remain above water and covering a large area in strategic locations (Wood et al. 2016; Becker et al. 2020; Hartill et al. 2020).

### 11.3.2 New Trends in Socioeconomic Research

The inclusion of economic benefits, in addition to ecological benefits, in research plans has become increasingly important in studies aimed at better allocating resources for reef development. Lima et al. (2019a) published an article that provided an overview of all literature published up to 2018 on trends, not only at the ecological level but also with regard to socioeconomic research on artificial reefs. This article focused on several aspects, such as a systematic review of the literature on artificial reef research worldwide since the early 1960s, focusing on long-term research trends, particularly in ecology and socioeconomics, as well as research trends in this field. The authors also mentioned that artificial reef science was incorporating increasingly sophisticated techniques in ecology and related areas; however, regarding the socioeconomic aspects of artificial reefs, there was still a scarcity of new techniques.

Another important contribution from social science research is that these more recent studies highlighted the importance of mitigating the impact of recreational diving on natural reefs and diverting divers to artificial reefs (Oliveira et al. 2015; Firth et al. 2023). For example, studies in the Persian Gulf indicate that fishing and tourism (such as diving and surfing) could be included as concurrent activities that justify investment in artificial reefs (Bartholomew et al. 2022).

The trend over time reflects the gradual recognition of the human perspective in artificial reef studies. In situ monitoring has proved to be crucial in making field observations essential for assessments. Earlier, to make these assessments, it was necessary to observe from a high vantage point, use binoculars at a certain distance (to avoid being intrusive), or be in a reef area aboard a boat with a crew (which was somewhat intrusive).

More recently, it has become possible to monitor human behavior without being intrusive. With the development and ubiquity of the Internet, as well as the use of various Industry 4.0 technologies, it is now possible to expand our potential for reef monitoring. This is particularly evident in the use of applications and devices that require fewer human and material resources for monitoring artificial reefs (Boyle et al. 2022; Ricci and Bohnenstiehl 2022).

### 11.3.3 Mentoring Students and Research Internships

Community education in the area of the environment is fundamental to the development and progress of any society, especially for communities in coastal areas. It is widely agreed that investing in the university education of young adults is worthwhile (Sparkman et al. 2012). By the end of their courses, these young people are exposed to education that includes an appreciation of the intricacies of the environment, and they are eager to apply their knowledge. Such knowledge is important for inhabitants of coastal areas to navigate their life path with greater consideration and appreciation for their environment.

Final-year social and natural science students have already acquired some knowledge related to their majors and developed technical skills they are eager to use. Teamwork skills are also important (Grant et al. 2015). Some university-affiliated research centers, such as CinTurs at the University of Algarve (Portugal), annually offer positions (i.e., internships) during the second semester for students who wish to apply some of their previously acquired knowledge and are motivated to develop additional skills related to projects launched by researchers and/or their professors (Figure 11.2). This is believed to be a crucial phase in the lives of students transitioning to working life, where they face academic prospects and a potential job that may be unrelated to their studies. This type of internship aims to make this transition by consolidating their knowledge, helping them work in groups and better communicate in science, and allowing them to better establish themselves in the scientific world (Shivni et al. 2021; Kilty et al. 2021).

These projects are sometimes a prelude to participating in funded projects or simply developing trendy ideas in certain fields of knowledge. They allow students to have direct contact between their course learning and its practical application on a current topic. Often, these experiences will

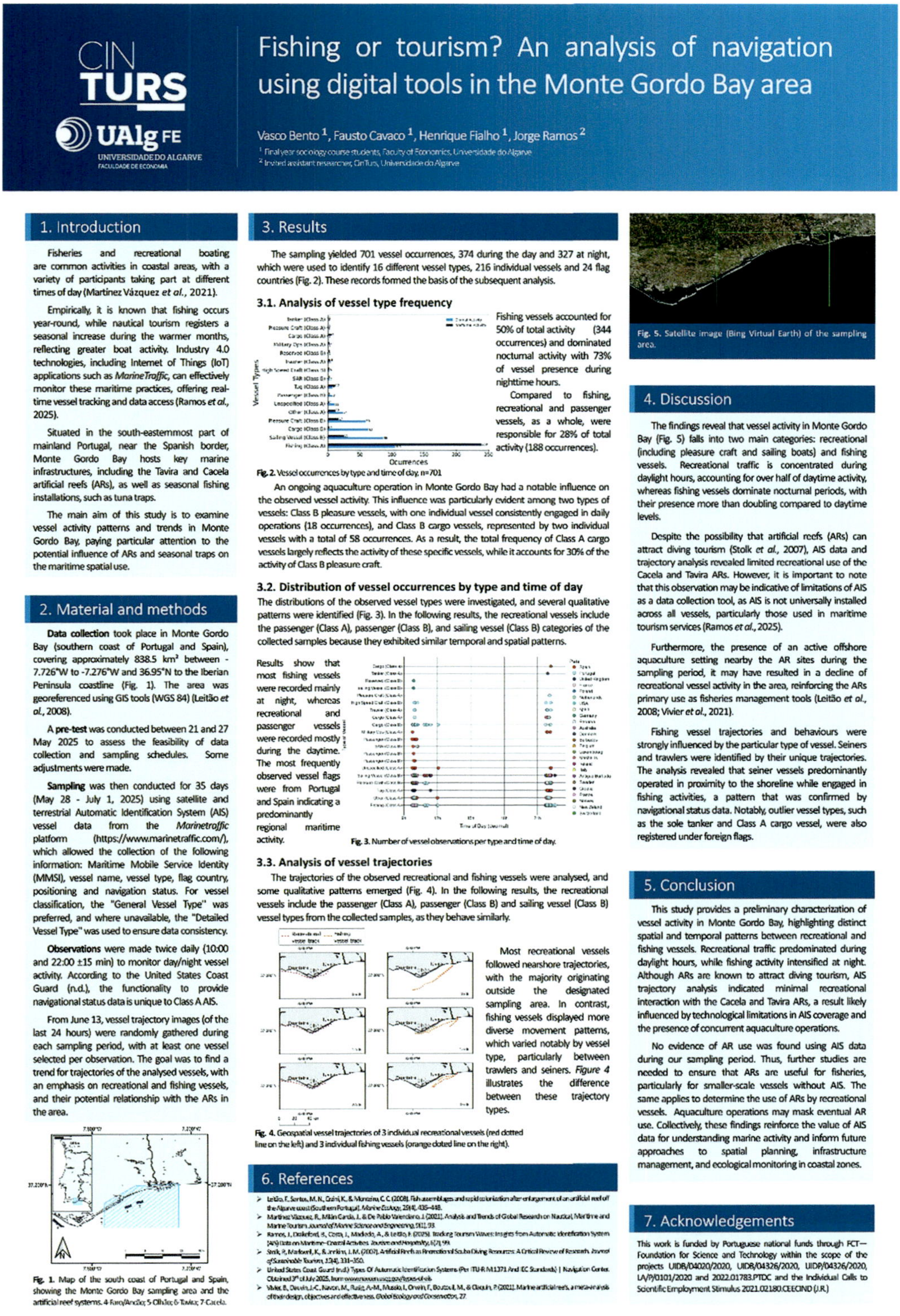


**FIGURE 11.2** Scientific poster on fishing, coastal tourism, and artificial reefs. Teamwork project carried out by third-year sociology students (academic year 2024/2025). The students developed various skills during a four-month integrated research internship.

lead to the eventual publication of their work in internationally accessible repositories (e.g., Ito et al. 2022). It is also important to consider artificial reefs implemented as part of a region's underwater heritage. For example, in the Algarve, internships are oriented around several research questions that fall within various fields of knowledge. Specifically, in the social sciences, a recurring issue is the ability to assess the usefulness of reefs for local communities, based on monitoring methods supported by digital tools (Ramos et al. 2024). In other words, the sampling effort in this type of approach is simplified to conduct this monitoring without a significant expenditure of resources (whether human, financial, or material). Furthermore, the results to date have been quite satisfactory, as students complete these four-month internships with their academic and group work skills significantly strengthened.

## 11.4 CORAL REEFS, DIVING, AND TOURISM

Artificial reef deployment attempts to complement—or even substitute—the functions of natural reefs. They complement places where there are no natural reefs to, for example, increase biodiversity. They substitute them when there is a risk of natural reefs being damaged. Coral reefs are a particular example of natural reefs that suffer damage frequently. Among the academic community, there is a consensus that coral reefs are incredibly beautiful, harbor immense biodiversity, and, in general, enhance a multitude of ecosystem services. However, it is noted that most coral reefs are located in tropical zones and, owing to atmospheric phenomena typical of these latitudes, some of these coral reef areas are subject to disruption by storms with some frequency. Oppositely, rising sea temperatures have also led to coral bleaching phenomena (Lirman and Schopmeyer 2016). In both situations above, there has been some interest from the scientific community in coral recovery through the promotion of projects involving the transplantation of coral fragments created as the result of storms (Hein et al. 2019) or the transplantation of corals more resistant to higher temperatures or bleaching phenomena (Shinzato et al. 2011).

These phenomena have also attracted social science researchers to study the enormous problems arising from the mass tourism occurring in natural coral reefs areas in certain destinations (Shakeela and Becken 2015; Eider et al. 2023). Since the famous case sparked by the 2000 film "The Beach," where mass tourism became widespread and caused the near-total depletion of corals in Maya Bay on Koh Phi Phi Leh Island (Thailand), the restoration of coral reefs from coral fragments has been driven by several factors (e.g., Ceccarelli et al. 2020). In a way, public awareness via a popular film helped generate an increase in tourism, which was not properly accompanied by adequate management measures. As a result, the aforementioned site was closed for several years, during which time the coral reefs were able to recover. More recently, this paradisiacal beach reopened with the implementation of a series of management measures to adequately cope with tourist demand (Koh and Fakfare 2020).

Based on the scientific knowledge that coral reefs have greater biodiversity, projects designed to expand coral reef areas have been developed for some time (Sebens 1994; Kittinger et al. 2016). Experiments have been conducted with metal lattice structures attached to different types of corals, sometimes stimulating them with small electrical discharges to encourage faster growth (Boström-Einarsson et al. 2020). This type of initiative also attracts many tourist divers who place a high value on conservation and nature enhancement activities (Halim 2017; Eider et al. 2023). For example, in Bali (Indonesia) there are initiatives where tourist divers can enjoy the underwater landscapes while also contributing to the creation and improvement of ecosystems where corals are planted and marked by the tourists themselves (Naja et al. 2021).

## 11.5 THE ROLE OF NEW TECHNOLOGIES IN SOCIOECONOMIC RESEARCH

With the development of Industry 4.0 from the mid-2010s onwards, there has been a notable increase in the way technology is used in environmental management. The ubiquitous availability of the Internet has led to facilitating information access and sharing via cloud computing. Thus, the connection between different information devices (i.e., transmitters and receivers) has become established. This has been chiefly achieved through the IoT (Nguyen and Simkin 2017; Clarysse et al. 2022) and the ability of 3D printing to design biomimetic corals (e.g., Leonard et al. 2022; Levy et al. 2022). Furthermore, the ability to produce objects with highly complex shapes and using sophisticated biomaterials (Dunn et al. 2022) has facilitated artificial reef design and deployment in addressing coral reef enhancement.

There are also other types of developments in this context. This is chiefly through augmented reality, where it is possible to take guided tours in a diving environment to visit artificial reefs (e.g., sunken vessels) from any physical space, just as if one were in the environmental situation of natural reefs (Yuval and Treibitz 2024).

At the same time, Industry 4.0 technologies, in a broad sense, have become more democratized, enabling high-quality research to be conducted using relatively low-cost conventional equipment. Access to faster information transfer and analysis at an affordable price has greatly expanded research possibilities in various fields of knowledge. Ecotourism, in the field of social sciences, can benefit from some of these technologies to improve service quality and enhance experiences. However, there are technologies like robot tour guides, as well as augmented reality and virtual reality, that are sometimes perceived as disruptive to service quality and tourist experiences (Seyitoğlu et al. 2025).

The application of artificial intelligence has perhaps been the technology that appears most frequently in literature associated with artificial reefs, but there are other advances in technology that have also recently received attention (Bao et al. 2025). For instance, 3D printing provides the possibility to produce innovative natural reef mimics (Luo et al. 2022) when applied to artificial reefs. For example, there are even concepts for 3D printed artificial coral reefs on the streets of some cities, indicating that people are inordinately drawn to the beauty of coral reefs and their underwater realm.

Quick response (QR) codes from devices such as mobile phones or tablets allow access to online questionnaires that are used to collect information from respondents anywhere in the world. Subsequently, these data can be stored through cloud computing. Thus, the availability of these data has become widespread in recent years, enabling researchers to conduct scientific studies at a relatively low cost while having a high influence on associated coastal communities (e.g., Firth et al. 2023).

## 11.6 THE FUTURE OF THE SOCIAL SCIENCES IN ARTIFICIAL REEF RESEARCH

The challenges facing humanity today are immense. Scientists, while serving the larger purpose of gaining scientific knowledge, are often challenged to produce solutions to problems (Figure 11.3). In this context, all scientifically derived information, including that information related to artificial reefs, can have a significant impact, not only today but also on the future of our socioeconomic understanding of artificial reefs (Gabrys and Yusoff 2012).

Studies are already beginning to emerge on the creation of artificial reefs with the aim of higher carbon sequestration and economic circularity, thereby contributing to climate change mitigation (Carral et al. 2019; Tanaya et al. 2025). The term "blue economy" has been widely used in the last decade and has been adapted to many activities related to the coast and the sea. Including artificial reefs in this endeavor seems promising (Picken 2025).

Those artificial reefs deployed to protect the coastline from the negative effects of coastal erosion and associated phenomena have been recurrent in recent decades and perhaps provide one of the most common ecosystem services. However, this type of service, along with the provision of green structures, has recently gained some interest (e.g., Harris 2009; Silva et al. 2016; da Silva et al. 2022). Breakwaters made from biomaterials such as oyster shells and incorporated into concrete mixes have been used to break the effects of waves and to give new life to waste derived from biological materials (Safak et al. 2020; Rupasinghe et al. 2024).

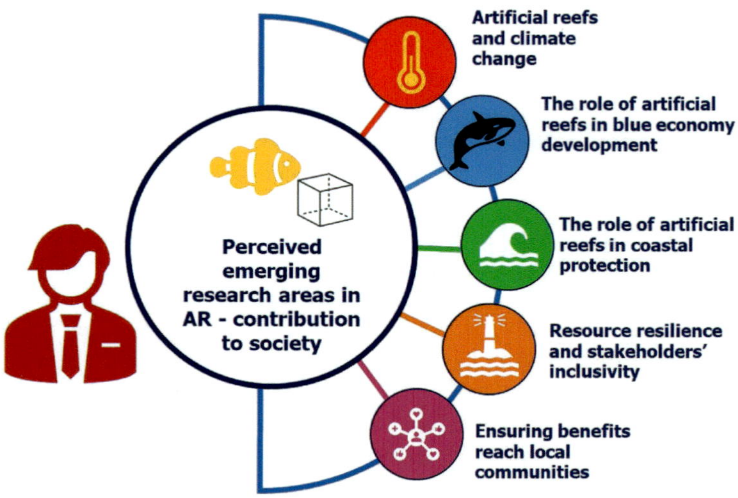


**FIGURE 11.3** Established or emerging research strands in the area of artificial reefs related to their contribution to society.

The creation of reef infrastructure that enables the development of community activities such as fishing or tourism, while also integrating into the cultural aspects of these communities, represents a major step forward. Similarly, public awareness-raising and scientific dissemination initiatives can provide added value in making local communities more resilient, encouraging tourism, and triggering more balanced socioeconomic systems (Kittinger et al. 2016; Higgins et al. 2022).

Any coastal community with the capacity to have useful reef infrastructure nearby will potentially generate community wealth that will tend to be distributed among the population (Schuett et al. 2016). One way to leverage this type of asset will undoubtedly involve educational initiatives. One objective of a reef deployment project would be to raise environmental awareness among young people, providing them with knowledge about the potential of how this type of structure can mitigate environmental problems. Thus, involving people from other generations to help with environmental issues can expand the range of possible solutions. In this sense, it is noteworthy (or recommended) that one can even imagine the utility of future theme parks that incorporate different uses of artificial reefs, articulating the reefs' purposefully designed responses to environmental and social challenges.

## 11.7 CONCLUSION

The topic of artificial reefs has gone through several phases over the past few decades. Before their modern rediscovery after World War II, fishers likely believed that certain structures possessed miraculous properties. Some of our ancient ancestors may have thought they had the ability to "create" life in places previously deprived of or disadvantaged by it. In the second half of the 20th century, the "panacea" concept emerged, suggesting that artificial reefs composed of discarded materials could be the solution for transforming unwanted waste into structures that fostered aquatic life.

Although interest has fluctuated over time, artificial reefs have at last been demonstrated to have immense potential benefits to communities in coastal areas, with some cases being more successful than others. This fluctuating interest is related to trends linked to increased funding at certain times, as well as the use of new materials or techniques in the construction of reef structures. Additionally, artificial reefs have the potential to serve as ecological stimuli, reduce the carbon footprint, and promote greater circularity in the economy. The combination of some of these outcomes means that there is a continuing interest in reef structures and that there is a constant community working on their potential, always aiming for a promising future in this field.

The most obvious interest in the usefulness of artificial reefs relates to the proper colonization of the structures. These interests are important not only for those researching in biological sciences but also for those in the fields of architecture and engineering of the structures themselves, their forms and materials. By encompassing these interests, there is also a whole range of social sciences that can be captivated by the development of artificial reefs. This chapter addresses this perspective: the usefulness of reefs for humans in their various economic and recreational activities and the potential of the structures in the context of valuing ecosystem services.

## 11.8 ACKNOWLEDGMENT

Efforts to produce this chapter were financed by Portuguese national funds through the FCT—Foundation for Science and Technology, under project: *Centro de Investigação em Turismo, Sustentabilidade e Bem Estar* (UID/04020/2025) with DOI https://doi.org/10.54499/UID/04020/2025, and the Individual Call to Scientific Employment Stimulus (2021.02180.CEECIND) and Tenure track (2023.10993.TENURE.037) to J.R.

## LITERATURE CITED

Aabel, J. P., Cripps, S. J., Jensen, A. C., and Picken, G. (1997). *Creating artificial reefs from decommissioned platforms in the North Sea: Review of knowledge and proposed programme of research*. Dames & Moore Group. Report of Dames and Moore, RF Rogaland Research, Southampton Oceanography Centre and Cordah to the Exploration and Production Forum.

Al Ismaili, S., Al Masroori, H., and Dutta, S. (2024). Contribution of artificial reefs on fisheries productivity in the Sultanate of Oman from the fishermen perspective. *Ocean and Coastal Management*, *254*, 107201.

Anderson, R. B. (2023). Time, seawalls, and money: Anthropologies of rising seas and eroding coasts. *Environment and Society*, *14*(1), 23–42.

Baine, M. (2001). Artificial reefs: A review of their design, application, management and performance. *Ocean and Coastal Management*, *44*(3–4), 241–259.

Bao, H., Nikolaeva, A., Xia, J., and Ma, F. (2025). Evolution trends and future prospects in artificial marine reef research: A 28-year bibliometric analysis. *Sustainability*, *17*(1), 184.

Bartholomew, A., Burt, J. A., and Firth, L. B. (2022). Artificial reefs in the Arabian Gulf: Benefits, challenges and recommendations for policy-makers. *Regional Studies in Marine Science*, *56*, 102723.

Becker, A., Taylor, M., McLeod, J., and Lowry, M. (2020). Application of a long-range camera to monitor fishing effort on an offshore artificial reef. *Fisheries Research*, *228*, 105589.

Blount, C., Komyakova, V., Barnes, L., Smith, M. L., Zhang, D., Reeds, K.,. . . and Needham, E. (2021). Using ecological evidence to refine approaches to deploying offshore artificial reefs for recreational fisheries. *Bulletin of Marine Science*, *97*(4), 665–698.

Bohnsack, J. A., and Sutherland, D. L. (1985). Artificial reef research: A review with recommendations for future priorities. *Bulletin of Marine Science*, *37*(1), 11–39.

Bortone, S. A. (2015). CARAH (International Conference on Artificial Reefs and Related Aquatic Habitats): An historical perspective of accomplishments. *Journal of Applied Ichthyology*, *31*, 3–14.

Bortone, S. A., Martin, T., and Bundrick, C. M. (1994). Factors affecting fish assemblage development on a modular artificial reef in a northern Gulf of Mexico estuary. *Bulletin of Marine Science*, *55*(2–3), 319–332.

Boström-Einarsson, L., Babcock, R. C., Bayraktarov, E., Ceccarelli, D., Cook, N., Ferse, S. C., . . . and McLeod, I. M. (2020). Coral restoration–A systematic review of current methods, successes, failures and future directions. *PloS One*, *15*(1), e0226631.

Boyle, K. S., Hightower, C. L., Nelson, T. R., and Powers, S. P. (2022). Use of passive acoustic monitoring to estimate fishing effort on artificial reefs in Alabama during the recreational red snapper fishing season. *Fisheries Research*, *249*, 106262.

Brochier, T., Brehmer, P., Mbaye, A., Diop, M., Watanuki, N., Terashima, H.,. . . and Auger, P. (2021). Successful artificial reefs depend on getting the context right due to complex socio-bio-economic interactions. *Scientific Reports*, *11*(1), 16698.

Calandra, M., Wencélius, J., Moussa, R. M., Gache, C., Berthe, C., Waqalevu, V.,. . . and Lecchini, D. (2022). Local perceptions of socio-ecological drivers and effects of coastal armoring: The case of Moorea, French Polynesia. *Population and Environment, 43*(3), 423–443.
Carral, L., Alvarez-Feal, C., Rodríguez-Guerreiro, M. J., Vargas, A., Arean, N., and Carballo, R. (2019). Methodology for positioning a group of green artificial reef based on a database management system, applied in the estuary of Ares-Betanzos (Nw Iberian Peninsula). *Journal of Cleaner Production, 233*, 1047–1060.
Ceccarelli, D. M., McLeod, I. M., Boström-Einarsson, L., Bryan, S. E., Chartrand, K. M., Emslie, M. J.,. . . and Bay, L. K. (2020). Substrate stabilisation and small structures in coral restoration: State of knowledge, and considerations for management and implementation. *PloS One, 15*(10), e0240846.
Chong, L., Siders, Z. A., Lorenzen, K., Ahrens, R. N., and Camp, E. V. (2024). Global synthesis of effects and feedbacks from artificial reefs on socioecological systems in recreational fisheries. *Fish and Fisheries, 25*(2), 303–319.
Clarysse, B., He, V. F., and Tucci, C. L. (2022). How the Internet of Things reshapes the organization of innovation and entrepreneurship. *Technovation, 118*, 102644.
Creque, S. M., Raffenberg, M. J., Brofka, W. A., and Dettmers, J. M. (2006). If you build it, will they come? Fish and angler use at a freshwater artificial reef. *North American Journal of Fisheries Management, 26*(3), 702–713.
Cruz, F., Valente, I. B., Maslov, D., Miranda, T., and Pereira, E. B. (2025). Engineering-driven approach for the structural design of geometrically complex modular artificial reefs. *Applied Sciences, 15*(11), 5907.
da Silva, G. V., Hamilton, D., Strauss, D., Murray, T., and Tomlinson, R. (2022). Sediment pathways and morphodynamic response to a multi-purpose artificial reef–New insights. *Coastal Engineering, 171*, 104027.
Dunn, K., Haeusler, M. H., Yu, K. D., Zavoleas, Y., Undery, G., and Firth, C. S. (2022). Building bio-shelters using Industry 4.0 data-driven hybrid fabrication technologies: Techniques for customisable complex artificial reef design. In: Ramos, J. H. P. (ed.). *Impact of artificial reefs on the environment and communities* (pp. 196–221). IGI Global Scientific Publishing, Hershey, PA.
Eider, D., Partelow, S., Albrecht, S., Adrianto, L., and Kluger, L. C. (2023). SCUBA tourism and coral reefs: A social-ecological network analysis of governance challenges in Indonesia. *Current Issues in Tourism, 26*(7), 1031–1050.
Elrick-Barr, C. E., Zimmerhackel, J. S., Hill, G., Clifton, J., Ackermann, F., Burton, M., and Harvey, E. S. (2022). Man-made structures in the marine environment: A review of stakeholders' social and economic values and perceptions. *Environmental Science and Policy, 129*, 12–18.
Fabi, G., Spagnolo, A., Bellan-Santini, D., Charbonnel, E., Çiçek, B. A., García, J. J. G.,. . . and Santos, M. N. D. (2011). Overview on artificial reefs in Europe. *Brazilian Journal of Oceanography, 59*, 155–166.
Firth, L. B., Farnworth, M., Fraser, K. P., and McQuatters-Gollop, A. (2023). Make a difference: Choose artificial reefs over natural reefs to compensate for the environmental impacts of dive tourism. *Science of the Total Environment, 901*, 165488.
Gabrys, J., and Yusoff, K. (2012). Arts, sciences and climate change: Practices and politics at the threshold. *Science as Culture, 21*(1), 1–24.
Gannon, J. E., Danehy, R. J., Anderson, J. W., Merritt, G., and Bader, A. P. (1985). The ecology of natural shoals in Lake Ontario and their importance to artificial reef development. In: D'itri, F. M. (ed.). *Artificial reefs: Marine and freshwater applications* (pp. 113–140). Lewis Publishers Inc., Chelsea, MI.
Gibson Banks, K., Curtis, J. M., Williams, J. A., Wetz, J. J., and Stunz, G. W. (2021). Designing cost-effective artificial reefs: Fine-scale movement and habitat use of Red Snapper around a nearshore artificial reef complex. *North American Journal of Fisheries Management, 41*(6), 1850–1862.
Glarou, M., Zrust, M., and Svendsen, J. C. (2020). Using artificial-reef knowledge to enhance the ecological function of offshore wind turbine foundations: Implications for fish abundance and diversity. *Journal of Marine Science and Engineering, 8*(5), 332.
Grant, B. L., Liu, X., and Gardella, J. A. (2015). Supporting the development of science communication skills in STEM university students: Understanding their learning experiences as they work in middle and high school classrooms. *International Journal of Science Education, Part B, 5*(2), 139–160.
Halim, H. S. (2017). Scrutinizing coastal ecotourism in Gili Trawangan, Indonesia. *International Journal of Marine Science, 7*(25), 247–259.
Harris, L. E. (2009). Artificial reefs for ecosystem restoration and coastal erosion protection with aquaculture and recreational amenities. *Reef Journal, 1*(1), 235–246.
Hartill, B. W., Taylor, S. M., Keller, K., and Weltersbach, M. S. (2020). Digital camera monitoring of recreational fishing effort: Applications and challenges. *Fish and Fisheries, 21*(1), 204–215.

Hein, M. Y., Birtles, A., Willis, B. L., Gardiner, N., Beeden, R., and Marshall, N. A. (2019). Coral restoration: Socio-ecological perspectives of benefits and limitations. *Biological Conservation, 229*, 14–25.
Herbert, R. J., Collins, K., Mallinson, J., Hall, A. E., Pegg, J., Ross, K.,. . . and Clements, T. (2017). Epibenthic and mobile species colonisation of a geotextile artificial surf reef on the south coast of England. *PloS One, 12*(9), e0184100.
Higgins, E., Metaxas, A., and Scheibling, R. E. (2022). A systematic review of artificial reefs as platforms for coral reef research and conservation. *PLoS One, 17*(1), e0261964.
Holland, S. M. (1991). Demographic indicators of the relative need for artificial reefs in Florida. *Coastal Management, 19*(2), 219–238.
Islam, G. M. N., Noh, K. M., Sidique, S. F., Noh, A. F. M., and Ali, A. (2014). Economic impacts of artificial reefs on small-scale fishers in Peninsular Malaysia. *Human Ecology, 42*(6), 989–998.
Ito, P., Costa, J., and Ramos, J. (2022). *Pilot monitoring to investigate the usefulness of artificial reefs using vessel tracking technology and cloud computing: A case study in southern Portugal.* International Institute of Fisheries Economics & Trade, Oregon State University, Corvallis, OR.
Jensen, A. (2002). Artificial reefs of Europe: Perspective and future. *ICES Journal of Marine Science, 59*(suppl), S3-S13.
Kilty, T., Burrows, A., Welsh, K., Kilty, K., McBride, S., and Bergmaier, P. (2021). Transcending disciplines: Engaging college students in interdisciplinary research, integrated STEM, and partnerships. *Journal of Technology and Science Education, 11*(1), 146–166.
Kirkbride-Smith, A. E., Wheeler, P. M., and Johnson, M. L. (2016). Artificial reefs and marine protected areas: A study in willingness to pay to access Folkestone Marine Reserve, Barbados, West Indies. *PeerJ, 4*, e2175.
Kittinger, J. N., Bambico, T. M., Minton, D., Miller, A., Mejia, M., Kalei, N.,. . . and Glazier, E. W. (2016). Restoring ecosystems, restoring community: Socioeconomic and cultural dimensions of a community-based coral reef restoration project. *Regional Environmental Change, 16*(2), 301–313.
Koh, E., and Fakfare, P. (2020). Overcoming "over-tourism": The closure of Maya Bay. *International Journal of Tourism Cities, 6*(2), 279–296.
Leonard, C., Hédouin, L., Lacorne, M. C., Dalle, J., Lapinski, M., Blanc, P., and Nugues, M. M. (2022). Performance of innovative materials as recruitment substrates for coral restoration. *Restoration Ecology, 30*(7), e13625.
Levy, N., Berman, O., Yuval, M., Loya, Y., Treibitz, T., Tarazi, E., and Levy, O. (2022). Emerging 3D technologies for future reformation of coral reefs: Enhancing biodiversity using biomimetic structures based on designs by nature. *Science of the Total Environment, 830*, 154749.
Lim, C. P., Matsuda, Y., and Shigemi, Y. (1995). Co-management in marine fisheries: The Japanese experience. *Coastal Management, 23*(3), 195–221.
Lima, J. S., Zalmon, I. R., and Love, M. (2019a). Overview and trends of ecological and socioeconomic research on artificial reefs. *Marine Environmental Research, 145*, 81–96.
Lima, J. S., Zappes, C. A., Di Beneditto, A. P. M., and Zalmon, I. R. (2018). Artisanal fisheries and artificial reefs on the southeast coast of Brazil: Contributions to research and management. *Ocean & Coastal Management, 163*, 372–382.
Lima, J. S., Zappes, C. A., Di Beneditto, A. P. M., and Zalmon, I. R. (2019b). Ethnoecology and socioeconomic around an artificial reef: The case of artisanal fisheries from southeastern Brazil. *Biota Neotropica, 19*(2), e20180620.
Lirman, D., and Schopmeyer, S. (2016). Ecological solutions to reef degradation: Optimizing coral reef restoration in the Caribbean and Western Atlantic. *PeerJ, 4*, e2597.
Luo, D., Wang, B., Bao, D. W., and Ward, S. (2022). Opportunities for further development of 3D-printed floating artificial reefs. *Journal of Aquaculture and Marine Biology, 11*(2), 58–63.
Macdonald, J. M. (1994). Student note: Artificial reef debate: Habitat enhancement or waste disposal? *Ocean Development & International Law, 25*, 87–118.
Milon, J. W. (1988). A nested demand shares model of artificial marine habitat choice by sport anglers. *Marine Resource Economics, 5*(3), 191–213.
Milon, J. W. (1991). Social and economic evaluation of artificial aquatic habitats. In: Seaman Jr., W., and Sprague, L. M. (eds.). *Artificial habitats for marine and freshwater fisheries* (pp. 237–270). Academic Press, San Diego, CA.
Milon, J. W., Holland, S. M., and Whitmarsh, D. J. (2000). Social and Economic Evaluation Methods. In: Seaman, W. (ed.). *Artificial reef evaluation: With application to natural marine habitats* (Vol. 1, pp. 165–194). CRC Press, New York.

Monteiro, C. C., and Santos, M. N. (2000). Portuguese artificial reefs. In: Jensen, A. C., Collins, K. J., and Lockwood, A. P. M. (eds.). *Artificial reefs in European seas* (pp. 249–261). Springer, Dordrecht, Netherlands.
Murray, J. D. (1994). A policy and management assessment of US artificial reef programs. *Bulletin of Marine Science, 55*(2–3), 960–969.
Na, W. B. (2023). Artificial reefs research. *Journal of Marine Science and Engineering, 11*(1), 166.
Naja, D. A., Suprayogi, S., Marfai, M. A., and Mardiatno, D. (2021). A study on the social network analyses of dive centers and sustainable tourism development in Pemuteran Bali, Indonesia. *Geo Journal of Tourism and Geosites, 36*, 603–615.
Nakamura, M. (1985). Evolution of artificial fishing reef concepts in Japan. *Bulletin of Marine Science, 37*(1), 271–278.
Nguyen, B., and Simkin, L. (2017). The Internet of Things (IoT) and marketing: The state of play, future trends and the implications for marketing. *Journal of Marketing Management, 33*(1–2), 1–6.
Oliveira, M. T., Ramos, J., and Santos, M. N. (2015). An approach to the economic value of diving sites: Artificial versus natural reefs off Sal Island, Cape Verde. *Journal of Applied Ichthyology, 31*, 86–95.
Otake, S. (2020). Design and creation of fishing grounds in Japan with artificial reefs. In: Bortone, S. A., and Otake, S. (eds.). *Modern fisheries engineering: Realizing a healthy and sustainable marine ecosystem* (pp. 67–74). CRC Press, Boca Raton, FL
Picken, F. (2025). Tourism and the blue economy. *Tourism Geographies, 27*(3–4), 538–546.
Pickering, H., Whitmarsh, D., and Jensen, A. (1999). Artificial reefs as a tool to aid rehabilitation of coastal ecosystems: Investigating the potential. *Marine Pollution Bulletin, 37*(8–12), 505–514.
Polak, O., and Shashar, N. (2013). Economic value of biological attributes of artificial coral reefs. *ICES Journal of Marine Science, 70*(4), 904–912.
Pondella, D. J., Claisse, J. T., and Williams, C. M. (2022). Theory, practice, and design criteria for utilizing artificial reefs to increase production of marine fishes. *Frontiers in Marine Science, 9*, 983253.
Ramm, L. A., Florisson, J. H., Watts, S. L., Becker, A., and Tweedley, J. R. (2021). Artificial reefs in the Anthropocene: A review of geographical and historical trends in their design, purpose, and monitoring. *Bulletin of Marine Science, 97*(4), 699–728.
Ramos, J., Drakeford, B., Madiedo, A., Costa, J., and Leitão, F. (2024). A Bayesian approach to infer the sustainable use of artificial reefs in fisheries and recreation. *Sustainability, 16*(2), 810.
Ramos, J., Santos, M. N., Whitmarsh, D., and Monteiro, C. C. (2006). Patterns of use in an artificial reef system: A case study in Portugal. *Bulletin of Marine Science, 78*(1), 203–211.
Ramos, J., Santos, M. N., Whitmarsh, D., and Monteiro, C. C. (2007). Stakeholder perceptions regarding the environmental and socio-economic impacts of the Algarve artificial reefs. *Hydrobiologia, 580*(1), 181–191.
Ramos, J., Tuaty-Guerra, M., Almeida, M., Raposo, A. C., Gaudêncio, M. J., Silva, A. D., Rodrigues, N., Leandro, S. M., and Caetano, M. (2021). An artificial reef at the edge of the deep: An interdisciplinary case study. *Ocean and Coastal Management, 210*, 105729.
Randall, J. E. (1962). An analysis of the fish populations of artificial and natural reefs in the Virgin Islands. *Caribbean Journal of Science, 3*, 1.
Rendle, E. J., and Rodwell, L. D. (2014). Artificial surf reefs: A preliminary assessment of the potential to enhance a coastal economy. *Marine Policy, 45*, 349–358.
Ricci, S. W., and Bohnenstiehl, D. R. (2022). Monitoring visitation at North Carolina artificial reef sites using high spatiotemporal resolution PlanetScope imagery. *Regional Studies in Marine Science, 55*, 102511.
Riera, E., Mauroy, B., Francour, P., and Hubas, C. (2024). Establishing complexity targets to enhance artificial reef designs. *Scientific Reports, 14*(1), 22060.
Rupasinghe, M., San Nicolas, R., Lanham, B. S., and Morris, R. L. (2024). Sustainable oyster shell incorporated artificial reef concrete for living shorelines. *Construction and Building Materials, 410*, 134217.
Safak, I., Norby, P. L., Dix, N., Grizzle, R. E., Southwell, M., Veenstra, J. J.,. . .and Angelini, C. (2020). Coupling breakwalls with oyster restoration structures enhances living shoreline performance along energetic shorelines. *Ecological Engineering, 158*, 106071.
Saharuddin, A. H., Ali, A., Lokman, M. H., and Salihin, W. (2012). Recent development and management of artificial reefs (ARs) in Malaysia. In: *Proceedings of MTS-IEEE oceans 2012* (pp. 1–23). Yeosu, Korea.
Salaün, J., Pioch, S., and Dauvin, J. C. (2022a). Artificial reef stakeholders in France: A social-network analysis. In: Ramos J. H. P. (ed.). *Impact of artificial reefs on the environment and communities* (pp. 21–47). IGI Global Scientific Publishing, Hershey, PA.
Salaün, J., Pioch, S., and Dauvin, J. C. (2022b). Socio-ecological analysis to assess the success of artificial reef projects. *Journal of Coastal Research, 38*(3), 624–638.

Samples, K. C. (1989). Assessing recreational and commercial conflicts over artificial fishery habitat use: Theory and practice. *Bulletin of Marine Science, 44*(2), 844–852.
Schuett, M. A., Ding, C., Kyle, G., and Shively, J. D. (2016). Examining the behavior, management preferences, and sociodemographics of artificial reef users in the Gulf of Mexico offshore from Texas. *North American Journal of Fisheries Management, 36*(2), 321–328.
Seaman, W. (2000). *Artificial reef evaluation: with application to natural marine habitats*. CRC Press.
Seaman, W. (2022). *Structure in the sea: The science, technology and effects of purpose-built reefs and related surfaces*. Elsevier.
Sebens, K. P. (1994). Biodiversity of coral reefs: What are we losing and why? *American Zoologist, 34*(1), 115–133.
Seyitoğlu, F., Atsız, O., and Akyuz, M. (2025). The role of technology in tourist guiding: Future career, service quality, marketing and tourist experiences. *Consumer Behavior in Tourism and Hospitality, 20*(1), 57–74.
Shakeela, A., and Becken, S. (2015). Understanding tourism leaders' perceptions of risks from climate change: An assessment of policy-making processes in the Maldives using the social amplification of risk framework (SARF). *Journal of Sustainable Tourism, 23*(1), 65–84.
Shani, A., Polak, O., and Shashar, N. (2012). Artificial reefs and mass marine ecotourism. *Tourism Geographies, 14*(3), 361–382.
Shinzato, C., Shoguchi, E., Kawashima, T., Hamada, M., Hisata, K., Tanaka, M.,. . . and Satoh, N. (2011). Using the Acropora digitifera genome to understand coral responses to environmental change. *Nature, 476*(7360), 320–323.
Shivni, R., Cline, C., Newport, M., Yuan, S., and Bergan-Roller, H. E. (2021). Establishing a baseline of science communication skills in an undergraduate environmental science course. *International Journal of STEM Education, 8*(1), 47.
Silva, R., Chávez, V., Bouma, T. J., van Tussenbroek, B. I., Arkema, K. K., Martínez, M. L.,. . . and Pereira, P. (2019). The incorporation of biophysical and social components in coastal management. *Estuaries and Coasts, 42*(7), 1695–1708.
Silva, R., Mendoza, E., Mariño-Tapia, I., Martínez, M. L., and Escalante, E. (2016). An artificial reef improves coastal protection and provides a base for coral recovery. *Journal of Coastal Research*, (75), 467–471.
Sparkman, L., Maulding, W., and Roberts, J. (2012). Non-cognitive predictors of student success in college. *College Student Journal, 46*(3), 642–652.
Stolk, P., Markwell, K., and Jenkins, J. (2005). Perceptions of artificial reefs as scuba diving resources: A study of Australian recreational scuba divers. *Annals of Leisure Research, 8*(2–3), 153–166.
Sutton, S. G., and Bushnell, S. L. (2007). Socio-economic aspects of artificial reefs: Considerations for the Great Barrier Reef Marine Park. *Ocean and Coastal Management, 50*(10), 829–846.
Tanaya, T., Iwamura, S., Okada, W., and Kuwae, T. (2025). Artificial structures can facilitate rapid coral recovery under climate change. *Scientific Reports, 15*(1), 9116.
Taylor, M. D., and Suthers, I. M. (2021). The socio-ecological system of urban fisheries in estuaries. *Estuaries and Coasts, 44*(7), 1744–1751.
Tessier, A., Francour, P., Charbonnel, E., Dalias, N., Bodilis, P., Seaman, W., and Lenfant, P. (2015). Assessment of French artificial reefs: Due to limitations of research, trends may be misleading. *Hydrobiologia, 753*(1), 1–29.
Thierry, J. M. (1988). Artificial reefs in Japan—A general outline. *Aquacultural Engineering, 7*(5), 321–348.
Tickell, S. C. Y., Sáenz-Arroyo, A., and Milner-Gulland, E. J. (2019). Sunken worlds: The past and future of human-made reefs in marine conservation. *BioScience, 69*(9), 725–735.
Vona, R., Basile, V., Capobianco, N., and Loia, F. (2022). The role of oil and gas offshore platform reconversion in creating artificial reefs: A multi-business and socio-economic perspective. In: Ramos, J. H. P. (ed.). *Impact of artificial reefs on the environment and communities* (pp. 232–256). IGI Global Scientific Publishing, Hershey, PA.
Watanuki, N., and Gonzales, B. J. (2006). The potential of artificial reefs as fisheries management tools in developing countries. *Bulletin of Marine Science, 78*(1), 9–19.
Westerberg, V., Jacobsen, J. B., and Lifran, R. (2013). The case for offshore wind farms, artificial reefs and sustainable tourism in the French mediterranean. *Tourism Management, 34*, 172–183.
Whitmarsh, D., Santos, M. N., Ramos, J., and Monteiro, C. C. (2008). Marine habitat modification through artificial reefs off the Algarve (southern Portugal): An economic analysis of the fisheries and the prospects for management. *Ocean and Coastal Management, 51*(6), 463–468.
Wood, G., Lynch, T. P., Devine, C., Keller, K., and Figueira, W. (2016). High-resolution photo-mosaic time-series imagery for monitoring human use of an artificial reef. *Ecology and Evolution, 6*(19), 6963–6968.
Yuval, M., and Treibitz, T. (2024). Releasing a dataset of 3D models of artificial reefs from the northern red-sea for 3D printing and virtual reality applications. *Remote Sensing Applications: Society and Environment, 36*, 101305.

# 12 The "Gap" in Artificial Reef Assessments

## *The Need for a Combination of Social and Ecological Approaches*

*Jessica Salaün*, Sylvain Pioch, and Jean-Claude Dauvin*
* Corresponding author: jess.salaun18@gmail.com

## 12.1 INTRODUCTION

### 12.1.1 The Two Dimensions of Artificial Reefs

Artificial reefs are intentionally deployed structures designed to provide habitat, shelter, or substrate for marine organisms (Baine 2001; Hammond et al. 2020; Vivier et al. 2021). While their ecological role has been well established, artificial reefs also generate significant social benefits, ranging from local economic development to regional cultural and recreational values (Sutton and Bushnell 2007; Kirkbride-Smith 2014; Salaün et al. 2022). They are not only habitats for fish, as they are emblematic resource management options for stakeholders and communities. Artificial reef projects are typically designed around four commonly recognised goals (Charbonnel 2005):

1. Protection—aiming to preserve habitats and biodiversity, often by creating obstacles to fishing, while ensuring no risks to human safety.
2. Production—stimulating species growth and reproduction that benefit fisheries and coastal economies.
3. Recreation—enhancing tourism and the perceived value of local territories alongside ecological gains.
4. Restoration—improving habitat quality while advancing scientific knowledge and environmental stewardship.

Despite their dual nature, performance assessments of artificial reefs often overlook the social dimension, which is essential for a more realistic evaluation (Milon et al. 2000; Ramos et al. 2007; Lee et al. 2018).

### 12.1.2 Current Assessment Approaches

Accurate and repeatable assessment is central to determining whether an artificial reef meets its intended outcomes. Yet, most existing study designs focus almost exclusively on ecological metrics. Three main approaches dominate the studies conducted to date:

- Colonisation and faunal development, which assess the presence and growth of benthic fauna and fish communities (Folpp et al. 2011; Paxton et al. 2020),


DOI: 10.1201/9781003676775-12

- Habitat quality comparisons, evaluating whether artificial reefs replicate the structure and function of natural reefs (Page et al. 2007; Hallier and Gaertner 2008; Pioch et al 2018b; Taormina et al. 2022; Al Kindi et al. 2025),
- Design effectiveness comparisons, analysing the relative ecological performance of different reef materials, forms, or configurations (Dafforn et al. 2015; Riera et al. 2018; Vivier et al. 2021; Hayek et al. 2023).

While some assessment includes social-economic indicators (Ramos 2007; Kirkbride-Smith 2014; Lima et al. 2019; Brochier et al. 2021), systematic analyses of how people use, manage, and benefit from artificial reefs remain rare (Milon et al. 2000; Ramos et al. 2007; Lee et al. 2018).

An analysis of monitoring practices across all artificial reefs deployed in France (49 sites) from the first deployment in 1968 to 2021 highlights the extent of this assessment bias (Tessier et al. 2015; Salaün 2022). When examining the various factors monitored across these sites, the analysis showed that in the English Channel, assessments of the structural integrity, fish faunal abundance, or richness were conducted at all monitored sites (100%), while benthic invertebrate communities were included in the assessment at 66% of the monitored sites. By contrast, no monitoring in this region addressed hydrodynamic conditions, yields of commercially important species, or human activities. In the Atlantic Ocean off France, a similar pattern emerged: monitoring structural integrity and fish fauna was conducted at 55% of sites, and benthic invertebrates were included in the analysis at 44% of the monitored sites (Salaün 2022). Hydrodynamic conditions, commercial fauna, and human activities were considered at only 22–33% of sites. In the Mediterranean Sea, 59% of sites were monitored for structural integrity and fish fauna, but other factors could not be compared due to insufficient information (Cépralmar 2015; Salaün 2022). These analyses, based on the French experiences, confirm that while ecological factors such as structural stability and biological community descriptors are widely tracked, social indicators are almost entirely absent from current monitoring study designs.

### 12.1.3 Why the Social Dimension Matters

By definition, artificial reefs are human interventions (FAO 2015). Their success should therefore be measured not only by ecological performance but also by their ability to meet social objectives such as raising awareness and increasing knowledge (Bideci and Cater 2019). A reef that supports abundant marine life but fails to reduce user conflicts, support livelihoods, or foster local engagement may ultimately lose legitimacy and funding (McAfee et al. 2021; Salaün 2022).

This issue gains urgency in the context of current policy. With the global target of protecting 30% of marine areas by 2030 taken under the Convention on Biological Diversity, artificial reefs delivering both ecological and social benefits could be recognised as Other Effective Area-Based Conservation Measures (Brodie et al. 2025). Such recognition would likely increase artificial reef deployment worldwide, but only if credible, transparent, and standardised evaluation methods are adopted (Alves-Pinto et al. 2021). Without robust social-ecological assessment methods, artificial reef projects risk expansion with no accountability, undermining both conservation and community outcomes.

The main objective of this chapter is to develop approaches that integrate ecological and social dimensions into the assessment of artificial reef performances. Previous evaluations have largely focused on ecological performance, giving limited attention to the social benefits that artificial reefs provide to coastal communities. This chapter introduces a more balanced and operational assessment perspective through two complementary methodological approaches applied at different scales: (1) a site-specific assessment, linking each stakeholder's objectives to specific indicators, and (2) a cross-site comparison, applying standardised indicators to compare multiple projects. Together, these approaches address the current gap in artificial reef evaluations and support more adaptive, inclusive, and effective management strategies.

## 12.2 DESIGNING SITE-SPECIFIC ASSESSMENTS BY CAPTURING STAKEHOLDERS' EXPECTATIONS

### 12.2.1 Clarifying Objectives through Stakeholder Expectations

A common limitation in many artificial reef projects lies in the formulation of overly broad or imprecise general, overarching objectives (e.g., enhancing biodiversity or supporting fisheries) that are insufficient to concretely guide assessments (Salaün 2022). Without breaking these general objectives into specific objectives accompanied by clear success criteria, it becomes impossible to determine what reef attributes should be monitored to assess whether a project has achieved its objectives and overall project goals (Becker et al. 2018).

Assessment frameworks for artificial reefs have been proposed by Seaman and Jensen (2000) and later by Claudet (2006), which outline steps in a sequence from general objectives to specific objectives, followed by success criteria and monitoring protocols. Building on their logic, the disaggregation of general objectives such as "protection," "production," "restoration," and "recreation" into specific, context-relevant goals is considered. This disaggregation step is crucial in enabling effective assessment. Here, we propose to perform this step by grounding the definition of specific objectives in stakeholder interests.

Behind a shared general objective, artificial reefs serve multiple functions for diverse stakeholders, including fishers, authorities, divers, scientists, and tourism operators, with each stakeholder holding different expectations. By identifying stakeholders' interests early in the project design, it becomes possible to define specific objectives that reflect both ecological and social priorities (based on the translation process theory defined by Callon 1986). For instance, a reef deployed with "production" goals may align with fishers' interests in enhancing populations of species with commercial value. Simultaneously, however, local policymakers may also be interested in the project to develop a harbour and territorial infrastructures. Similarly, scientists may be interested in the same project to test scientific hypotheses and improve knowledge. These interests give structure to the specific objectives and lay the groundwork for selecting appropriate success criteria.

The selection of stakeholder-related interests may occur at various stages of the project lifecycle. Ideally, this process is initiated during the early phases of project planning to establish the initial objectives. However, since stakeholder priorities and project circumstances can shift over time, this step may be repeated to ensure continued alignment of objectives with evolving expectations. Stakeholders can be identified through a review of project documentation, and these stakeholders may be selected for their current participation in the project, their dependence on its outcomes, and their expertise in the thematic areas the project addresses. In some cases, "snowball" selection of stakeholder may occur, whereby initial participants recommend additional relevant stakeholders that can help broaden the range of individuals to be included in the study. Once identified, stakeholders' interests may be examined through semi-structured interviews or surveys designed to elicit information about their understanding of project objectives, their motivations and concerns, and their expectations regarding project outcomes and collaborative processes. Alternatively, the survey or interview processes may be conducted directly by the project manager or in collaboration with external partners such as Non-Governmental Organizations (NGOs), researchers, or other intermediaries. Working with external individuals can help create a neutral and trustworthy environment, allowing stakeholders to express their views more freely and without fear of judgement or influence from others. Subsequently, all points of view are then analysed by the thematic content method (Fallery and Rodhain 2007) to translate these expectations into specific objectives and success criteria. Stakeholder expectations may vary widely as their concerns may include differing domains such as ecology, economy, education, legitimacy, access, or social value. The resulting evaluation framework is therefore both multidimensional (social and ecological) and context-driven (Figure 12.1).

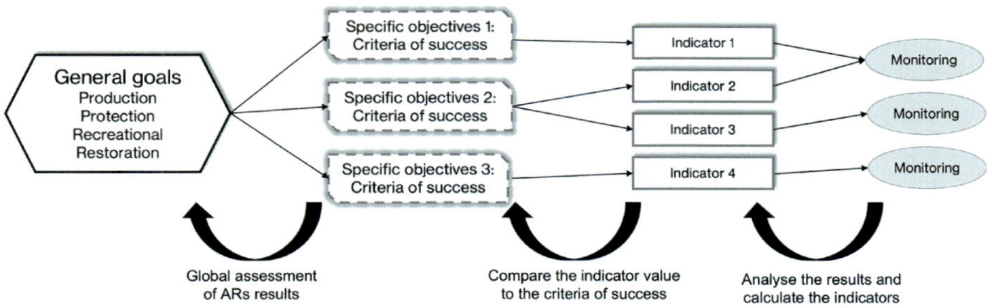


**FIGURE 12.1** The site-specific assessment framework adapted to artificial reefs (adapted from Claudet 2006).

### 12.2.2 Network Analysis Approach to Provide Indicators

To operationalise the site-specific assessment approach, we propose the use of additional indicators (defined as variables that can track a system's state or transformation; Bavoux and Chapelon 2014) derived from a network analysis. This approach offers a robust method for representing and analysing complex systems by describing the relationships between components, whether human or non-human. Network analysis not only reveals their structural ecological characteristics but also identifies diverse indicators suited to characterise both ecological and social dimensions.

This method has been well-established in both sociology and ecology. In the social sciences, Social Network Analysis (SNA), developed in the 1970s, introduced mathematical and conceptual analyses to explore interdependencies within and across organisations (Wasserman and Faust 1994; Eloire et al. 2011). In ecology, network-based approaches such as food-web models have long been used to characterise species interactions and to examine system-level properties like resilience, redundancy, and productivity (Newman 2003).

In line with a balanced social-ecological perspective, the construction of two complementary network types is proposed here. Social networks represent the interactions among human individual, where nodes refer to stakeholders (e.g., fishers, managers, and researchers), and links indicate information flow, financial transactions, technical resources, or skills. Ecological networks, by contrast, represent the trophic structure of the ecosystem. These are typically derived from biomass and energy flow models such as Ecopath with Ecosim, where nodes refer to species or functional groups, and links indicate feeding relationships and material transfers (Polovina 1984; Christensen and Pauly 1992). Such modelling approaches were used for the first time for artificial reefs in the Bay of Biscay (Salaün et al. 2023) and in Cherbourg Harbour (Raoux et al. 2024) off France.

The combination of these two network types allows for a comprehensive understanding of how artificial reefs are embedded in both social and ecological systems. SNA helps identify patterns of collaboration, knowledge exchange, or conflict among stakeholders, while ecological network analysis illuminates changes in biomass distribution, trophic dependencies, or energy flow. While each network is analysed separately due to the distinct nature of their relationships, their joint interpretation enables a holistic assessment of artificial reef performance. This dual-structure approach ensures analytical clarity while enabling the integration of both domains. Moreover, network analysis yields indicators across multiple levels, from individual species or humans to entire network configurations. These indicators can be used to assess a variety of specific objectives. For instance, the indicator of interaction density in a social network serves to measure the frequency and the strength of relationships among stakeholders. It can be used to evaluate the presence and robustness of collaborations between individuals, or it can bring focus to a specific type of interaction such as communication flows, which serve to assess the degree of knowledge exchange between individuals. Similarly, ecological indicators such as total system throughput reflect overall ecosystem activity and can be used to assess the contribution an artificial reef makes to the production and transfer of biomass across trophic levels. These indicators can serve either as absolute measures when the criterion of success is defined as a quantitative benefit or as comparative tools to detect changes over time such as increases or decreases in specific parameters. In the latter case, before-and-after analyses are conducted, requiring the modelling of both social and ecological networks prior to and following artificial reef deployment. Such comparative approaches enable the identification of measurable trends and the assessment of whether artificial reefs achieve their specific objectives.

### 12.2.3 Example of Application of Site-Specific Assessments

To illustrate how the site-specific assessments enable the verification of an artificial reef project meeting its objectives, the assessment was applied to two contrasting sites: Capbreton (Bay of Biscay) and Cherbourg (English Channel). The first study site (i.e., Capbreton) is in the Bay of Biscay, 2.2 km offshore at 20 m depth on a sandy bottom enriched by seasonal upwelling. Since 1999, several types of artificial reefs have been deployed at Capbreton and have been managed by the Atlantique Landes Récifs association in partnership with local authorities with the broad aim of supporting fisheries and local marine activities (Salaün 2022). The second site (i.e., Cherbourg) is located in a harbour at a shallow depth of only 5 m. The artificial reefs at Cherbourg were installed in 2015 as part of the European RECIF project and were intended to enhance biodiversity and production in the English Channel ecosystem while being part of the harbour development project (Vivier 2021). The objectives for artificial reef deployment at both sites were general (production, protection, and/or restoration).

Following the approaches described previously, the first step in the assessment consisted of defining the specific objectives of the artificial reef project, and then assessing whether these objectives were fulfilled using indicators derived from social and ecological network analyses. To this end, stakeholder interviews and ecological monitoring were conducted for the Capbreton and Cherbourg artificial reef projects. A total of 31 in-person or phone interviews were conducted with stakeholders involved in the Capbreton artificial reef project, and 12 with those engaged in Cherbourg between July 2019 and December 2020 (Salaün et al. 2022). These interviews aimed to identify stakeholder interests, clarify their roles and relationships, and map the structure of the social network before and after artificial reef deployment. In parallel, trophic network models were developed using ecological data compiled from field monitoring and the literature, focusing on 23 functional groups representing the ecosystem before and after artificial reef deployment. These groups included commercially important species, species of cultural or ecological value, and reef-associated taxa ranging from primary producers such as phytoplankton and detritus to top predators such as marine mammals and diving seabirds. Each functional group was parameterised on the basis of biomass, diet composition, and additional ecological parameters such as its Production-to-Biomass ratio, Consumption-to-Biomass ratio, and Ecotrophic Efficiency. Detailed methodological information and specific datasets are available in Salaün et al. (2023) for the reefs at Capbreton and Raoux et al. (2024) for the reefs at the Cherbourg Harbour. Comparative analyses of models constructed before and after reef deployment were then performed using social and ecological indicators.

The analysis of stakeholders' interests helped define the specific objectives of each artificial reef project. For Capbreton, these included reducing conflicts among different fishing practices (small-scale net fishing and trawling), promoting economic and recreational uses, supporting fishing activities, and improving ecological functions (Salaün 2022). For Cherbourg, the specific objectives were to promote harbour development, raise ecological awareness, support fisheries beyond the site, and enhance biodiversity (Vivier 2021). To measure progress, a combination of social indicators (actor diversity, interaction intensity, conflict presence, communication activities, and involvement of fishing organisations in the management of artificial reefs) and ecological indicators (i.e., species richness, biomass, and total system throughput) was used. While these are more specific than the original objectives (production, protection, and/or restoration), they were insufficiently precise to define a quantitative threshold of success. Therefore, holistic and classical criteria of success were included, such as increases or decreases in conventional ecological characters like taxonomic richness or biomass, or descriptions of the network properties. When success criteria are established in advance, setting more precise targets, a greater range of network indicators (e.g., flux density, degree, betweenness centrality, and maturity of the system) can be applied.

The assessment results (i.e., a comparison of indicator values with the success criteria) showed that, in Capbreton, social objectives were fully met, with more diverse and connected stakeholder networks and no reported conflicts in the area (Figure 12.2). Ecological indicators also increased, indicating

| **Site-specific assessments**<br>**Original and general goal** | Specific objectives | Indicator and criteria of success in brackets | Indicator values: Before | Indicator values: After | Assessment |
|---|---|---|---|---|---|
| **Capbreton** | | | | | |
| **Protection and production** | • Ensure public interest | *Diversity and number of actors (increase)* | 12<br>29 | 14<br>62 | ✓<br>✓ |
| | • Promote harbour activity (economic) and diving activity | *Number of interactions (increase)* | 44 | 157 | ✓ |
| | • Reduce conflicts in fisheries activity | *Presence of trawlers in the area (absence)* | absence | | ✓ |
| | • Support fishermen activity | *Involvement of fishing organisations (presence)* | 3 | 4 | ✓ |
| | • Enhance ecological functionalities such habitat | *Taxonomic fishes, macro-decapod and benthic fauna richness, biomass (gC.m⁻²) and Total system throughput (gC.m⁻².Year⁻¹) (increase)* | 113<br>5<br>379 | 142<br>6.33<br>413 | ✓<br>✓<br>✓ |
| **Cherbourg** | | | | | |
| **Restoration and production** | • Promote ecological actions | *Communication and manifestation to promote ARs (presence)* | absence | | ✗ |
| | • Promote harbour activity and ensure its extension | *Extension of harbour activity (yes)* | yes | | ✓ |
| | • Enhance knowledge about effect of the design and materials on ecological colonisation | *Valorizstion through publications (yes)* | yes | | ✓ |
| | • Support fishermen activity outside the harbour | *Involvement of fishing organisations (presence)* | 0 | 0 | ✗ |
| | • Enhance ecological functionalities | *Taxonomic fishes, macro-decapod and benthic fauna richness, biomass (gC.m⁻²) and Total system throughput (gC.m⁻².Year⁻¹) (increase)* | 156<br>5.88<br>450 | 62<br>6.52<br>509 | ✗<br>✓<br>✓ |

**FIGURE 12.2** Application of the site-specific assessment framework: evaluation of Capbreton and Cherbourg artificial reefs.

functional and biodiversity gains. In Cherbourg, most social objectives were achieved through increased outreach and harbour activity, but ecological outcomes were mixed: functional gains in biomass and throughput contrasted with a drop in species richness, suggesting partial fulfilment of biodiversity goals (Figure 12.2). Overall, the approach confirmed that clarifying objectives and selecting relevant indicators (whether derived from network analysis or other monitoring methods) make it possible to verify artificial reef performance in a rigorous, measurable way for both ecological and social dimensions.

The main limitation of this approach lies in its reliance on interviews and, consequently, the memory and subjective perceptions of participants. Because several years had passed since the initial stages of these projects, the original interests of stakeholders were sometimes difficult to recall, and the objectives may have changed over time, replacing or reshaping earlier intentions. This required conducting multiple interviews to cross-check information and ensure consistency. This methodology was applied solely to sites off France. This limited the diversity of examples concerning the types of interests, objectives, and indicators that could emerge in other contexts, as these may differ substantially across socio-ecological systems. However, studies conducted on other artificial structures (Tunca et al. 2022) and marine seawalls (Dodds et al. 2025) have highlighted complementary social interests to those identified in our study, such as well-being or equity, that could be added to the potential lists of interests.

Nevertheless, this site-specific nature also constitutes a major strength, as it allows the method to be adapted to diverse environmental conditions, management frameworks, and stakeholder configurations. When applied iteratively throughout the project life cycle, this integrated socio-ecological approach provides a structured basis for adaptive management. It supports decision-making by identifying when adjustments or additional actions are needed to achieve the specific objectives. In doing so, it ensures that artificial reef management remains responsive to changing ecological conditions and stakeholder expectations. In addition to assessing artificial reef performance, the approach also presents significant potential for application during the planning phase of new projects. By identifying stakeholders and clarifying their expectations at an early stage, it contributes to the definition of precise and shared objectives, thereby supporting the design, implementation, and subsequent evaluation of artificial reef projects.

## 12.3 COMPARING ARTIFICIAL REEF PERFORMANCES BETWEEN SITES

The previous assessment approach is highly site-specific, reflecting local objectives and expectations. While effective for evaluating individual projects, this specificity makes it difficult to compare performance between sites, as each operates under different ecological conditions and stakeholder priorities. To address this limitation, we propose a standardised framework capable of translating diverse site-level results into a common performance metric.

### 12.3.1 The Adaptation of the Rapid Assessment Method

To evaluate and compare artificial reef performance across different sites, the MERCIe methodology was employed (i.e., Méthode d'Évaluation Rapide de la Compensation des Impacts écologiques). This is a rapid assessment method (Boileau et al. 2022) originally designed in the United States and developed in the 1990s to quantify ecological losses from development projects, which are to be offset with ecological gains generated by compensatory restoration measures to achieve a "no net loss" of biodiversity within the mitigation hierarchy (Fennessy et al. 2007; Levrel et al. 2015; Jacob et al. 2018). In France, this method has been adapted within the disciplines of environmental geography and marine spatial planning (Mechin and Pioch 2016; Pioch et al. 2018a) to provide a rapid yet robust evaluation of the environmental state before and after an intervention. The original MERCIe method is designed to evaluate ecological losses by comparing the state of the ecosystem before and after an intervention. The initial state of the site is compared to the negative impact of the intervention, resulting in ecological losses such as a decrease in biodiversity. These variations are expressed as a "delta impact" (Mechin and Pioch 2016), which becomes the compensatory target for achieving ecological equivalence in nearby locations. The method first characterises the ecosystem through three components (i.e., physico-chemical conditions, habitat quality, and species composition) using measurable indicators. It then quantifies potential gains and losses as the mean score of these components, multiplied by the surface area affected, thereby producing an adjusted measure of compensation requirements (Mechin and Pioch 2016; Jacob 2017; Pioch et al 2018b).

To adapt the MERCIe Framework to artificial reef projects, the spatial unit was defined as the authorised concession area, encompassing the seafloor surface covered by all artificial reefs deployed within the concession area. The surface factor used in the original method was removed, as all gains and losses were measured within the same area. In this modification, the ecological dimension is supplemented by a social dimension. Gains and losses are calculated separately for the ecological and social dimensions, with each expressed as a "delta of benefits" (either social or ecological) corresponding to the mean score of the relevant indicators within that dimension. This framework is designed to detect changes in performance rather than absolute levels, assessing how artificial reefs enhance ecological and social functions over time. The selected indicators therefore capture functional trends rather than project-specific objectives.

Artificial reefs are deployed in diverse socio-ecological contexts and serve different, site-specific objectives. To allow meaningful comparison across sites, it is necessary to identify indicators of performance that are adaptable and not dependent on fixed thresholds or locally specific goals. The selection of ecological indicators was informed by existing literature on artificial reef performance, including studies addressing ecological functions, management practices, and rapid marine assessment methods (Bas 2017; Jacob 2017; Mechin et al. 2023). For the ecological dimension, we propose eight indicators: (1) presence of spawning zones, (2) presence of feeding zones, (3) presence of resting, refuge, or migratory habitats for juveniles and reproductive species (nursery function), (4) presence of invasive species, (5) degree of habitat fragmentation, (6) structural persistence of reef units, (7) frequency of biological data acquisition, and (8) biological growth. The social indicators, as part of the social dimension, are designed to characterise the human-related attributes of artificial reefs. As the MERCIe framework was initially developed for ecological assessments only, no predefined social metrics were available. The social indicators were therefore defined on the basis of the range of stakeholder interests identified previously (economic, scientific, cultural, educational, relational, etc.), ensuring that the selected indicators reflect the diversity of benefits expected by users and managers. The eight indicators designed for this purpose are (1) presence of a specific manager, (2) existence of economic activities linked to the reef, (3) provision of physical protection functions, including erosion control, protection from trawling, and design elements ensuring human safety, (4) implementation of communication, education, and public awareness initiatives, (5) strength and quality of relationships among stakeholders, (6) presence of scientific studies, (7) frequency of data acquisition on stakeholder satisfaction, and (8) growth in the number of participants involved in reef-related activities. When comparing sites, it is essential that all sites be assessed using the same variables for the selected indicators. For example, biological growth (8) may be measured through biomass or species diversity, depending on the available data. To evaluate ecological functions (1–4), most of which are initially absent in areas where artificial reefs are deployed, the analysis focuses on verifying that the reefs effectively provide these functions. These indicators are evaluated using in situ observations (through diving or video surveys) and, when possible, complementary analyses such as gut content examination to confirm trophic enhancement.

Then, each indicator is scored on a scale from 0 to 1, where a score of 1 reflects an optimal condition or positive change (e.g., a confirmed spawning area on the site, no fragmentation, or observed biological growth), a score of 0.5 indicates a partial condition or stability (such as spawning areas nearby but not directly on the site, low fragmentation, or stable biological growth), and a score of 0 denotes absence or negative change (such as no spawning area, high fragmentation, or a decline in biological growth). The indicators are measured for two states: before and after artificial reef deployment (Figure 12.3). The measurements of the initial ecological condition are established

**Ecological indicators**

1. presence of spawning zones,
2. presence of feeding zones,
3. presence of resting, refuge, or migratory habitats for juveniles and reproductive species (nursery function),
4. presence of invasive species,
5. degree of habitat fragmentation,
6. structural persistence of the reef units,
7. frequency of biological data acquisition,
8. biological growth measured through changes in biomass, abundance, or species observations

**Social indicators**

1. designation of a specific manager
2. existence of economic activities,
3. provision of physical protection functions,
4. implementation of communication, education, and public awareness initiatives,
5. strength and quality of relationships between stakeholders,
6. presence of scientific studies,
7. frequency of data acquisition on stakeholders satisfaction,
8. growth in the number of actors involved in reef-related activities

| Ecological assessment | Mean of ecological indicators values **before** artificial reefs deployment | Mean of ecological indicators values **after** artificial reefs deployment | Ecological performance **Δ = after–before** |
|---|---|---|---|
| Etretat | 0.50 | 0.68 | 0.18 |
| Cherbourg | 0.62 | 0.90 | 0.28 |
| Croisic | 0.60 | 0.73 | 0.13 |
| Oléron | 0.77 | 0.82 | 0.05 |
| Capbreton | 0.72 | 0.77 | 0.05 |
| Gruissan | 0.50 | 0.63 | 0.13 |
| Agde | 0.60 | 0.73 | 0.13 |
| Carry-le-Rouet | 0.77 | 0.87 | 0.10 |
| Marseille | 0.72 | 0.82 | 0.10 |
| Vallauris | 0.73 | 0.78 | 0.05 |

| Social assessment | Mean of social indicators values **before** artificial reefs deployment | Mean of social indicators values **after** artificial reefs deployment | Social performance **Δ = after–before** |
|---|---|---|---|
| Etretat | 0.48 | 0.05 | -0.43 |
| Cherbourg | 0.35 | 0.42 | 0.07 |
| Croisic | 0.52 | 0.40 | -0.12 |
| Oléron | 0.52 | 0.75 | 0.23 |
| Capbreton | 0.43 | 0.72 | 0.28 |
| Gruissan | 0.22 | 0.28 | 0.07 |
| Agde | 0.27 | 0.77 | 0.50 |
| Carry-le-Rouet | 0.65 | 0.90 | 0.25 |
| Marseille | 0.62 | 0.75 | 0.13 |
| Vallauris | 0.40 | 0.48 | 0.08 |

**FIGURE 12.3** Ecological and social indicators and results of social and ecological performance assessments for ten artificial reef sites distributed across coastal regions off France (a check mark (✓) indicates that the objective was fulfilled, while a cross (✗) indicates that it was not fulfilled).

from site-specific studies conducted before reef deployment or from comparable surveys in the surrounding area, with the biological growth indicator presumed stable prior to deployment. In both ecological and social assessments, the performance score for each site is calculated as the mean of the corresponding indicator scores, without assigning greater weight to any particular indicator. Performance in each dimension is expressed as the difference between the scores obtained after reef deployment and those recorded beforehand, representing the "delta social/ecological state." A positive delta indicates that the artificial reefs contribute to improving the ecological or social state of the system, whereas a negative delta reflects losses. This approach ensures that both dimensions are evaluated consistently, enabling a direct comparison and integrated interpretation of results across sites.

### 12.3.2 Example of Application across Sites in Metropolitan France

The adapted MERCIe approach was applied to ten artificial reef sites distributed along several coastal regions of metropolitan France. These included the Cherbourg Harbour site in the northern Cotentin, the Étretat offshore site in the eastern English Channel, three sites in the Bay of Biscay (Le Croisic, Capbreton, and Oléron), and five sites in the Mediterranean Sea (Gruissan, Agde, Marseille, Carry-le-Rouet, and Vallauris) (Salaün 2022). Sites were selected to reflect differences in governance intensity and intervention levels, but not with regard to ecological conditions.

For the ecological dimension, the assessment relied on existing studies, including site-specific pre-deployment impact assessments and broader regional surveys such as national marine strategy datasets. These studies provided evidence on the presence/absence of ecological functions (indicators 1–4), the degree of habitat fragmentation (5), the structural persistence of reef units (6), and the biological growth of fish assemblages based on species diversity (8). The frequency of biological data acquisition (7) was determined by verifying whether studies had been conducted within a five-year interval, in accordance with the recommendations of the national marine strategy. As the analysis was conducted in 2021, the reference period included data up to that year. Subsequently, the ecological performance was calculated as the difference between scores determined before and after deployment (Δ ecological state). The results (Figure 12.3) show that positive ecological performance values were recorded at all sites, demonstrating the capacity of the reefs to improve functional habitat for marine fauna (Figure 12.3). However, the magnitude of improvement varied with initial conditions. For example, the Île d'Oléron site already had high ecological values prior to artificial reef deployment, consequently limiting relative gains in the environment due to the reef.

For the social dimension, human activity (indicators 1–4) was assessed using spatial data from national marine strategy datasets. The frequency of data acquisition on stakeholder satisfaction (7) was determined using the same five-year interval applied to the ecological indicators (until 2021). The growth in the number of individuals involved in reef-related activities (8) was estimated using SNA, where nodes represented organisations rather than individuals. Then, the social performance was similarly calculated as the difference between scores before-deployment and after-deployment (Δ social state). While most sites indicated positive social gains, Étretat and Le Croisic had negative values attributed to them, indicating a decline in social outcomes despite ecological improvements (Figure 12.3).

To explore patterns among the ten study sites, a Principal Component Analysis (PCA) was conducted on four variables describing both the ecological and social dimensions of performance: (1) a score of social performance after deployment, (2) the Δ social state, (3) a score of ecological performance after deployment, and (4) the Δ ecological state. The first two principal components, representing the main gradients of overall performance and the balance between social and ecological dynamics, together explain 83.62% of the total variance. This analysis provides a reduced, two-dimensional representation that highlights similarities and contrasts among the study sites based on their socio-ecological performance. The PCA results were then used as the basis for a Hierarchical Ascendant Classification, which grouped sites according to their performance and

Δ state (Figure 12.4). Three groups emerged from this analysis: (1) artificial reef sites exhibiting improvement and high post-deployment scores in both ecological and social dimensions, (2) artificial reef sites displaying limited or declining social change with low post-deployment scores, and (3) the intermediate case of Cherbourg.

To describe each group and better understand the common characteristics that could influence their performance, the different components of the socio-ecological system defined by Ostrom (2009) are referred to here as the governance system, actors, resource system, resource units, and the interactions among them. Following the socio-ecological System Framework, these components were characterised for each group using data from existing studies, including site-specific pre-deployment impact assessments, broader regional surveys such as national marine strategy datasets, and interviews with site managers.

The first group consists of sites sharing high ecological gains and strong post-deployment social scores. This group includes Oléron, Capbreton, Agde, Carry-le-Rouet, Marseille, and Vallauris. All six sites displayed strong social performance, with post-deployment scores above 0.48 and positive social gains, as well as clear ecological improvements such as increased biodiversity and functional productivity. These results reflect the presence of numerous and diverse stakeholders, forming well-developed networks where information circulates efficiently. Public institutions, municipalities, and environmental associations often act as central participants, ensuring stable governance and continuity of engagement. The characteristics of the artificial reef sites in this group are heterogeneous: while some sites extend over large implantation areas of around 1 km², others sites cover smaller surfaces of about 0.1 km². Despite these differences, all sites showed high levels of ecological productivity, as measured by species richness values. Governance is generally structured through prefectural decrees, with activity bans as the dominant rule. Most sites are embedded within broader

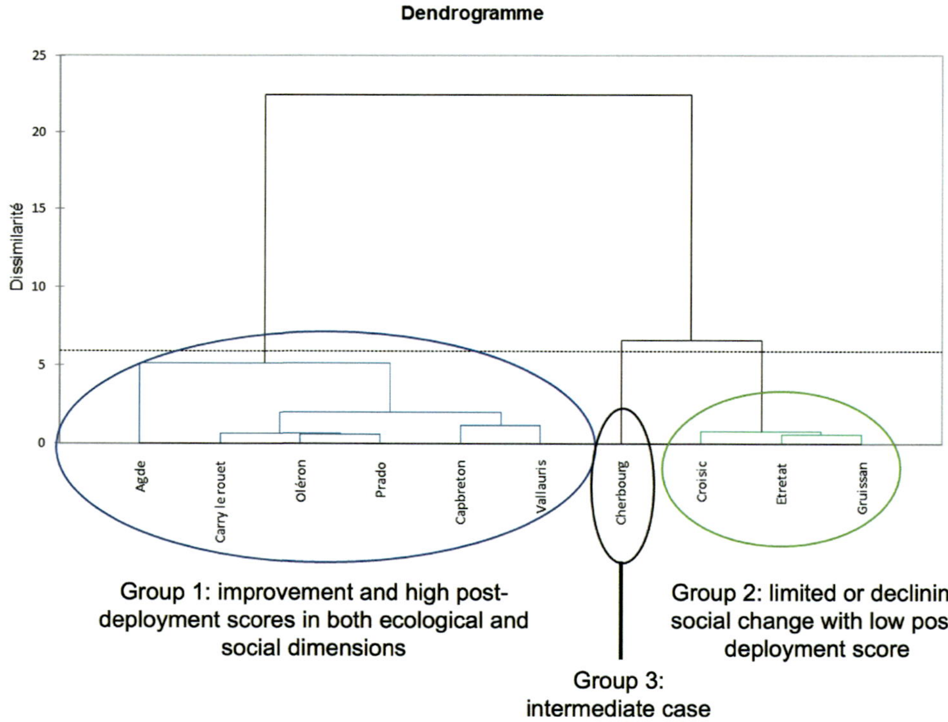


**FIGURE 12.4** Hierarchical Ascendant Classification (HAC) of the ten artificial reef sites showing three groups based on social and ecological scores.

territorial management frameworks such as marine parks, marine protected areas, or Natura 2000 sites. Funding sources are also diversified, drawing on at least two types of organisations. Interactions between human individuals and artificial reefs are frequent, including (when allowed) fishing activities, biological monitoring, and diving practices. This group is therefore characterised by strong actor diversity, dense relational networks, formalised governance, integration into larger management schemes, and sustained interactions with the artificial reefs and their ecosystem. These features explain the ability of a reef to improve both ecological and social performance.

The second group consists of sites with positive ecological outcomes but low or negative social gains. This group includes Étretat, Le Croisic, and Gruissan. The outcome reflects a limited and poorly connected set of stakeholders, where the municipality managed the sites in the absence of designated managers or broader participation. Information exchange was minimal, governance rules differed between sites, and no organisation was tasked with enforcing use regulations or monitoring outcomes. For example, a fishing reserve was established at Étretat that allowed limited fishing, all activities were prohibited at Croisic (e.g., fishing, diving, or any form of exploitation), and there were no restrictions at Gruissan. Despite these differences, all three sites lacked robust governance structures and financial partners, which resulted in limited interaction between local individuals and the artificial reefs. This situation can be explained by a complete restriction of access to the site at Le Croisic, and by activity limitations such as restrictions on fishing and diving at Gruissan. This group of artificial reef sites illustrates how artificial reef deployments can result in clear ecological benefits but remain socially underperforming when actor diversity, governance mechanisms, and stakeholder engagement are low.

The Cherbourg site can be seen as an intermediate case between the first and second groups: although its social score was very low, it had the highest ecological performance among all sites. This specificity is partly explained by the fact that the artificial reefs were only recently deployed and by the reef's primary objective of advancing scientific knowledge. As a result, most of the participants involved belong to the scientific community, which reduced the diversity and number of stakeholders compared to other sites in the first group. The ecosystem here is well-documented with a rich dataset available from extensive monitoring efforts. Governance, however, is largely restricted to the permit holder, with little opportunity for broader stakeholder consultation. Access rules are strict, with all activities prohibited except for scientific studies. Consequently, interactions with the artificial reef site are limited to research activities, with minimal public outreach or communication to local communities. Thus, Cherbourg has a profile distinct from other sites with artificial reefs: a highly successful ecological project but with constrained social integration due to its narrow user base, governance structure, and restricted use.

This comparative analysis revealed that artificial reef projects that delivered both ecological and social benefits shared three consistent features. Firstly, they involve a high diversity of stakeholders and foster dense interaction networks, promoting collaboration and mobilising varied expertise. Secondly, they operate under governance that imposes clear usage rules to facilitate enforcement. Thirdly, they are embedded within broader territorial or marine planning frameworks, aligning artificial reef deployment objectives with larger-scale environmental and socio-economic strategies.

Overall, differences between sites were driven more by variations in social performance than by ecological performance. These findings indicate that relying solely on ecological metrics can mask critical shortcomings. Sites that perform ecologically but underperform socially may result in user conflicts, reduced compliance with regulations, eroded stakeholder engagement, and a loss of political or financial support (Bennett and Dearden 2014). Incorporating social performance assessments enables managers to anticipate and address these risks, adapting governance strategies to maintain both ecological functions and stakeholder commitment over the long term.

Conducting comparative assessments across sites is particularly challenging because it requires collecting comparable datasets for each location while maintaining accuracy and methodological consistency. For instance, measuring the growth of biological communities poses two main difficulties: first, ensuring that the same indicators are used; and second, applying identical sampling methods. In practice, some datasets include biomass values while others provide only species richness; some focus exclusively on commercially important species, whereas others include all observed taxa. Therefore, before undertaking a comparative analysis, it is essential to clearly define which ecological components will be assessed—for example, total fish assemblages, specific indicator species, or particular trophic groups. To address data inconsistencies, we recommend using the most specific and comparable indicators available for each site. While this site-comparative approach can provide valuable insights into the performance of artificial reefs, the results should be interpreted with caution. This method does not directly measure the effectiveness of artificial reefs, as no universal threshold of "success" exists; even small positive trends may represent meaningful progress in certain contexts.

However, it offers a first overview of social and ecological performance across sites, allowing comparisons between artificial reef sites that can help identify which dimensions and indicators are underperforming. Such insights can then guide targeted management actions aimed at improving performance, drawing on lessons learned from better-performing sites. Beyond these specific results, this comparative and integrative approach also highlights the potential applicability of the methodology, suggesting that it could serve as a protocol for the evaluation of a larger number of artificial reef projects at the national scale (as demonstrated through this application) and potentially at the global level.

## 12.4 CONCLUSION

This chapter demonstrates that assessing artificial reefs solely through ecological indicators provides an incomplete understanding of an artificial reef's effectiveness. Across ten French sites evaluated here, all artificial reefs were associated with measurable ecological gains, yet three projects failed to meet social expectations. Without incorporating this social dimension, these shortcomings would have remained undetected, concealing vulnerabilities that threaten the long-term resilience of these social-ecological systems.

The contribution of this chapter lies in demonstrating the importance of the methodological integration of both ecological and social assessments into a single, coherent framework. By combining site-specific evaluations (anchored in local objectives and stakeholder expectations) with cross-site comparisons using standardised indicators, the approach ensures both contextual relevance and comparability. The use of network analysis and rapid assessment methods captures the structure and dynamics of a system in a way that is both rigorous and operational for managers (i.e., directly translatable into practical data aligned with artificial reef objectives). Thus, this improved assessment procedure can provide early-warning signals of emerging governance or engagement issues. Low social performance can undermine ecological success over time through user conflicts, reduced compliance with rules, loss of stakeholder engagement, and eventual funding or political withdrawal. By closely aligning social and ecological indicators, this approach enables managers to identify such risks early and adjust governance strategies accordingly.

Ultimately, successful artificial reefs are those that balance ecological gains with social value in attaining their objectives. This success is due to the integration of biodiversity enhancement and functional improvements with stakeholder satisfaction, equitable benefit sharing, and stable governance. Artificial reefs are not simply marine engineering projects; they are also socio-ecological investments. In the future, evaluating them as such is essential for ensuring lasting ecological and social returns.

## 12.5 ACKNOWLEDGEMENTS

This research was conducted as part of a PhD project co-funded by the National Association for Technical Research and Atlantique Landes Récifs. We thank all interview participants for their time and the reviewers for their valuable comments.

## REFERENCES


Al Kindi A, Marshell A, Dutta S. 2025. Evaluating the effectiveness of artificial reefs in enhancing fisheries productivity and biodiversity in the Sea of Oman. *Estuarine, Coastal and Shelf Science* 318:109249. https://doi.org/10.1016/j.ecss.2025.109249

Alves-Pinto H, Geldmann J, Jonas H, Maioli V, Balmford A, Latawiec AE, Strassburg B. 2021. Opportunities and challenges of other effective area-based conservation measures (OECMs) for biodiversity conservation. *Perspect Ecol Conserv.* 19(2):115–120. https://doi.org/10.1016/j.pecon.2021.01.004

Baine M. 2001. Artificial reefs: A review of their design, application, management and performance. *Ocean Coast Manage.* 44:241–259. https://doi.org/10.1016/S0964-5691(01)00048-6

Bas A. 2017. *Analyse de la compensation écologique comme instrument d'internalisation et de lutte contre l'érosion de la biodiversité marine: illustration par l'éolien en mer* [PhD dissertation]. Brest, France: Université de Bretagne Occidentale. 255 p.

Bavoux JJ, Chapelon L. 2014. *Dictionnaire d'analyse spatiale.* Paris, France: Armand Colin. 607 p.

Becker A, Taylor MD, Folpp H, Lowry MB. 2018. Managing the development of artificial reef systems: The need for quantitative goals. *Fish.* 19(4):740–752. https://doi.org/10.1111/faf.12288

Bennett NJ, Dearden P. 2014. Why local people do not support conservation: Community perceptions of marine protected area livelihood impacts, governance and management in Thailand. *Marine Policy* 44:107–116.

Bideci C, Cater C. 2019. Dive and novelty seeking in experimental artificial reefs. *Journal on Tourism & Sustainability* 3(1):22–31.

Boileau J, Calvet C, Pioch S, Moulherat S. 2022. Ecological equivalence assessment: The potential of genetic tools, remote sensing and metapopulation models to better apply the mitigation hierarchy. *Journal of Environmental Management* 305:114415.

Brochier T, Brehmer P, Mbaye A, Diop M, Watanuki N, Terashima H, . . . & Auger P. 2021. Successful artificial reefs depend on getting the context right due to complex socio-bio-economic interactions. *Scientific Reports* 11(1):16698.

Brodie JF, Deith MC, Burns P, Goetz S, Cunningham C, Hill JK, Mohd-Azlan J. 2025. The contribution of other effective area-based conservation measures (OECMs) to protecting global biodiversity. *Nature Communications* 16(1):7886.

Callon M. 1986. Éléments pour une sociologie de la traduction: la domestication des coquilles St-Jacques et des marins pêcheurs dans la baie de St. Brieuc. *Ann Sociol.* 36:169–208.

Cépralmar. 2015. *Guide pratique d'aide à l'élaboration, l'exploitation et la gestion des récifs artificiels en Languedoc-Roussillon.* OiEau. 121 p.

Charbonnel E. 2005. Les récifs artificiels comme outils de gestion des ressources littorales: éléments de synthèse et de réflexion. In: *Séminaire international d'échanges d'expériences.* Nantes, France.

Christensen V, Pauly D. 1992. ECOPATH II—A software for balancing steady-state ecosystem models and calculating network characteristics. *Ecol Model.* 61:169–185. https://doi.org/10.1016/0304-3800(92)90016-8

Claudet J. 2006. *Aires marines protégées et récifs artificiels: méthodes d'évaluation, protocoles expérimentaux et indicateurs* [PhD dissertation]. Perpignan, France: Université de Perpignan. 266 p.

Dafforn KA, Glasby TM, Airoldi L, Rivero NK, Mayer-Pinto M, Johnston EL. 2015. Marine urbanization: An ecological framework for designing multifunctional artificial structures. *Front Ecol Environ.* 13:82–90. https://doi.org/10.1890/140050

Dodds KC, Vozzo ML, Mayer-Pinto M, Dafforn KA, Bishop MJ. 2025. Stakeholders support marine eco-engineering, but what are the perceived benefits and who should pay? *People and Nature* 7:653–667. https://doi.org/10.1002/pan3.10795

Eloire F, Penalva Icher É., Lazega E. 2011. Application de l'analyse des réseaux complets à l'échelle inter-organisationnelle. *Terrains Travaux.* 2:77–98.

Fallery B, Rodhain F. 2007. *Quatre approches pour l'analyse de données textuelles: Lexicale, linguistique, cognitive, thématique.* Proceedings of the XVI ème Conférence de l'Association Internationale de Management Stratégique AIMS (Montréal, Canada), pp. 1–16.

FAO. 2015. *Practical guidelines for the use of artificial reefs in the Mediterranean and the Black Sea.* Rome, Italy: General Fisheries Commission for the Mediterranean. 84 p.

Fennessy S, Jacobs A, Kentula M. 2007. An evaluation of rapid methods for assessing the ecological condition of wetlands. *Wetlands.* 27:543–560. https://doi.org/10.1672/0277-5212(2007)27[543:AEORMF]2.0.CO;2

Folpp H, Lowry M, Gregson M, Suthers IM. 2011. Colonization and community development of fish assemblages associated with estuarine artificial reefs. *Braz J Oceanogr.* 59:55–67. https://doi.org/10.1590/S1679-87592011000100006

Hallier JP, Gaertner D. 2008. Drifting fish aggregation devices could act as an ecological trap for tropical tuna species. *Mar Ecol Prog Ser.* 353:255–264. https://doi.org/10.3354/meps07180

Hammond M, Bond T, Prince J, Hovey RK, McLean DL. 2020. An assessment of change to fish and benthic communities following installation of an artificial reef. *Reg Stud Mar Sci.* 39:101408. https://doi.org/10.1016/j.rsma.2020.101408

Hayek M, Salgues M, Souche JC, De Weerdt K, Pioch S. 2023. How to improve the bioreceptivity of concrete infrastructure used in marine ecosystems? Literature review for mechanisms, key factors, and colonization effects. *Journal of Coastal Research* 39(3):553–568.

Jacob C. 2017. *Approche géographique de la compensation écologique en milieu marin: analyse de l'émergence d'un système de gouvernance environnementale* [PhD dissertation]. Montpellier, France: Université Paul Valéry—Montpellier III. 240 p.

Jacob C, Thorin S, Pioch S. 2018. Marine biodiversity offsetting: An analysis of the emergence of an environmental governance system in California. *Marine Policy* 93:128–141.

Kirkbride-Smith AE. 2014. *The economic, social and conservation benefits of recreation-orientated artificial reefs* [Doctoral dissertation]. University of Hull.

Lee MO, Otake S, Kim JK. 2018. Transition of artificial reefs (ARs) research and its prospects. *Ocean Coast Manage.* 154:55–65. https://doi.org/10.1016/j.ocecoaman.2018.01.001

Levrel H, Frascaria-Lacoste N, Hay J, Martin G, Pioch S. 2015. *Restaurer la nature pour atténuer les impacts du développement: analyse des mesures compensatoires pour la biodiversité.* Paris, France: Quae. 320 p.

Lima JS, Zalmon IR, Love M. 2019. Overview and trends of ecological and socioeconomic research on artificial reefs. *Marine Environmental Research* 145:81–96.

McAfee D, Reinhold SL, Alleway HK, Connell SD. 2021. Environmental solutions fast-tracked: Reversing public scepticism to public engagement. *Biological Conservation* 253:108899.

Mechin A, Pioch S. 2016. *Une méthode expérimentale pour évaluer rapidement la compensation en zone humide, la méthode MERCIe: principes et applications.* Rapport technique, ONEMA. 86 p.

Mechin A, Pioch S, Cluchier A. 2023. Offset sizing tools: A review of practices used in the field and their operationality. *Journal of Environmental Management* 346:118990.

Milon J, Whitmarsh D, Holland S. 2000. Social and economic evaluation methods. In: Seaman W, Jensen A, editors. *Artificial reef evaluation: With application to natural marine habitats.* Boca Raton, FL: CRC Press. p. 166–190. https://doi.org/10.1201/9781420036633.ch8

Newman ME. 2003. The structure and function of complex networks. *SIAM Rev.* 45:167–256. https://doi.org/10.1137/S003614450342480

Ostrom E. 2009. A general framework for analyzing sustainability of social-ecological systems. *Science* 325(5939):419–422. https://doi.org/10.1126/science.1172133

Page HM, Dugan JE, Schroeder DM, Nishimoto MM, Love MS, Hoesterey JC. 2007. Trophic links and condition of a temperate reef fish: Comparisons among offshore oil platform and natural reef habitats. *Mar Ecol Prog Ser.* 344:245–256. https://doi.org/10.3354/meps06929

Paxton AB, Shertzer KW, Bacheler NM, Kellison GT, Riley KL, Taylor JC 2020. Meta-analysis reveals artificial reefs can be effective tools for fish community enhancement but are not one-size-fits-all. *Frontiers in Marine Science* 7:2. https://doi.org/10.3389/fmars.2020.002

Pioch S, Pinault M, Brathwaite A, Méchin A, Pascal N. 2018b. *Handbook II: Methodology for scaling mitigation and compensatory measures in tropical marine ecosystems.* Paris: International Coral Reef Initiative, IFRECOR ed. https://icriforum.org/wp-content/uploads/2019/12/HandBook%202%20-%20V5.pdf

Pioch S, Relini G, Souche JC, Stive M, de Monbrison D, Nassif S, Simard F, Allemand D, Saussol P, Spieler R. 2018a. Enhancing eco-engineering of coastal infrastructure with eco-design: Moving from mitigation to integration. *Ecol Eng.* 120:574–584. https://doi.org/10.1016/j.ecoleng.2017.06.051

Polovina JJ. 1984. Model of a coral reef ecosystem. *Coral Reefs* 3:1–11. https://doi.org/10.1007/BF00306135

Ramos J. 2007. *Socio-economic implications of artificial reef deployment: A Portuguese case study* [Doctoral dissertation]. Portsmouth: Department of Economics, University of Portsmouth.

Ramos J, Santos MN, Whitmarsh D, Monteiro CC. 2007. Stakeholder perceptions regarding the environmental and socio-economic impacts of the Algarve artificial reefs. *Hydrobiologia* 580:181–191. https://doi.org/10.1007/s10750-006-0453-8

Raoux A, Salaün J, Pezy JP, Vivier B, Navon M, Deloor M, Claquin P, Pioch S, Niquil N, Dauvin JC. 2024. Trophic modelling, a key framework to provide adapted metrics to feedback ecological performance of artificial structures. *Reg Stud Mar Sci.* 80:103890. https://doi.org/10.1016/j.rsma.2023.103890

Riera E, Lamy D, Goulard C, Francour P, Hubas C 2018. Biofilm monitoring as a tool to assess the efficiency of artificial reefs as substrates: Toward 3D printed reefs. *Ecological Engineering* 120:230–237. https://doi.org/10.1016/j.ecoleng.2018

Salaün J. 2022. *Vers une approche socio-écologique d'évaluation de la performance de projets d'aménagements côtiers avec des récifs artificiels: étude de cas sur trois régions biogéographiques maritimes en France métropolitaine* [PhD dissertation]. Montpellier, France: Université Paul Valéry—Montpellier III. 347 p.

Salaün J, Pioch S, Dauvin JC. 2022. Social-ecological analysis illustrates real objectives of artificial reefs projects: French case studies. *J Coast Res*. 38:624–638. https://doi.org/10.2112/JCOASTRES-D-21-00050.1

Salaün J, Raoux A, Pezy JP, Pioch S, Dauvin JC. 2023. Trophic modelling approach of artificial reefs ecosystem changes in the Bay of Biscay. *Reg Stud Mar Sci*. 63:103100. https://doi.org/10.1016/j.rsma.2023.103100

Seaman W, Jensen A. 2000. Purposes and practices of artificial reef evaluation. In: Seaman W, Jensen A, editors. *Artificial reef evaluation: With application to natural marine habitats*. Boca Raton, FL: CRC Press. p. 1–19. https://doi.org/10.1201/9781420036633.ch1

Sutton SG, Bushnell SL. 2007. Socio-economic aspects of artificial reefs: Considerations for the Great Barrier Reef Marine Park. *Ocean & Coastal Management* 50(10):829–846.

Taormina B, Claquin P, Vivier B, Navon M, Pezy JP, Raoux A, Dauvin JC. 2022. A review of methods and indicators used to evaluate the ecological modifications generated by artificial structures on marine ecosystems. *J Environ Manage*. 310:114646. https://doi.org/10.1016/j.jenvman.2022.114646

Tessier A, Francour P, Charbonnel E, Dalias N, Bodilis P, Seaman W, Lenfant P. 2015. Assessment of French artificial reefs: Due to limitations of research, trends may be misleading. *Hydrobiologia* 753(1):1–29.

Tunca S, Lizaso JLS, Ünal V. 2022. A stakeholder analysis of the artificial reefs in Edremit Bay, Turkey: Social, Economics, and Management Aspects. In: *Impact of artificial reefs on the environment and communities*. IGI Global Scientific Publishing. p. 48–68.

Vivier B. 2021. *Rôle de la colonisation d'infrastructures marines sur la production marine et la biodiversité des écosystèmes côtiers en Manche* [PhD dissertation]. Caen, France: Université de Caen Normandie. 378 p.

Vivier B, Dauvin JC, Navon M, Rusig AM, Mussio I, Orvain F, Claquin P. 2021. Marine artificial reefs in the world: A literature analysis of their designs, objectives and effectiveness. *Glob Ecol Biogeogr*. 27:e01538. https://doi.org/10.1111/geb.1538

Wasserman S, Faust K. 1994. *Social network analysis: Methods and applications*. Cambridge, UK: Cambridge University Press. 857 p.

# 13 Artificial Reefs
## *A Perspective of What We Learned and What We Should Do Next*

*Juliano Silva Lima, Igor David da Costa, and Ilana Rosental Zalmon**
* Corresponding author: ilana@uenf.br

## 13.1 INTRODUCTION

The deployment of artificial reefs constitutes one of the oldest and most versatile anthropogenic strategies for modulating aquatic ecosystems. Artificial reefs are defined as structures designed to mimic, restore, or enhance the ecological functions originally performed by natural reefs in marine, estuarine, and freshwater environments (Bortone et al. 2011; Giglio et al. 2024). Although the contemporary concept of artificial reefs incorporates advanced principles of ecological engineering, spatial planning, and environmental governance, their origins can be traced to millennia-old empirical practices. As early as the 18th century, coastal communities in Japan, as well as indigenous peoples across the Americas, used natural materials such as wooden logs, stones, and bundled vegetation to attract and concentrate fish schools, thereby providing some of the earliest documented examples of intentional structural enrichment of aquatic habitats (Lima et al. 2019a).

The conceptual and technological advancement of artificial reefs became more pronounced in the post–World War II period, particularly in Western countries. Initially, decommissioned naval vessels (and subsequently, obsolete offshore oil platforms) were deliberately converted into artificial habitats, demonstrating a pivotal shift from empirical methods to systematic strategies grounded in applied science (Bortone 2006; Lee et al. 2018). Since then, artificial reef deployments have progressively incorporated elements of coastal engineering, ecological monitoring, and adaptive planning, emerging as multifunctional structures with diverse applications.

While the historical impetus for artificial reef deployments centered on enhancing fish biomass, their role has expanded substantially in recent decades. Artificial reefs are now recognized not only as instruments to support sustainable fisheries but also as platforms for biodiversity conservation, coastal zone stabilization, the promotion of underwater tourism, and the support of aquaculture operations (Dafforn et al. 2015; Becker et al. 2018). Their capacity to aggregate and sustain populations of economically important species has reinforced their strategic importance in fisheries management, particularly in regions with intense anthropogenic pressures (Sánchez-Caballero et al. 2024; Sheng et al. 2025).

Ecologically, artificial reefs play a critical role in the recovery of degraded habitats by providing substrate essential for benthic organism settlement and shelter for juvenile stages of various species. Thus, artificial reefs contribute to the functional resilience of coastal ecosystems (Vivier et al. 2021; Frehse et al. 2025). Furthermore, artificial reef installations can aid in mitigating erosive processes by acting as physical barriers against the hydrodynamic forces of currents and waves (a function increasingly relevant with the advent of human-induced climate change), resulting in increasing vulnerability (Pardo et al. 2025; Korniejenko et al. 2025). Recent studies indicate that well-designed artificial reefs not only attract individual organisms but also enhance reproductive processes and net biomass production, underscoring their dual value as productive and conservation-oriented management options (Da Costa et al. 2022; Watson et al. 2025).


DOI: 10.1201/9781003676775-13

The socioeconomic dimension of artificial reefs has also gained increasing attention, particularly concerning underwater tourism and coastal recreation. The establishment of diving (both snorkeling and SCUBA) parks and the intentional sinking of vessels as recreational attractions have spurred local economic growth (especially through artisanal fisheries) while simultaneously promoting environmental education and public engagement in ocean conservation (Chen et al. 2013; Ramos 2024). When properly regulated, these activities represent a unique opportunity to harmonize marine conservation with sustainable economic development, generating ecological, social, and economic benefits in a synergistic manner (Firth et al. 2016; Ramos et al. 2021).

A critical understanding of artificial reef functions, limitations, and potential is thus essential to maximize the benefits of their application, especially in light of escalating anthropogenic pressures and the global pursuit of innovative nature-based solutions (NbS). Accordingly, the objective of this chapter is to provide a comprehensive and up-to-date circumspection of the current state and future prospects for artificial reef development and management. Included here is a depiction of their broad functional spectrum, along with a conceptual re-orientation of their function that positions them as strategic components in multiple environmental, economic, and social management scenarios. To this end, we reviewed scientific publications spanning the period from 1962 to 2025, encompassing multiple dimensions (e.g., ecological, technological, and socioeconomic), and investigated how artificial reef deployments have changed from concrete modules, primarily designed to enhance fisheries production, into biogenic multifunctional infrastructures toward sustainable targeted fisheries management and habitat restoration. This change has occurred through the application of innovative methods, such as using new architected materials, and the adoption of NbS to increase the resilience of coastal cities to climate change.

## 13.2 BACKGROUND

### 13.2.1 Global Overview of Artificial Reef Research

Artificial reef research conducted to date presents a consolidated, yet geographically uneven, global landscape in terms of thematic approaches and scientific maturity (Ramm et al. 2021). Since the first systematic records in Japan and the United States during the 1960s, artificial reef science has undergone considerable expansion, establishing new centers of the scientific application of artificial reefs in Asia (notably China and Japan), Australia, Europe, and, more recently, in countries such as Brazil and Colombia (Lima et al. 2019a). Areas with relatively large numbers of reefs include the eastern seaboard of the USA, Western Europe, countries bordering the Mediterranean Sea, and the coasts of Japan, Korea, and China. However, this expansion has not occurred uniformly, as some countries have led the innovation in reef construction and monitoring techniques while others have concentrated efforts on foundational ecological studies or localized socioeconomic evaluations. These differences in approach reflect the distinct environmental, social, and economic contexts of each region (Lima and Zalmon 2020).

In the United States, a multifaceted approach is evident, particularly in the reuse of decommissioned oil platforms through the Rigs-to-Reefs program, coupled with systematic studies on the ecological and socioeconomic functions of artificial reefs (Love et al. 2019). These initiatives employ advanced surveying technologies that include remotely operated vehicles (ROVs) and acoustic surveys—especially in California and the Gulf of Mexico. This has helped enable detailed mapping of associated biological communities and the assessment of processes such as the recruitment of commercially valuable recreational and commercially important species (Ajemian et al. 2015). The east coast of the United States also has had a large focus on recreational fishing (Hughes 2015) with several studies on various aspects of artificial reef design and their effects on fish populations (Bohnsack and Sutherland 1985; Baine 2001).

In China, Japan, and South Korea, research has been largely technology-driven, with the application of ecological engineering methods to develop new materials and reef structures, including eco-concrete substrates and biogenic formations (Huang et al. 2016; Kim et al. 2014). These efforts have been complemented by highly complex field experiments aimed at optimizing the attraction and production functions of these installations (Xu et al. 2017; Yun and Kim 2018). The concept of combining aquaculture-based enhancement with artificial reefs was first conceived in China in 1979 and termed "marine ranching" (Qin et al. 2020), generating substantial direct economic and ecological benefits (Zhou et al. 2019; Pan et al. 2022). In Southeast Asia, particularly in Malaysia and the Philippines, studies have focused on the socioeconomic dimensions of artificial reef deployment, emphasizing co-management strategies and local community participation (Latun and Abdullah 1990; Raja et al. 1994; Delos Reyes and Martens 1995; Macusi et al. 2017; Jani, 2020).

In Australia, most reefs were deployed to enhance faunal communities or fisheries, particularly recreational fishing (Arlinghaus et al. 2015; Ramm et al. 2021). Furthermore, Australian researchers have compared biological communities associated with natural reefs and artificial reefs, seeking to evaluate the conservation potential and biodiversity support of artificial reefs in contexts of coastal degradation (Champion et al. 2015; Smith et al. 2017). In Europe (especially in France, Italy, Portugal, and the United Kingdom), a wide range of research topics has been investigated. These studies include applications toward local small-scale fisheries, which are crucial from a social and economic perspective (Harmelin 2000; Fabi et al. 2011; Reis et al. 2021). Later, the studies advanced toward nutrient flux analyses in Mediterranean artificial reef zones (Cresson et al. 2014) as well as the examination of public perceptions regarding the recreational and tourism uses of artificial reefs (Evans et al. 2017; Ramos et al. 2021). More recently, these studies have adopted a more emerging method by employing Nature Inclusive Design to enhance biodiversity and ecosystem functioning.

In the Middle East, Turkish environmental managers use artificial reefs mainly to prevent illegal trawling and create new fishing opportunities (Lök et al. 2002). Concomitantly, Israeli managers have conducted studies on marine communities inhabiting artificial reef structures for diving and recreational management. These efforts include coral manipulation on artificial reefs aimed at restoring marine habitats with several 3D-printed reefs in the Gulf of Eilat to attract corals, fish, and invertebrates that support the regrowth of natural coral reefs (Berman et al. 2023; Levy et al. 2024; Yuval et al. 2024). In contrast, African research efforts, though more limited, have explored the multispecies attraction potential of fish aggregating devices and shipwrecks (Lechanteur and Griffiths 2001). In South America, there has been a notable increase in scientific output on artificial reefs in Brazil. While initially modest, studies by Brazilian researchers have matured to encompass integrated ecological, ethnoecological, and socioeconomic evaluations, particularly within estuarine and vulnerable coastal zones (Lima et al. 2018; Lima and Zalmon 2022; Rocha et al. 2024).

Despite the significant growth in the number and scope of artificial-reef-related studies, several structural limitations remain. The predominance of short-term monitoring (often less than two years in duration) limits the detection of long-term effects of ecological processes such as biological succession, trophic shifts, and population stability (Lee et al. 2018; Vivier et al. 2021). Furthermore, scientific integration is hindered by institutional barriers, methodological disparities, budget constraints, and logistical challenges associated with large-scale, multinational monitoring efforts (Becker et al. 2018; Lima et al. 2020a). The lack of robust collaborative projects continues to restrict the advancement of artificial reef investigations toward truly comparative and globally integrated standards. Subsequently, this has limited the generalization of findings to account for environmental variability and to formulate guidelines for diverse social contexts (Frehse et al. 2025).

### 13.2.2 Materials and Criteria for Construction of Artificial Reefs

The selection of materials for artificial reef construction represents one of the most critical factors influencing the ecological, structural, and socioeconomic success of these environmental interventions. Historically, material choice has been largely dictated by local availability and costs of acquisition and deployment. Consequently, this has led to the widespread use of conventional materials such as concrete, natural rock, tires, and repurposed metal structures that include decommissioned vessels and offshore platforms (Baine 2001; Werner et al. 2024; Korniejenko et al. 2025). Concrete, in particular, has become the most commonly used material due to its durability, chemical stability, moldability into diverse shapes, and suitability for colonization by a wide range of benthic organisms (Firth et al. 2016). Natural rock provides a structurally complex reef material that promotes habitat heterogeneity. In contrast, tires, though initially attractive due to their low cost, have proven problematic due to their limited stability and potential release of toxic compounds over time. This has led to regulatory restrictions and a phase-out of tires for use in artificial reefs in various jurisdictions (Aleksandrov et al. 2002; Vivier et al. 2021).

Conventional artificial substrates such as concrete blocks are considered inadequate substitutes for natural reef habitats (Sánchez-Caballero et al. 2021). With an increasing emphasis on sustainable and ecologically compatible alternatives in artificial reef construction, a broader variety of materials is becoming prominent as artificial reef materials. According to Hickling et al. (2022), Nature-Inclusive Design structures integrate habitat features that mimic discrete microhabitats or directly duplicate the full diversity of natural reef complexity (Evans et al. 2021). Furthermore, biogenic materials, such as mollusk shells and oyster reef remnants, have been utilized to replicate the natural characteristics of marine substrates and to encourage the settlement of larvae from ecologically and commercially important species (Vivier et al. 2021). Recent research has also explored the use of biodegradable polymers, organic-origin structures, and modified concrete enriched with surface-roughening additives, all aimed at enhancing larval settlement and increasing the diversity of colonizing species (Dafforn et al. 2015; Becker et al. 2018). Innovative alternative materials such as eco-concrete (i.e., concrete that uses less energy to produce with fewer carbon emissions) embedded with microtextures have demonstrated significant potential to improve habitat biomimicry, which favors organism recruitment and enhances the ecological functionality of artificial reefs (Firth et al. 2016; Huang et al. 2016). More recently, Levy et al. (2024) used biomimetic tiles fabricated with 3D design, printing, and molding to duplicate the rugosity of natural reef surfaces with important implications for reef restoration strategies.

However, the selection of materials must take ecological factors into account. Moreover, economic factors also play a decisive role in project feasibility, particularly in the context of limited funding or in scenarios where there is demand for rapid fisheries enhancement or tourism development (Lima et al. 2019b; Korniejenko et al. 2025). While conventional concrete offers ecological and structural advantages in artificial reef construction, this material incurs higher production and deployment costs compared to the reuse of existing structures such as sunken ships or decommissioned metal oil platforms (Bull and Love 2019). Nonetheless, the repurposing of materials requires rigorous environmental safety assessments to avoid introducing contaminants such as heavy metals, hydrocarbons, or other toxic agents that could impair water quality and marine organism health (Dafforn et al. 2015; Becker et al. 2018).

Furthermore, increasingly refined ecological criteria serve as guidelines for both material selection and structural design. These guidelines often take into account chemical compatibility with the local environment, the structural complexity required to support species diversity and system resilience, and the ability to harmoniously integrate with the hydrodynamic conditions of the deployment site (Firth et al. 2016; Vivier et al. 2021). Studies have shown that substrates with greater surface roughness (i.e., rugosity), micro-relief variability, and cavity heterogeneity promote early colonization and foster more complex trophic networks (Dafforn et al. 2015; Becker et al. 2018). These reef attributes consequently can reinforce the role of artificial reefs as biodiversity enhancers and providers of vital ecosystem services.

Beyond material properties, construction criteria extend to oceanographic and engineering considerations that ensure long-term performance and ecological integration (Bugnot et al. 2021). Minimum deployment requirements typically include sufficient unit weight and stability to resist local wave action, current velocities, and storm events. Additionally, deployments should include adequate vertical relief and spacing to optimize water circulation and habitat availability. Furthermore, deployment sites should be selected on the basis of bathymetry, substrate type, and proximity to natural reefs to prevent ecological redundancy or degradation (Ramm et al. 2021). Engineering standards often recommend redundancy in structural strength, the avoidance of sharp or entangling features, and modularity to facilitate adaptive design and phased construction. Oceanographic parameters (i.e., prevailing currents, sediment transport dynamics, and water quality) are crucial to ensure that reefs not only remain structurally stable but also deliver the intended ecological services, from enhancing fish biomass to promoting coral recruitment. Ultimately, compliance with international conventions (e.g., London Convention and Protocol/UNEP 2009) and national regulatory frameworks provides the baseline for safe, effective, and sustainable reef deployment.

The shift from opportunistic reef construction (e.g., rubber tires or discarded waste) toward purpose-built, ecologically compatible materials reflects advances in technology, engineering practice, and environmental awareness (Ramm et al. 2021; Komyakova et al. 2022). This paradigm shift underscores the critical role of science-based criteria in ensuring that artificial reefs deliver not only immediate fisheries benefits but also long-term contributions to marine biodiversity conservation and ecosystem resilience (Lima and Zalmon 2020).

### 13.2.3 Methods for Evaluating Artificial Reefs

The evaluation of artificial reefs is fundamental to assessing their ecological and socioeconomic effectiveness, as well as their engineering and oceanographic aspects. This process requires the application of diverse and complementary methodological approaches. Historically, assessment techniques have aligned with theoretical and technological advancements in marine ecology, applied social sciences, and the growing demands of coastal ecosystem management (Becker et al. 2018; Hylkema et al. 2021). To accurately determine whether artificial reefs are fulfilling their intended objectives, it is essential to employ assessment methodologies that simultaneously account for the biological dynamics of associated assemblages and the perceptions of, and impacts on, human populations that result from the deployment of these structures (Islam et al. 2014; Lima and Zalmon 2020).

From an ecological perspective, non-destructive observational methods are predominant, given their application to monitor reef-associated fauna without compromising the integrity of the environment. For example, underwater visual censuses conducted by divers using SCUBA enable the direct observation of fish and macroinvertebrate identification, abundance, species diversity, and size (Oricchio et al. 2016; Davis and Smith 2017). According to Ramm et al. (2021), cost appeared to play an important role in determining the non-destructive monitoring method chosen, with the prevalence of visual census declining in higher-income countries and being replaced by visual methods employing more expensive technology, such as ROVs and Baited Remote Underwater Videos.

Although less frequently employed today, destructive methods are still utilized in specific studies that require organism collection for detailed laboratory analyses, including biomass estimation, dietary assessment, and genetic investigations (Ribeiro et al. 2008; Lima et al. 2020b). Such methods that involve the use of trawls, traps, and manual collection are particularly relevant for examining assemblage trophic structure, leading to an understanding of the influence that artificial reefs have on local food web dynamics (Campbell et al. 2010; Cresson et al. 2014). These methods are also cost-effective and can be used in areas where the water is turbid or has limited clarity due to eutrophication (Wu et al. 2019), while also involving local artisanal or commercial fishers.

From a socioeconomic perspective, evaluation techniques are generally traditional methods that include semi-structured interviews and targeted questionnaires for artisanal fishers, diving tourism operators, and environmental managers and have become vital for accessing local perceptions and assessing the effectiveness of artificial reefs (Knights et al. 2024). This information helps determine the distribution of generated benefits and identify potential conflicts that may arise during implementation and management processes (Hylkema et al. 2021). Some of these studies incorporate ecosystem services analysis and apply a socioecological framework (Ramos and Santos 2015; Lima et al. 2019b; Costa et al. 2024).

The evaluation of artificial reefs has significantly progressed beyond the traditional focus on fish-catch data, incorporating methodologies capable of providing data to evaluate the complex social, cultural, and economic impacts associated with the deployment of these structures (Chen et al. 2013; Macusi et al. 2017). In the last five years, some improvements in management and analytical methodologies include optimizing image-based protocols to monitor macro-epibenthic communities colonizing artificial structures (Taormina et al. 2020); using underwater drones to record surveys; implementing multimetric indices to evaluate the performance and management of artificial reefs (Lima et al. 2020a); applying a Bayesian approach to infer the sustainable use of artificial reefs in fisheries and recreation (Ramos et al. 2024); and employing environmental DNA (eDNA) analysis to determine fine-scale spatial fish distribution (Miyajima-Taga et al. 2024).

## 13.3 MAIN FOCUS OF THE CHAPTER

This chapter presents a global overview of artificial reef research, the materials and criteria used in artificial reef construction, and the methodologies employed in their evaluation. Concomitantly, this chapter lays the foundation for a review of the environmental and socioeconomic issues associated with the deployment of these structures. In examining these issues, the chapter introduces emerging trends and near-future perspectives mainly discussed in the most recent CARAH (i.e., Conference on Artificial Reefs and Related Aquatic Habitats held in Santa Marta, Colombia, March 2025) that are shaping the scientific, technological, and policy benefits of artificial reefs as strategic management options for ecological conservation, sustainable resource management, and coastal spatial planning.

### 13.3.1 Benefits Related to Artificial Reefs

Among the numerous benefits attributed to artificial reefs, the most prominent is the strengthening of both artisanal and commercial fisheries—the main, original objective of these structures. By affecting both the aggregation and effective production of biomass, artificial reefs can enhance the availability of fishery resources for traditional communities, contributing to food security, diversification of fishing strategies, and the economic stability of historically vulnerable social groups (Ramos et al. 2011; Chen et al. 2013; Islam et al. 2014; Evans et al. 2017; Reye and García 2022). Studies conducted by Lima et al. (2020b) demonstrated that the deployment of artificial reefs has led to significant increases in the abundance of target species, particularly those of commercial interest, thereby improving the efficiency of local fishing activities. Additionally, Da Costa et al. (2022) highlighted that these artificial habitats also function as important reproductive and recruitment areas for economically significant species. These attributes reinforce the role of artificial reefs as not merely aggregation points but as functional habitats that contribute to the ecological and socioeconomic sustainability of fisheries.

Another widely recognized environmental benefit of artificial reefs is their contribution to the restoration of degraded habitats. In areas affected by erosion, destruction of natural reefs, or the degradation of benthic substrates, artificial reefs provide alternative surfaces for the settlement of benthic invertebrates and the establishment of diverse fish assemblages (Fabi et al. 2002; Watson et al. 2025). Recent projects have shown that, when appropriately designed and implemented, artificial reefs can restore lost ecological functions such as water filtration, shelter provision, and trophic resource availability for multiple species (Dafforn et al. 2015; Reye and García 2022). In the realm of coastal protection, numerous initiatives have demonstrated the potential of artificial reefs to reduce wave energy and mitigate shoreline erosion (Firth et al. 2016; Watson et al. 2025). Structures designed with ecological engineering principles not only provide habitat for marine species but also reduce the hydrodynamic impacts of currents and waves, thus enhancing the resilience of coastal systems in the face of climate change and sea-level rise.

The implementation of artificial reefs also generates both direct and indirect employment opportunities in coastal communities. Among others, these opportunities are associated with the construction, monitoring, and maintenance of the structures and associated activities such as fishing, tourism, and service provision (Campbell et al. 2010; Bortone et al. 2011). Recent data indicate that well-structured reef projects can create employment for fishers, diving guides, and tour operators (Becker et al. 2018). The promotion of ecotourism and coastal recreation has also emerged as a strategic function of artificial reefs, with direct implications for income generation and the cultural appreciation of marine environments. Artificial reefs designed for recreational diving have become major tourist attractions, fostering environmental awareness and encouraging the conservation of local ecosystems (Chen et al. 2013; Ramos et al. 2024). In addition to boosting sectors such as hospitality, gastronomy, and transport, diving activities can be integrated into environmental education campaigns, thereby enhancing societal engagement in ocean conservation efforts (Ramos 2024).

Moreover, artificial reefs serve as valuable platforms for scientific research and environmental education. Studies on ecological succession, population connectivity, trophic dynamics, and ecosystem services have benefited from the existence of these structures, which act as "natural laboratories" within controlled and replicable marine settings (Evans et al. 2017; McLean et al. 2022). Participatory monitoring initiatives involving schools, universities, and non-governmental organizations have further strengthened scientific literacy and fostered environmental awareness among younger generations, contributing to broader efforts in marine conservation.

### 13.3.2 Emerging Challenges Related to Artificial Reefs

Despite the well-documented potential benefits of artificial reefs, their expanding deployment presents a series of emerging challenges that warrant critical attention. One of the foremost concerns remains the inappropriate use of materials, which can severely compromise both environmental quality and ecological functionality (Lima and Zalmon 2020; Ramm et al. 2021). Historically, materials such as tires, scrap metal, and industrial waste have been widely employed in reef projects, often without proper risk assessment regarding the release of toxic substances or the accelerated degradation of these materials (Aleksandrov et al. 2002; Vivier et al. 2021). These shortcomings not only pose threats to local biodiversity but also hinder public acceptance of artificial reef initiatives and elevate the costs associated with environmental remediation and structural maintenance (Lemasson et al. 2024). In response, recent trends have favored inert and environmentally compatible materials such as marine-grade concrete, biogenic substrates, and treated steel, aligned with more advanced design and ecology.

A key frontier in artificial reef development involves the creation of structures that mimic the ecological complexity and heterogeneity of natural reef systems (Higgins et al., 2022). Rather than relying solely on rigid, geometrically simple substrates, new approaches emphasize embedding ecological principles into design, incorporating rugosity, cavities, and microhabitats that support diverse species assemblages (Perkol-Finkel et al. 2005; Maslov et al. 2024). These "nature-inclusive" designs represent a paradigm shift in which artificial reefs are no longer conceived as passive fishing devices but as ecological infrastructures capable of enhancing resilience and ecosystem functioning (Higgins et al. 2022).

Additive manufacturing technologies offer unprecedented opportunities for precision, customization, and scalability in artificial reef construction (Pan et al. 2022). Through 3D printing, intricate surface textures and structural configurations can be reproduced, enhancing larval settlement and replicating the ecological fingerprint of natural substrates (Korniejenko et al. 2025). Recent research indicates that the nature-like design method applied to artificial reefs (i.e., varying in scale, complexity, and function) broadens the possibilities for how intelligent, human-made underwater structures can effectively support nearshore ecosystems (Maslov et al. 2024). Despite their promise, these improvements in structure and design remain untested with regard to material durability, cost-effectiveness, and ecological performance in large-scale applications.

Recent studies demonstrate that artificial reefs can exhibit densities, biomass, and species richness among fishes comparable to those of natural reefs. Nevertheless, other research highlights the limitations of such metrics and stresses the importance of evaluating community composition and ecosystem functions rather than focusing solely on abundance (Bracho-Villavicencio et al. 2023; Paxton et al., 2024). Without this broader perspective, biodiversity gains may be artificially inflated as a result of attraction effects or edge phenomena. Despite the emphasis on fisheries enhancement and biomass increases, systematic assessments of the wider ecological impacts of artificial reefs remain scarce (Higgins et al. 2022; Komyakova et al. 2022). Notably, these structures can alter local hydrodynamics, influence sedimentation patterns, and selectively favor certain functional groups, with cascading consequences for food web dynamics (Dafforn et al. 2015; Vivier et al. 2021). To determine whether artificial reefs deliver genuine conservation outcomes beyond short-term increases in target species, the establishment of long-term monitoring is essential. In line with this need, international guidelines recommend the definition of environmental baselines, explicit conservation targets, and the implementation of short-, medium-, and long-term monitoring programs (Higgins et al. 2022). Results of long-term monitoring can be linked to management objectives, encompassing evaluations of water quality, biofouling, sediment cycling, and potential biological risks (Lemasson et al. 2024).

Simultaneously, artificial reefs can provide habitat and disrupt the environment (Komyakova et al. 2022). While they can serve as refuges and expand feeding opportunities for marine organisms, they may also facilitate the spread of invasive species and alter competitive balances within ecosystems (Sheehy and Vik 2010). Artificial structures, by virtue of their physical differences from natural habitats, may serve as "stepping stones" that connect otherwise isolated populations of exotic species, thereby expanding a species' geographic range and disrupting native communities (Lima and Zalmon 2020). The spread of sun corals (*Tubastraea* spp.) in the South Atlantic and Red Lionfish (*Pterois volitans*) in the Caribbean are emblematic cases demonstrating how well-intended but poorly planned artificial reefs can inadvertently favor invasive species expansion by altering trophic dynamics, outcompeting native taxa, and undermining ecosystem integrity (Sheehy and Vik 2010). Castro et al. (2021) indicated that the increase in artificial reefs through the sinking of ships has resulted in negative consequences to the conservation of marine ecosystems. They also indicated that early detection of these consequences could be dampened through early detection of the spread and impact of invasive species.

An additional emerging issue is the capacity of artificial structures to retain marine debris, particularly macro- and microplastics (Da Costa et al. 2023). Due to their physical configuration and positioning within the water column, artificial reefs can become retention points for solid waste originating from coastal activities, including fishing operations and maritime traffic (Dafforn et al. 2015). This unintended function not only diminishes the aesthetic and recreational value of reef environments but also introduces serious ecological risks. The ingestion of plastic debris by marine organisms may result in direct mortality or sublethal effects, such as metabolic and reproductive dysfunctions (Firth et al. 2016; Salaün et al. 2023; Da Costa et al. 2023). As such, management strategies, including the design of reef structures that minimize debris retention and the implementation of ongoing cleaning and monitoring programs, are essential to ensuring the long-term sustainability of artificial reef initiatives.

Beyond their ecological impacts, artificial reef deployments can also have significant social and economic implications (Chong et al. 2023). The presence of reefs can increase fisheries, tourism, and coastal protection but may simultaneously exacerbate conflicts among stakeholders competing for limited marine space (Elrick-Barr et al. 2022; Sapehee et al. 2025). In some cases, reef deployment increases capture efficiency. This can lead to declines in target species abundance and intensify pre-existing tensions between artisanal fishers and tourism operators, or between traditional communities and large-scale maritime infrastructure developments (Ramos et al. 2011; Islam et al. 2014). These conflicts highlight the need for participatory approaches during the planning stages of reef projects. Proper planning should include preliminary socioeconomic assessments and broad stakeholder consultations to ensure the recognition of the rights, needs, and expectations of affected groups (Lima et al. 2018). While numerous studies have examined the biological benefits of artificial reefs, the assessment of their social outcomes and potential drawbacks remains limited. It is therefore essential that artificial reef deployments should be perceived not only as management options for fisheries enhancement but also as social assets that provide tangible benefits to coastal communities (Chong et al. 2023; Ramos 2024; Knights et al. 2024).

Emerging research has begun to reframe artificial reefs as multifunctional, socio-ecological infrastructures rather than single-purpose, fish-aggregating devices (Komyakova et al. 2022). This perspective advances typologies that explicitly link ecological targets (e.g., complexity, connectivity, functional diversity) with engineering properties, including stability, maintainability, and modularity (McLean et al. 2022). A new conceptual structure proposes a stepwise, objective-driven design that connects siting, morphology, and material selection to measurable outcomes. This system should also introduce decision-support criteria to compare alternatives under local environmental constraints (Pan et al. 2022). To operationalize these concepts, new artificial reef typologies emphasize complexity thresholds tuned to focal taxa and ecological processes while encouraging adaptive, modular reef units whose structural properties can be adjusted to meet ecological performance criteria over time (Bugnot et al. 2021).

At the governance level, international guidance indicates how to integrate permitting, cleaning, and monitoring requirements into artificial reef classification and lifecycle planning (Ramm et al. 2021). These measures distinguish legitimate habitat creation from waste disposal practices and establish reef typologies to be in compliance with long-term stewardship (Bugnot et al. 2021). Adaptive management, coupled with the establishment of co-management forums, offers essential strategies for mitigating conflicts and ensuring equitable and sustainable use of coastal spaces (Knights et al. 2024). Socioeconomic evaluations, however, remain underdeveloped, and critical gaps persist in assessing the full spectrum of socioeconomic benefits with drawbacks associated with reef deployments (Sapehee et al. 2025). Addressing these gaps is fundamental to ensuring that artificial reefs are recognized not only for their biological contributions but also as social assets that deliver tangible benefits to coastal communities (Ramos et al. 2007).

The global proliferation of artificial reef projects further underscores the need for coherent legal frameworks and robust governance mechanisms. International conventions, such as the London Convention and Protocol/UNEP (2009), have established guidelines to prevent the misuse of waste materials in reef construction, yet national policies often remain fragmented or weakly enforced (Ramm et al. 2021). Effective regulation should integrate environmental safeguards, technical standards for design and deployment, and accountability mechanisms to ensure that reefs deliver long-term public goods rather than short-lived or private benefits (Frehse et al. 2025). This paradigm shift increasingly positions artificial reefs as experimental platforms for habitat restoration, blue carbon strategies, and coastal resilience initiatives (Firth et al. 2016; Salaün et al. 2023).

## 13.4 TRENDS AND ADVANCES IN ARTIFICIAL REEF RESEARCH

The field of artificial reef research is currently undergoing notable expansion in deployments and theoretical-methodological sophistication. This expansion reflects the growing recognition of artificial reefs as strategic instruments for marine conservation, sustainable fisheries management, and coastal adaptation to environmental change. It also reflects the academic consolidation of a domain that progressively integrates ecological, technological, and socioeconomic knowledge in a critical and interdisciplinary manner. In this context, an analysis of contemporary trends reveals consistent developments in four major areas: the growth of scientific output incorporating socioeconomic methods (Increasing Approaches), the adoption of new methodologies and analyses (Different Approaches), the reformulation of investigative frameworks (Last Approaches), and the development of projects based on original designs and innovative materials (Original Approaches).

### 13.4.1 Increasing Approaches

Research on artificial reefs has expanded beyond ecological monitoring to systematically incorporate socioeconomic and participatory perspectives, integrating local stakeholder knowledge into artificial reef planning and management (Elrick-Barr et al. 2022; Chong et al. 2023). Studies such as those by Lima and Zalmon (2022) and Costa et al. (2024) illustrate how traditional ecological knowledge of artisanal fishers is increasingly valued as a legitimate source for interpreting colonization processes, recruitment, and resource dynamics. This trend is also evident in research from Ramos et al. (2021) in Portugal and Reye and García (2022) in Mexico, which highlights the role of artificial reefs in promoting local development and strengthening participatory management in coastal areas. Beyond fisheries, new lines of research explore the relevance of artificial reefs for coastal recreation, thus expanding reef multifunctionality (Rani et al. 2025). This is exemplified by Belhassen et al. (2017) with studies on artificial coral reefs and Ramos et al. (2024) with an important review of the tourism industry.

The primary contribution of these expanded approaches lies in their ability to capture multiple dimensions of ecological and socioeconomic success. These successes range from functional biodiversity to tangible benefits for coastal communities and providing robust decision-alternatives in adaptive management programs (Taormina et al. 2022; Knights et al. 2024). However, the full realization of the socioeconomic benefits of artificial reefs requires strengthening specific public policies and developing participatory governance models. Recent literature emphasizes that the absence of a clear regulatory system, marine spatial planning instruments, and formal co-management mechanisms significantly limits the contribution of artificial reefs to the sustainable development of coastal regions (Evans et al. 2017; Dafforn et al. 2015; Sapehee et al. 2025).

Successful examples of artificial reef deployments demonstrate that integrating fishers, tourism operators, environmental authorities, and researchers in collective decision-making forums not only increases the social legitimacy of artificial reef projects but also enhances their long-term ecological and socioeconomic efficiency (Reye and García 2022; Elrick-Barr et al. 2022). Oppositely, addressing potential negative impacts is essential for advancing understanding in this field. A recent socioecological analysis by Chong et al. (2023) shows that the semi-permanent nature and apparent popularity of the artificial reefs with stakeholders are indicative of their potential to function as "social-ecological traps" through negative feedback. This negative feedback should be accounted for when considering reef deployments in fishery management.

### 13.4.2 Different Approaches

This contemporary trend naturally leads to methodological and analytical strategies that overcome the limitations of previous studies. For example, the role of nutrient cycling in influencing reef productivity was examined on the southern coast of Portugal by Falcão et al. (2007) and indicated that there were marked chemical changes in the ecosystem during the two-year study period. Their results reinforced the importance of these artificial structures in rehabilitating sandy coastal areas through trophic chain enhancement. According to Claisse et al. (2014), secondary (i.e., heterotrophic or animal) production is a main pathway of energy flow through an ecosystem as it makes energy available to consumers, including humans. Its estimation can play a valuable role in the examination of linkages between ecosystem functions and services.

In this attraction/production context, Smith et al. (2016) made a pivotal contribution by introducing metrics such as "local production," "new production," and "biomass flow" to assess the ecological functionality of artificial reefs. These metrics revealed that while these habitats can be highly productive, they do not always lead to a net increase in fisheries production. In response, other integrative methodologies have been developed. Da Costa et al. (2022), for example, assessed the reproductive contribution of artificial reefs based on age classes, gonadal stages, and larval presence, thus moving beyond the reductive view of artificial reefs as mere adult aggregation zones.

This broader strategy underscores the importance of understanding not only abundance but also the population dynamics and life cycles of associated species.

Layman and Allgeier (2020) proposed that the measurement of enhanced primary productivity (or other accelerated ecosystem processes) is an underutilized metric to evaluate an artificial reef deployment as part of a broader coastal management program. Moreover, the growing incorporation of high-resolution monitoring technologies is revolutionizing the understanding of the ecological processes associated with artificial reefs. Passive acoustic monitoring, as demonstrated by Leone and Warren (2023) in Long Island (New York, USA), has made it possible to identify spatial and temporal patterns of artificial reef association by fish, marine mammals, and even humans (e.g., recreational activities), thus providing a dynamic perspective on organism–habitat interactions. Simultaneously, the characterization of artificial-reef-associated microbiota, as investigated by Chen et al. (2024a), has revealed that specific microbial communities perform essential ecological functions, influencing processes such as nutrient bioaccumulation, biogeochemical cycling, and pathogen resistance.

The research and monitoring efforts associated with artificial reefs have also been transformed via the integration of emerging technologies that enable continuous, precise, and less invasive ecological assessments. The use of underwater drones, autonomous underwater vehicles (AUVs), and passive acoustic monitoring systems has become increasingly common. These systems allow real-time tracking of reef use patterns by various species, benthic cover variation, and subtle trophic flows (Miyajima-Taga et al. 2021; Leone and Warren 2023).

Complementing these methodological advances in sampling, studies by Rocha et al. (2024) indicate that even subtle changes in construction variables (i.e., surface roughness, microhabitat heterogeneity, and hydrodynamic design) can influence the ecological effectiveness of artificial reefs (Werner et al. 2024; Pardo et al. 2025). These findings reinforce the need to incorporate ecological engineering principles into the planning and design processes to enhance the functional compatibility of artificial reefs with local environmental dynamics.

Nevertheless, recognizing the benefits of artificial reefs must be balanced with a critical analysis of the ecological risks associated with their deployment. For example, it is possible that artificial reefs may act as ecological traps to non-native species (Sheehy and Vik 2010). Furthermore, the accumulation of macro- and microplastics in artificial reefs, as evidenced by Da Costa et al. (2023), presents a growing threat as debris can degrade habitat quality and introduce contaminants into marine food webs. Taormina et al. (2022) indicated the lack of reliable methods concerning some biological ecosystem components impacted by artificial structures. In this context, evaluations may require the modification or creation of indicators to achieve a more meaningful characterization of the ecological disturbances generated by these structures.

### 13.4.3 Last Approaches

Contemporary research increasingly prioritizes methods that integrate structural, functional, and socioeconomic analyses to comprehensively understand the trends in the application of artificial reefs over time. Recent studies have shown that artificial reefs, rather than merely aggregating biomass, can transform trophic structures and modulate ecosystem functions. Paxton et al. (2020) demonstrated that artificial habitats sustain high densities of large reef predators. Ostensibly, artificial habitats provide suitable substrates and shelter, thereby promoting substantial changes in local trophic networks and reinforcing the role of artificial reefs as dynamic components of marine ecology. Similarly, Salaün et al. (2023) and Wang et al. (2024), through trophic modeling, revealed that artificial reefs affect not only the species composition of fish assemblages but also the energy efficiency and structural resilience of ecosystems in the face of environmental disturbances. Similarly, Lima et al. (2020b), in a two-decade study on the southeastern Brazilian coast, found that progressive structural complexity and strengthened trophic networks are key factors in the ecological maturation of artificial reefs. These studies (i.e., Lima et al. 2020b, Salaün et al. 2023, and Wang et al. 2024) proposed a different maturity index as a new indicator to assess the maturation of artificial reefs.

Equally important is the development of multimetric indices such as the Artificial Reef Multimetric Indices (ARMIr and ARMIe), which synergistically integrate ecological and socioeconomic indicators into a composite metric, thereby enabling holistic performance evaluations of artificial reefs (Lima et al. 2020a). Parallel metrics have also been proposed to critically address the classical "attraction versus production" dilemma, offering more precise insights into the actual biomass enhancement effects of artificial reefs (Smith et al. 2016).

In parallel, there is growing concern regarding the establishment of quantitative parameters to guide artificial reef design and assessment. The concept of Complexity Target Points (CTPs) has been proposed to correlate specific structural features such as rugosity, spatial heterogeneity, and cavity volume with biodiversity metrics and ecosystem functionality (Salaün et al. 2023).

### 13.4.4 New and Original Approaches

Methodological trends in artificial reef research and management are unequivocally manifested in the systematic incorporation of new technology and monitoring methods. Optimized image-based protocols for macro-epibenthic monitoring (Taormina et al. 2020) and the use of underwater drones combined with eDNA analysis (Miyajima-Taga et al. 2024) have revolutionized biodiversity monitoring practices on artificial reefs. These techniques enable more comprehensive and sensitive assessments that are less invasive. These assessment attributes are crucial for application in habitats that are difficult to access or subject to collection restrictions. Moreover, these methods enhance the detection of cryptic, benthic species and subtle ecological dynamics.

Additionally, the expanded understanding of the ecological benefits of artificial reefs includes explicit consideration of the ecosystem services they provide. Beyond the traditional function of enhancing fish biomass, there is growing recognition of the potential for artificial reefs to mitigate climate change, particularly through carbon sequestration (Frehse et al. 2025). Recent studies demonstrate that artificial substrates, when integrated with restored natural habitats such as seagrass beds and oyster reefs, facilitate organic carbon accumulation in sediments, thereby contributing to a reduction in atmospheric $CO_2$ concentrations (Oanta et al. 2024). Including this service in artificial reef evaluation protocols enhances their strategic relevance within the "blue" economy and facilitates achieving global carbon neutrality goals.

Managing urban ocean life as natural infrastructure can offer an opportunity, particularly when resources dedicated to artificial infrastructure are redirected to conservation and restoration projects, while simultaneously reducing incentives that could potentially destroy coastal habitats and raise risks and costs to society (McCreless and Beck 2017; Bracho-Villavicencio et al. 2023). According to Airoldi et al. (2021), several of these "management" options have the potential to help bring nature back to the urban ocean while also addressing some of the critical needs of urban societies, such as climate adaptation, seafood production, clean water, and recreation, providing both human and environmental benefits in some of our most impacted ocean spaces (Frehse et al. 2025).

Amid this expanded functionality, innovations based on NbS have also emerged to encompass innovative designs for anchoring systems, coastal erosion control infrastructure, and coral reef restoration. Airoldi et al. (2021) described an array of evidence-based methods that include "greening" of gray infrastructure, restoring habitats, and developing biotechnologies. These strategies show that nature-based options can be more cost-effective and deliver better contingent benefits than conventional engineering solutions.

Also, to increase the resilience of coastal cities to climate-change effects, the implementation of intertidal artificial structures (i.e., textured, perforated panels installed on urban seawalls) can significantly increase species abundance and diversity, particularly through the formation of tide pools that maintain cooler temperatures (Bishop et al. 2022). These interventions promote favorable microhabitats in urbanized environments, reinforcing the role of artificial reefs as hybrid infrastructures that combine ecological conservation and coastal protection (Bugnot et al. 2021).

Parallel to this, research on advanced materials for artificial reefs has made remarkable progress, with increasing emphasis on sustainable and biodegradable options. Additive manufacturing technologies have been employed for ecological projects, especially those connected with artificial reefs, and this technology can effectively support ecological design and engineering, especially in restoration efforts (Korniejenko et al. 2025). Also, marine-grade geopolymers, organic-based composites, and modified eco-concretes are being developed to reduce the environmental impact of artificial reef construction while enhancing their functional integration with natural ecosystems. Research on artificial reefs for marine biodiversity protection has adopted the use of geopolymer binders due to their reduced carbon emissions and better performance in the marine environment (Zaidi et al. 2019; Ly et al. 2020; Rahman and Al-Ameri 2022). Also, novel methods include the use of 3D printing and the investigation of new morphologies to form marine habitats using additive manufacturing with eco-friendly materials to support a diversity of coral reef residents (Berman et al. 2023).

These eco-friendly materials not only emit less carbon dioxide during production but also exhibit greater chemical compatibility with marine biota, facilitating early colonization and the development of more resilient ecological communities. Such innovations align with initiatives to construct ecologically enhanced artificial reefs, such as the "Green Artificial Reef" proposed by Huang et al. (2016) and Carral et al. (2018, 2023). Similarly, experimental research conducted under laboratory conditions by Ronglan et al. (2024) shows the energy-dissipating capabilities of architected structures. These structures illustrate the potential growing, multifunctional use of artificial reefs in that they might act not only to enhance biodiversity but also as potential natural barriers against coastal erosion and extreme weather events. The adoption of standard monitoring units for the recruitment of reef fishes (i.e., SMURFS) to test different substrates to facilitate juvenile fish recruitment (Salinas and Cintra-Buenrostro 2024), along with urban intertidal "living wall" projects (Bishop et al. 2022) reaffirm that structural heterogeneity, ecological complexity, and material sustainability are central principles guiding the new generation of artificial reefs.

## 13.5 FUTURE PERSPECTIVES FOR THE DEVELOPMENT AND MANAGEMENT OF ARTIFICIAL REEFS

The future of artificial reefs extends far beyond their traditional role in enhancing fish stocks. They are now becoming critical infrastructure for marine conservation, coastal protection, and the socioeconomic empowerment of coastal communities (Knights et al. 2024). This paradigm shift reflects conceptual trends in the field. Additionally, it reflects the incorporation of emerging demands driven by climate change, global environmental degradation, and the urgent need for integrated and sustainable solutions for marine and coastal systems. These demands may be met by including the application of artificial intelligence for predictive modeling of ecological dynamics associated with artificial reefs.

One of the most prominent trends is the transition from a focus on fisheries production to a broader ecological vision, whereby artificial reefs are designed to support fundamental ecological processes. Multifunctionality is emerging as a guiding principle to simultaneously restore degraded habitats, promote ecological connectivity, enhance biodiversity, and increase ecosystem resilience to disturbances (Salaün et al. 2023; Paxton et al. 2020). Thus, artificial reefs are increasingly integrated into NbS strategies, in conjunction with natural habitats such as mangrove forests, seagrass beds, and oyster reefs to maximize the delivery of ecosystem services and mitigate the impacts of climate change (Carral et al. 2021, 2023; Ronglan et al. 2024).

Innovations in engineering and materials represent another promising avenue for the future application of artificial reefs. The use of bioactive substrates, modified bioconcrete, and biodegradable composites can be coupled with 3D printing technologies to construct structures with optimized microhabitats. This has expanded the capacity of artificial reefs to support larval recruitment, ecological succession, and functional stability (Wan et al. 2024; Chen et al. 2024b; Giglio et al. 2024). This technological innovation contributes not only to improved ecological performance but also to a reduction in the environmental impacts associated with the longevity of deployed structures.

In terms of monitoring and evaluation, there is a growing need to move beyond traditional metrics based solely on species abundance, biomass, or richness. Future assessments of artificial reefs must incorporate indicators of functional diversity, trophic stability, ecosystem services, and effectiveness in promoting ecological resilience (Smith et al. 2017; Da Costa et al. 2022). Technologies such as AUVs, multispectral sensors, passive acoustic monitoring, and eDNA analyses will be essential for enabling more comprehensive, continuous, and adaptive evaluations (Miyajima-Taga et al. 2024; Ramos et al. 2024).

As integral components of the "blue economy" agenda, artificial reef deployments are expected to increasingly facilitate sustainable activities such as ecological tourism, low-impact artisanal fisheries, and environmental education (Frehse et al. 2025). For example, marine ranching uses a variety of artificial reefs and habitat structures made from diverse materials and designs to support mariculture. Moreover, multifunctional artificial infrastructure provides new habitats that could be exploited for aquaculture, recreational, or restoration purposes. Airoldi et al. (2021), in an extensive review for a more sustainable urban ocean, state that multifunctional artificial reefs can merge coastal protection with recreational benefits (e.g., surfing) (ten Voorde et al. 2009). This includes outfitting floating bridges and breakwaters with aquaculture cages (Park et al. 2004) and wave energy converters (Zhao et al. 2019).

Additionally, the symbolic and cultural value of artificial reefs as local heritage should be further recognized in public policy and coastal management programs (Islam et al. 2014; Reye and García 2022). Participatory co-management incorporates local communities, economic sectors, and institutional stakeholders throughout the planning, implementation, and monitoring phases. This management process is crucial to ensure legitimacy, equity, and sustainability (Elrick-Barr et al. 2022; Chong et al. 2023; Knights et al. 2024).

Despite these positive trends, several critical information gaps remain before ensuring the future incorporation of artificial reefs into the management process. Foremost among these is the scarcity of long-term studies capable of determining cumulative effects such as changes in community functional composition, the role of artificial reefs as connectivity corridors, and their actual efficacy in ecological restoration (Lima and Zalmon 2022; McLean et al. 2022). The lack of standardized international methodologies for comparative evaluation of artificial reefs further hampers the synthesis of findings and the development of best practice guidelines (Taormina et al. 2022). The problems connected with potential "ocean sprawl" and the spread of invasive species should be taken into consideration when planning the research (Korniejenko et al. 2025).

Another pressing challenge involves the need to assess and mitigate emerging risks, including the facilitation of invasive species dispersal, the accumulation of macro- and microplastics, and the potential for artificial reefs to act as ecological traps (Paxton et al. 2024). Consequently, artificial reefs may attract organisms to habitats that subsequently fail to provide suitable conditions for their survival and reproduction (Bull and Love 2019; Da Costa et al. 2023). Risk assessment must therefore be integrated from the earliest stages of project planning and link to continuous and adaptive monitoring programs.

Furthermore, artificial reef governance must move toward more inclusive and adaptive models that incorporate principles of socio-environmental justice. The effective integration of local ecological knowledge into the design and management of artificial reefs remains limited and should be strengthened to ensure greater social and ecological project alignment (Lima et al. 2019a; Ramos 2024). Specific public policies that embed the deployment and management of artificial reefs into marine spatial planning and national conservation strategies are also urgently needed (Frehse et al.

2025). Finally, future artificial reef initiatives must be conceived not merely at the local level but planned at regional and global scales as integrated components of conservation networks, tailored to species migration dynamics, climate change scenarios, and the socioeconomic pressures on coastal and marine systems.

## 13.6 CONCLUSION

Overall, this chapter presents an up-to-date synthesis of artificial reef research, highlighting the conceptual evolution of artificial reefs from mere structures intended primarily to enhance fisheries to multifunctional ecosystem components aimed at marine conservation, sustainable resource management, and coastal resilience. The integration of ecological, technological, and socioeconomic perspectives into artificial reef development reinforces their role as strategic infrastructure for biodiversity conservation, fisheries and tourism development, and coastal ecosystem protection.

Moreover, we highlight that the future of artificial reef socioeconomic and management issues lies in the adoption of interdisciplinary approaches that embrace community participation, adaptive governance, and ecosystem-based management principles. The studies discussed herein advance the critical understanding of the challenges and opportunities associated with artificial reef deployment and operation across diverse environmental and cultural contexts. Ultimately, our conclusions potentially offer a valuable contribution to the scientific and practical plans guiding the future development and management of artificial reefs by emphasizing that project success depends on balancing ecological functionality, socioeconomic benefits, and cultural integration, thereby ensuring the long-term sustainability and resilience of coastal and marine ecosystems.

## 13.7 ACKNOWLEDGMENT

The authors thank the Research Foundation of Rio de Janeiro (FAPERJ) and the Coordination for the Improvement of Higher Education Personnel (CAPES) for the PhD scholarship and the internship abroad scholarship, respectively. This research was supported by FAPERJ (E-26/203.909/2024) and by the Brazilian National Council for Agency for Scientific and Technological Development-CNPq (301505/2022-5). This study was also partially financed by the CAPES (finance code 001).

## REFERENCES

Airoldi L, Beck MW, Firth LB, Bugnot AB, Steinberg PD, Dafforn KA. 2021. Emerging solutions to return nature to the urban ocean. *Annu Rev Mar Sci*. 13:445–477. https://doi.org/10.1146/annurev-marine-032020-020015.

Ajemian MJ, Wetz JJ, Shipley-Lozano B, Shively JD, Stunz GW. 2015. An analysis of artificial reef fish community structure along the northwestern Gulf of Mexico shelf: Potential impacts of "Rigs-to-Reefs" programs. *PLOS ONE*. 10(5):e0126354. https://doi.org/10.1371/journal.pone.0126354.

Aleksandrov BG, Minicheva GG, Strikalenko TV. 2002. Ecological aspects of artificial reef construction using scrap tires. *Russ J Mar Biol*. 28(2):120–126.

Arlinghaus R, Tillner R, Bork M. 2015. Explaining participation rates in recreational fishing across industrialised countries. *Fish Manag Ecol*. 22(1):45–55.

Baine M. 2001. Artificial reefs: A review of their design, application, management and performance. *Ocean Coastal Manage*. 44(3–4):241–259. https://doi.org/10.1016/S0964-5691(01)00048-5.

Becker A, Taylor MD, Folpp H, Lowry MB. 2018. Managing the development of artificial reef systems: The need for quantitative goals. *Fish Fish*. 19(4):740–752. https://doi.org/10.1111/faf.12288.

Belhassen Y, Rousseau M, Tynyakov J, Shashar N. 2017. Evaluating the attractiveness and effectiveness of artificial coral reefs as a recreational ecosystem service. *J Environ Manage*. 203:448–456. https://doi.org/10.1016/j.jenvman.2017.08.020.

Berman O, Weizman M, Ore A, Ner R, Parnas H, Shashar N, Tarazi E. 2023. Design and application of a novel 3D printing method for bio-inspired artificial reefs. *Ecol Eng*. 193:106892. https://doi.org/10.1016/j.ecoleng.2023.106892.

Bishop MJ, Vozzo ML, Mayer-Pinto M, Dafforn KA. 2022. Complexity–biodiversity relationships on marine urban structures: Reintroducing habitat heterogeneity through eco-engineering. *Philos Trans R Soc Lond B Biol Sci.* 377(1855):20210393. https://doi.org/10.1098/rstb.2021.0393.
Bohnsack JA, Sutherland DL. 1985. Artificial reef research: A review with recommendations for future priorities. *Bul Mar Sci.* 37(1):11–39.
Bortone SA. 2006. A perspective of artificial reef research: The past, present, and future. *Bull Mar Sci.* 78(1):1–8.
Bortone SA, Brandini FP, Fabi G, Otake S. 2011. *Artificial reefs in fisheries management.* Boca Raton, FL: CRC Press.
Bracho-Villavicencio C, Matthews-Cascon H, Rossi S. 2023. Artificial reefs around the world: A review of the state of the art and a meta-analysis of its effectiveness for the restoration of marine ecosystems. *Environ.* 10:121. https://doi.org/10.3390/environments10070121.
Bugnot AB, Mayer-Pinto M, Airoldi L, Dafforn KA, Johnston EL, Strain EMA, Morris RL, Naylor LA, Bishop MJ, Heery E, Aguilera MA, Cole VJ, Hanley ME. 2021. Current and projected global extent of marine built structures. *Nat Sustain.* 4(1):16–22. https://doi.org/10.1038/s41893-020-00595-1.
Bull AS, Love MS. 2019. Worldwide oil and gas platform decommissioning: A review of practices and reefing options. *Ocean Coastal Manage.* 168:274–306. https://doi.org/10.1016/j.ocecoaman.2018.10.024.
Burt J, Bartholomew A, Usseglio P, Bauman A, Sale PF. 2009. Are artificial reefs surrogates of natural habitats for corals and fish in Dubai, United Arab Emirates? *Coral Reefs* 28:663–675. https://doi.org/10.1007/s00338-009-0500-1.
Campbell MD, Patino R, Tolan J, Strauss R, Diamond SL. 2010. Sublethal effects of catch-and-release fishing: Measuring capture stress, fish impairment, and predation risk using a condition index. *ICES J Mar Sci.* 67(3):513–521. https://doi.org/10.1093/icesjms/fsp255.
Carral L, Cartelle Barros JJ, Carro Fidalgo H, Camba Fabal C, Munín Doce A. 2021. Greenhouse gas emissions and energy consumption of coastal ecosystem enhancement programme through sustainable artificial reefs in Galicia. *Int J Environ Res Public Health.* 18:1909. https://doi.org/10.3390/ijerph18041909.
Carral L, Rodriguez-Guerreiro M, Tarrio-Saavedra J, Alvarez-Feal JC, Fraguela Formoso J. 2018. Social interest in developing a Green modular artificial reef structure in concrete for the ecosystems of the Galician rías. *J Clean Prod.* 172:1881–1898.
Carral L, Tarrío-Saavedra J, Barros, JJC, Fabal CC, Ramil A, Álvarez-Feal C. 2023. Considerations on the programmed functional life (one generation) of a green artificial reef in terms of the sustainability of the modified ecosystem. *Heliyon.* 9(4):e14978. https://doi.org/10.1016/j.heliyon.2023.e14978.
Castro KL, Battini N, Giachetti CB, Trovant B, Abelando M, Basso NG, Schwindt E. 2021. Early detection of marine invasive species following the deployment of an artificial reef: Integrating tools to assist the decision-making process. *J Environ Manage.* 294:113333. https://doi.org/10.1016/j.jenvman.2021.113333.
Champion C, Suthers IM, Smith JA. 2015. Zooplanktivory is a key process for fish production on a coastal artificial reef. *Mar Ecol Prog Ser.* 541:1–14. https://doi.org/10.3354/meps11529.
Chen B, Wei Y, Yu K, Liang Y, Yu X, Liao Z, Bao Z. 2024a. The microbiome dynamics and interaction of endosymbiotic Symbiodiniaceae and fungi are associated with thermal bleaching susceptibility of coral holobionts. *Applied and Environmental Microbiology.* 90(4):e01939–23. https://doi.org/10.1128/aem.01939-23.
Chen JL, Chuang CT, Jan RQ, Liu LC, Jan MS. 2013. Recreational benefits of ecosystem services on and around artificial reefs: A case study in Penghu, Taiwan. *Ocean Coastal Manage.* 85:58–64. https://doi.org/10.1016/j.ocecoaman.2013.08.006.
Chen X, Che X, Zhou Y, Tian C, Li X. 2024b. A numerical simulation study and effectiveness evaluation on the flow field effect of trapezoidal artificial reefs in different layouts. *J Mar Sci Eng.* 12:3. https://doi.org/10.3390/jmse12010003.
Chong L, Siders ZA, Lorenzen K, Ahrens RNM, Camp EV. 2023. Global synthesis of effects and feedbacks from artificial reefs on socioecological systems in recreational fisheries. *Fish Fish.* 25:303–319. https://doi.org/10.1111/faf.12809.
Claisse JT, Pondella DJ II, Love M, Zahn LA, Williams CM, Williams JP, Bull AS. 2014. Oil platforms off California are among the most productive marine fish habitats globally. *Proc Natl Acad Sci USA.* 111(43):15462–15467. https://doi.org/10.1073/pnas.1411477111.
Costa LL, Lima JS, Costa ID, Zalmon IR. 2024. Can the Caribbean sharpnose shark Rhizoprionodon porosus Poey, 1861 be used as a 'flagship umbrella species' in an artificial reef complex to facilitate management actions and sustainability? *Mar Biol Res.* 20:114–122. https://doi.org/:10.1080/17451000.2024.2330382.
Cresson P, Ruitton S, Ourgaud M, Harmelin-Vivien M. 2014. Contrasting perception of fish trophic level from stomach content and stable isotope analyses: A Mediterranean artificial reef experience. *J Exp Mar Biol Ecol.* 452:54–62. https://doi.org/10.1016/j.jembe.2013.11.014.

Da Costa ID, Costa LL, Zalmon IR. 2023. Are fishes selecting the trash they eat? Influence of feeding mode and habitat on microplastic uptake in an artificial reef complex (ARC). *Sci Total Environ*. 904:166788. https://doi.org/10.1016/j.scitotenv. 2023.166788.

Da Costa ID, Santos JL, Costa LL, Lima JS, Zalmon IR. 2022. Reproductive potential and production role of artificial reefs in southeastern Brazil. *Estuar Coast Shelf Sci*. 258:107710. https://doi.org/10.1016/j.ecss.2021.107710.

Dafforn KA, Glasby TM, Airoldi L, Rivero NK, Mayer-Pinto M, Johnston EL. 2015. Marine urbanization: An ecological framework for designing multifunctional artificial structures. *Front Ecol Environ*. 13(2):82–90. https://doi.org/10.1890/140050.

Davis TR, Smith SDA. 2017. Proximity effects of natural and artificial reef walls on fish assemblages. *Reg Stud Mar Sci*. 9:17–23. https://doi.org/10.1016/j.rsma.2016.11.007.

Delos Reyes M, Martens R. 1995. Low-cost artificial reef program in the Philippines: An evaluation in the management of a tropical coastal ecosystem. *Beitr Paläont*. 20:1–6.

Elrick-Barr E, Zimmerhackel JS, Hill G, Clifton J, Ackermann F, Burton M, Harvey ES. 2022. Man-made structures in the marine environment: A review of stakeholders' social and economic values and perceptions. *Environ Sci Policy*. 129:12–18. https://doi.org/10.1016/j.envsci.2021.12.006.

Evans A, Lawrence PJ, Natanzi AS, Brooks P, Roughan D, Barrett N, Doherty P, Buggy M, McNally C. 2021. Replicating natural topography on marine artificial structures—A novel approach to eco-engineering. *Ecol Eng*. 160:106144. https://doi.org/10.1016/j.ecoleng.2020.106144.

Evans AJ, Garrod B, Firth LB, Hawkins SJ, Morris-Webb ES, Goudge H, Moore PJ. 2017. Stakeholder priorities for multi-functional coastal defense developments and steps to effective implementation. *Mar Policy*. 75:143–155. https://doi.org/10.1016/j.marpol.2016.10.006.

Fabi G, Luccarini F, Panfili M, Solustri C, Spagnolo A. 2002. Effects of an artificial reef on the surrounding soft-bottom community (central Adriatic Sea). *ICES J Mar Sci*. 68(5):916–927. https://doi.org/10.1093/icesjms/fsr043.

Fabi G, Spagnolo A, Bellan Santini D, Charbonnel E, Çiçek BA, García JG, Jensen AC, Kallianiotis A, Santos MN. 2011. Overview on artificial reefs in Europe. *Braz J Oceanogr*. 59(Special Issue 1):155–166. https://doi.org/10.1590/S1679-87592011000500017.

Falcão M, Santos MN, Vicente M, Monteiro CC. 2007. Biogeochemical processes and nutrient cycling within an artificial reef off southern Portugal. *Mar Environ Res*. 63:429–444. https://doi.org/10.1016/j.marenvres.2006.12.001.

Firth LB, Knights AM, Bridger D, Evans AJ, Mieszkowska N, Moore PJ, O'Connor NE, Sheehan EV, Thompson RC, Hawkins SJ. 2016. *Ocean sprawl: Challenges and opportunities for biodiversity management in a changing world*. Boca Raton, FL: CRC Press. https://doi.org/10.1201/9781315368597-5.

Frehse FA, Derviche P, Pereira FW, Hostim-Silva M, Vitule JRS. 2025. Artificial aquatic habitats: A systematic literature review and new perspectives. *Hydrobiologia*. 852:1997–2012. https://doi.org/10.1007/s10750-024-05385-0.

Giglio VJ, Aued AW, Cordeiro CAMM, Eggertsen L, Ferrari DS, Gonçalves LR, Hanazaki N, Luiz OJ, Luza AL, Mendes TC, Pinheiro HT, Segal B, Waecher LS, Bender MG. 2024. A global systematic literature review of ecosystem services in reef environments. *Ecosyst Serv*. 73:634–645. https://doi.org/10.1007/s00267-023-01912-y.

Harmelin J-G. 2000. Mediterranean marine protected areas: Some prominent traits and promising trends. *Environ Conserv*. 27(2):104–105. https://doi.org/10.1017/S037689290000014X.

Hickling S, Murphy J, Grasselli F. 2022. Enhanced nektonic communities and juvenile fish biodiversity around Reef Cubes®: Evidence from a Before–After–Gradient study in Torbay, UK. *Mar Environ Res*. 183:105742. https://doi.org/10.1016/j.marenvres.2022.105742.

Higgins E, Metaxas A, Scheibling RE. 2022. A systematic review of artificial reefs as platforms for coral reef research and conservation. *PLoS One*. 17(1):e0261964. https://doi.org/10.1371/journal.pone.0261964.

Huang X, Wang Z, Liu Y, Hu W, Ni W. 2016. On the use of blast furnace slag and steel slag in the preparation of green artificial reef concrete. *Constr Build Mater*. 112:241–246. https://doi.org/10.1016/j.conbuildmat.2016.02.088.

Hughes RM. 2015. Recreational fisheries in the USA: Economics, management strategies, and ecological threats. *Fish Sci*. 81(1):1–9. https://doi.org/10.1007/s12562-014-0815-x.

Hylkema A, Hakkaart QCA, Reid CB, Osinga R, Murk AJ, Debrot AO. 2021. Artificial reefs in the Caribbean: A need for comprehensive monitoring and integration into marine management plans. *Ocean Coast Manage*. 209:105672. https://doi.org/10.1016/j.ocecoaman.2021.105672.

Islam GMN, Noh KM, Sidique SF, Noh AFM, Ali A. 2014. Economic impacts of artificial reefs on small-scale fishers in Peninsular Malaysia. *Hum Ecol*. 42(6):989–998. https://doi.org/10.1007/s10745-014-9692-2.

Jani JM. 2020. Artificial reefs in Setiu: At a crossroad between tradition and modernity. *Malays J Soc Space*. 16(3):97–107. https://doi.org/10.17576/geo-2020-1603-08.
Kim D, Woo J, Yoon HS, Na WB. 2014. Wake lengths and structural responses of Korean general artificial reefs. *Ocean Eng*. 92:83–91. https://doi.org/10.1016/j.oceaneng.2014.09.040.
Knights AM, Lemasson AJ, Firth LB, Bond T, Claisse J, Coolen JWP, Copping A, Dannheim J, De Dominicis M, Degraer S, Elliott M, Fernandes PG, Fowler AM, Frost M, Henry L-A, Hicks N, Hyder K, Jagerroos S, Jones DOB, Love M,. . . Somerfield PJ. 2024. Developing expert scientific consensus on the environmental and societal effects of marine artificial structures prior to decommissioning. *J Environ Manage*. 336:119897. https://doi.org/10.1016/j.jenvman.2023.119897.
Komyakova V, Jaffrés JBD, Strain EMA, Cullen-Knox C, Fudge M, Langhamer O, Bender A, Yaakub SM, Wilson E, Allan BJM, Sella I, Haward M. 2022. Conceptualisation of multiple impacts interacting in the marine environment using marine infrastructure as an example. *Sci Total Environ*. 820:154748. https://doi.org/10.1016/j.scitotenv.2022.154748.
Korniejenko K, Oliwa K, Gądek S, Dynowski P, Źróbek A, Lin W-T. 2025. A review of additive manufacturing techniques in artificial reef construction: Materials, processes, and ecological impact. *Appl Sci*. 15(8):4216. https://doi.org/10.3390/app15084216.
Latun AR, Abdullah MP. 1990. *Artificial reefs in Malaysia: A country review paper*. Department of Fisheries Malaysia. 13 p.
Layman CA, Allgeier JE. 2020. An ecosystem ecology perspective on artificial reef production. *J Appl Ecol*. 57:2131–2141. https://doi.org/10.1111/1365-2664.13748.
Lechanteur, YARG. and Griffiths, CL. 2001. Fish associated with artificial reefs in False Bay, South Africa: A preliminary survey. *Afr. Zool*. 36(1):87–93. https://doi.org/10.1080/15627020.2001.11657118.
Lee MO, Otake S, Kim JK. 2018. Transition of artificial reefs (ARs) research and its prospects. *Ocean Coastal Manage*. 154:55–65. https://doi.org/10.1016/j.ocecoaman.2018.01.010.
Lemasson AJ, Somerfield PJ, Schratzberger M, McNeill CL, Nunes J, Pascoe C, Watson SCL, Thompson M, Couce E, Knights AM. 2024. A global meta-analysis of ecological effects from offshore marine artificial structures. *Nat Sustain*. 7:485–495. https://doi.org/10.1038/s41893-024-01311-z.
Leone M, Warren JB. 2023. Passive acoustic monitoring of artificial reefs south of Long Island, New York provides novel insights on fish, marine mammal, and post-pandemic human activity occurring in nearshore coastal habitats. *J Acoust Soc Am*. 154(4): A215. https://doi.org/10.1121/10.0023325.
Levy, N, Marques, JA, Simon-Blecher, N, Bourne, DG, Doniger, T, Benichou, JIC, Levy, O. 2024. Ecosystem transplant from a healthy reef boosts coral health at a degraded reef. *Nat Commun* 15:10033. https://doi.org/10.1038/s41467-024-54149-6.
Lima JS, Atalah J, Sanchez-Jerez P, Zalmon IR. 2020a. Evaluating the performance and management of artificial reefs using artificial reef multimetric index (ARMI). *Ocean Coastal Manage*. 198:105350. https://doi.org/10.1016/j.ocecoaman.2020.105350.
Lima JS, Sanchez-Jerez P, dos Santos LN, Zalmon IR. 2020b. Could artificial reefs increase access to estuarine fishery resources? Insights from a long-term assessment. *Estuar Coast Shelf Sci*. 242:106858. https://doi.org/10.1016/j.ecss.2020.106858.
Lima JS, Zalmon IR. 2020. Trends and obstacles in artificial reef research. In: Bortone AS, Otake S, editors. *Modern fisheries engineering: Realizing a healthy and sustainable marine ecosystem*. Boca Raton, FL: CRC Press; p. 11–30. https://doi.org/10.1201/9780429328039.
Lima JS, Zalmon IR. 2022. Impacts of artificial reefs on artisanal fishers: Southeastern Brazil. In: Ramos JHP, editor. *Impacts of artificial reefs on the Environment and Communities*. Hershey, PA: IGI Global; p. 87–109. https://doi.org/10.4018/978-1-6684-2344-8.ch005.
Lima JS, Zalmon IR, Love M. 2019a. Overview and trends of ecological and socioeconomic research on artificial reefs. *Mar Environ Res*. 145:81–96. https://doi.org/10.1016/j.marenvres.2019.01.010.
Lima JS, Zappes CA, Di Beneditto APM, Zalmon IR. 2018. Artisanal fisheries and artificial reefs on the southeast coast of Brazil: Contributions to research and management. *Ocean Coastal Manage*. 163:363–375. https://doi.org/10.1016/j.ocecoaman.2018.07.018.
Lima JS, Zappes CA, Zalmon IR. 2019b. Ethnoecology and socioeconomic aspects around an artificial reef: The case of artisanal fisheries from southeastern Brazil. *Biota Neotrop*. 19(3): e20180620. https://doi.org/10.1590/1676-0611-bn-2018-0620.
Lök A, Metin C, Ulaş A, Düzbastılar FO, Tokaç A. 2002. Artificial reefs in Turkey. *ICES J Mar Sci*. 59(suppl):S192–S195. https://doi.org/10.1006/jmsc.2002.1221.
London Convention and Protocol/UNEP. 2009. *London Convention and Protocol/UNEP: Guidelines for the placement of artificial reefs*. London: International Maritime Organization (IMO) and United Nations Environment Programme (UNEP). 100 p.

Love MS, Claisse JT, Roeper A. 2019. An analysis of the fish assemblages around 23 oil and gas platforms off California with comparisons with natural habitats. *Bull Mar Sci.* 95(4):477–514. https://doi.org/10.5343/bms.2018.0061.

Ly O, Yoris-Nobile AI, Sebaibi N, Blanco-Fernandez E, Boutouil M, Castro-Fresno D, Hall AE, Herbert RJ, Deboucha W, Reis B. 2020. Optimisation of 3D printed concrete for artificial reefs: Biofouling and mechanical analysis. *Constr Build Mater.* 272:121649. https://doi.org/10.1016/j.conbuildmat.2020.121649.

Macusi ED, Abreo NAS, Babaran RP. 2017. Local ecological knowledge (LEK) on fish behavior around anchored FADs: The case of tuna purse seine and ringnet fishers from southern Philippines. *Front Mar Sci.* 4:188. https://doi.org/10.3389/fmars.2017.00188.

Maslov D, Cruz F, Pinheiro M, Miranda T, Valente IB, Ferreira V, Pereira E. 2024. Functional conception of biomimetic artificial reefs using parametric design and modular construction. *J Mar Sci Eng.* 12:1682. https://doi.org/10.3390/jmse12091682.

Miyajima-Taga Y, Minamoto T, Yamanaka H. 2024. Fine-scale spatial distribution of fish community in high-rise artificial reefs investigated using an underwater drone and quantitative environmental DNA metabarcoding. *Mar Ecol Prog Ser.* 740:123–144. https://doi.org/10.3354/meps14631.

McCreless E, Beck MW. 2017. Rethinking our global coastal investment portfolio. *J Ocean Coast Econ.* 3:6.

McLean DL, Ferreira LC, Benthuysen JA, Miller KJ, Schläppy M-L, Ajemian MJ, Berry O, Birchenough SNR, Bond T, et al. 2022. Influence of offshore oil and gas structures on seascape ecological connectivity. *Glob Change Biol.* 28(10):16134. https://doi.org/10.1111/gcb.16134.

Miyajima-Taga Y, Minamoto T, Yamanaka H. 2024. Fine-scale spatial distribution of fish community in high-rise artificial reefs investigated using an underwater drone and quantitative environmental DNA metabarcoding. *Mar Ecol Prog Ser.* 740:123–144. https://doi.org/10.3354/meps14631.

Oanta GA, Arranz, AL, Barros, JJC, Otero, XL, Agras, GD, Couce, LC 2024. The EU's sustainable blue economy: The role of artificial reefs in achieving the climate neutrality goal. In: Carral L, Vega A, Carreño J, de Lara J, Lamas MI, Cartelle JJ, Tarrío J, Carballo R, Townsed P, editors. *Proceedings of the IV Iberoamerican Congress of Naval Engineering and 27th Pan-American Congress of Naval Engineering, Maritime Transportation and Port Engineering (COPINAVAL).* COPINAVAL ICNE 2022 2023. Springer Series on Naval Architecture, Marine Engineering, Shipbuilding and Shipping, vol 17. Springer, Cham. https://doi.org/10.1007/978-3-031-49799-5_61.

Oricchio FT, Pastro G, Vieira EA, Flores AAV, Gibran FZ, Dias GM. 2016. Distinct community dynamics at two artificial habitats in a recreational marina. *Mar Environ Res.* 122:85–92. https://doi.org/10.1016/j.marenvres.2016.09.010.

Pan Y, Tong H, Wei D, Xiao W, Xue D. 2022. Review of structure types and new development prospects of artificial reefs in China. *Front Mar Sci.* 9:853452. https://doi.org/10.3389/fmars.2022.853452.

Pardo JCF, Aune M, Harman C, Walday M, Skjellum SF. 2025. A synthesis review of nature positive approaches and coexistence in the offshore wind industry. *ICES J Mar Sci.* 82(4): fsad191. https://doi.org/10.1093/icesjms/fsad191.

Park RS, Lee MS, Kim SG, Lee SM, An SH, Yang YT. 2004. A study on the performance of floating twin-hull breakwaters including fish-cage. In: *Proceedings of the 14th international offshore polar engineering conference.* Cupertino, CA: Int Soc Offshore Polar Eng; p. 694–700.

Paxton AB, Foxfoot IR, Cutshaw C, Steward DA, Poussard L, Riley TN, Swannack TM, Piercy CD, Altman S, Puckett BJ, Storlazzi CD, Viehman TS. 2024. Evidence on the ecological and physical effects of built structures in shallow, tropical coral reefs: A systematic map. *Environ Evid.* 13:12. https://doi.org/10.1186/s13750-024-00336-3.

Paxton AB, Newton EA, Adler AM, Van Hoeck RV, Iversen Jr. ECT, Peterson CH, Silliman BR. 2020. Artificial habitats host elevated densities of large reef-associated predators. *PLOS ONE.* 15(8):e0237374. https://doi.org/10.1371/journal.pone.0237374.

Perkol-Finkel S, Shashar N, Benayahu Y. 2005. Can artificial reefs mimic natural reef communities? The roles of structural features and age. *Mar Environ Res.* 60(2):211–225. https://doi.org/10.1016/j.marenvres.2005.08.001.

Qin M, Yue C, Du Y. 2020. Evolution of China's marine ranching policy based on the perspective of policy tools. *Mar Policy.* 117:103941. https://doi.org/10.1016/j.marpol.2020.103941.

Rahman SK, Al-Ameri R. 2022. Marine geopolymer concrete—A hybrid curable self-compacting sustainable concrete for marine applications. *Appl Sci.* 12:3116. https://doi.org/3390/app1206311613. https://doi.org/10.54055/ejtr.v38i.3552.

Raja Noordin RO, Cheah EK, Sukarno W, Abdul Mutalib MH, Mobarak H, Raja Bidin RH, Che Omar MH. 1994. Design and construction of artificial reefs in Malaysia. *Bull Mar Sci.* 55:1050–1061.

Ramm LAW, Florisson JH, Watts SL, Becker A, Tweedley JR. 2021. Artificial reefs in the Anthropocene: A review of geographical and historical trends in their design, purpose, and monitoring. *Bull Mar Sci.* 97(4):699–728. https://doi.org/10.5343/bms.2020.0046.

Ramos J. 2024. How important are artificial reefs to the tourism industry? A review of the literature. *Eur J Tour Res.* 38:38.

Ramos J, Drakeford B, Madiedo A, Costa J, Leitão F. 2024. A Bayesian approach to infer the sustainable use of artificial reefs in fisheries and recreation. *Sustainability.* 16(2):810. https://doi.org/10.3390/su16020810.

Ramos J, Oliveira MT, Santos MN. 2011. Stakeholder perceptions of decision-making process on marine biodiversity conservation on Sal Island (Cape Verde). *Braz J Oceanogr.* 59:95–105. https://doi.org/10.1590/S1679-87592011000500012.

Ramos J, Santos MN. 2015. Facilitating fishing decisions in an artificial reef area off southern Portugal: A case study using generalized additive models. *J Appl Ichthyol* 31(S3):24–34. https://doi.org/10.1111/jai.12948.

Ramos J, Santos MN, Whitmarsh D, Monteiro CC. 2007. Stakeholder perceptions regarding the environmental and socio-economic impacts of the Algarve artificial reefs. *Hydrobiol.* 580(1):181–191.

Ramos J, Tuaty-Guerra M, Almeida M, Raposo AC, Gaudêncio MJ, Silva AD, Rodrigues N, Leandro SM, Caetano M. 2021. An artificial reef at the edge of the deep: An interdisciplinary case study. *Ocean Coastal Manage.* 210:105729. https://doi.org/10.1016/j.ocecoaman. 2021.105729.

Rani S, Kumar S, Kumar P, Abhiman, Dar SA, Kumar A. 2025. Artificial reefs: Solution for restoration of marine ecosystem and improving community resilience. *Int J Bioresour Stress Manag.* 16(5):1–11. https://doi.org/10.23910/1.2025.6014.

Reis B, van der Linden P, Sousa Pinto I, Almada E, Borges MT, Hall AE, Stafford R, Herbert RJH, Lobo-Arteaga J, Gaudêncio MJ, Tuaty-Guerra M, Ly O, Georges V, Audo M, Sebaibi N, Boutouil M, Blanco-Fernandez E, Franco JN. 2021. Artificial reefs in the Northeast Atlantic area: Present situation, knowledge gaps and future perspectives. *Ocean Coast Manage.* 209:105854. https://doi.org/10.1016/j.ocecoaman.2021.105854.

Reye JAS, García JO. 2022. Artificial reef contributions to sustainability: A case study on small-scale fisheries in Michoacan, Mexico. In: Ramos JHP, editor. *Impacts of artificial reefs on the Environment and Communities.* Hershey, PA: IGI Global; p. 70–86. https://doi.org/10.4018/978-1-6684-2344-8.ch004.

Ribeiro J, Monteiro CC, Monteiro P, Bentes L, Coelho R, Gonçalves JMS, Lino PG, Erzini K. 2008. Long-term changes in fish communities of the Ria Formosa coastal lagoon (southern Portugal) based on two studies made 20 years apart. *Estuar Coast Shelf Sci.* 76(1):57–68. https://doi.org/10.1016/j.ecss.2007.06.001.

Rocha SIB, Pires THS. Amadio S, Ropke CP, Deus CP. 2024. Fish attractors when resources abound: Prevalence of juveniles and lack of assemblage structure in a field experiment in the Amazon floodplain. *Biota Neotrop.* 24(1):e20231518. https://doi.org/10.1590/1676-0611-bn-2023–1518.

Ronglan E, Rubio AP, Silva AO, Fan D, Gair Jr. JL, Stathatou PM, Bastidas C, Strand E, Ferrandis JA, Gershenfeld N, Triantafyllou MS. 2024. Architected materials for artificial reefs to increase storm energy dissipation. *PNAS Nexus.* 3(3):pgae101. https://doi.org/10.1093/pnasnexus/pgae101.

Salaün M, Raoux A, Pezy J, Dauvin J, Pioch S. 2023. Structural and functional changes in Artificial Reefs ecosystem stressed by trophic modelling approach: Case study in the Bay of Biscay. *Reg Stud Mar Sci.* 65:103100. https://doi.org/10.1016/j.rsma.2023.103100.

Salinas S, Cintra--Buenrostro CE. 2024. "Another kid on the block": Testing the effects of artificial substrates on the recruitment of juvenile fishes in the northwestern Gulf of Mexico. *Mar Ecol.* 45(5):e12840. https://doi.org/10.1111/maec.12840.

SánchezCaballero CA, Borges-Souza JM, Abelson A. 2021. Can wrecks serve as exploitable surrogate habitats for degraded natural reefs? *Marine Environmental Research* 169:105399. https://doi.org/10.1016/j.marenvres.2021.105399.

SánchezCaballero CA, Borges-Souza JM, Saldierna-Martínez RJ, Abelson A. 2024. Assessing the conservation value of artificial and natural reefs via ichthyoplankton spatio-temporal dynamics. *Fishes.* 9(5):166. https://doi.org/10.3390/fishes9050166.

Sapehee NH, Samsudin H, Wan Talaat WIA, Othman N, Fikri MM, Hasri MN, Md Repin I, Abdul Latip AR. 2025. The socio-economic impact of artificial reefs on fishing communities in the east coast region of Malaysia. *Wellbeing Space Soc.* 6:100248. https://doi.org/10.1016/j.wss.2025.100248.

Sheehy DJ, Vik SF. 2010. The role of constructed reefs in non-indigenous species introductions. *Ecol. Eng.* 36(1):1–11. https://doi.org/10.1016/j.ecoleng.2009.09.012.

Sheng Q, Guo A, Zhang Y, Chu T. 2025. Effect of artificial fish nest placement on spawning site preference and reproductive efficiency in reservoir fish with adhesive and demersal eggs. *Sci Rep.* 15:3219. https://doi.org/10.1038/s41598-025-87211-4.

Smith JA, Cornwell WK, Lowry MB, Suthers IM. 2017. Modelling the distribution of fish around an artificial reef. *Mar Freshw Res.* 68(10):1955–1964. https://doi.org/10.1071/MF16019.

Smith JA, Lowry MB, Champion C, Suthers IM. 2016. A designed artificial reef is among the most productive marine fish habitats: New metrics to address 'production versus attraction'. *Mar Biol.* 163(9):188. https://doi.org/10.1007/s00227-016-2967-y.

Taormina B, Bald J, Want A, Lacroix D, Carlier A, Lejart M, Desroy N. 2020. Optimizing image-based protocol to monitor macroepibenthic communities colonizing artificial structures. *ICES J Mar Sci.* 77(1):274–284. https://doi.org/10.1093/icesjms/fsz249.

Taormina B, Claquin P, Vivier B, Navon M, Pezy JP, Raoux A, Dauvin JC. 2022. A review of methods and indicators used to evaluate the ecological modifications generated by artificial structures on marine ecosystems. *J Environ Manage.* 310:114646. https://doi.org/10.1016/j.jenvman.2022.114646.

ten Voorde M, do Carmo JSA, Neves MG. 2009. Designing a preliminary multifunctional artificial reef to protect the Portuguese coast. *J Coast Res.* 25(1):69–79. https://doi.org/10.2112/07-0850.1.

Vivier B, Dauvin JC, Navon M, Rusig AM, Mussio I, Orvain F, Boutouil M, Claquin P. 2021. Marine artificial reefs: A meta-analysis of their design, objectives and effectiveness. *Glob Ecol Conserv.* 27:e01538. https://doi.org/10.1016/j.gecco.2021.e01538.

Wan Y, Kong Q, Du H, Yang W. Zha W, Li W. 2024. Effectiveness of artificial reefs in enhancing phytoplankton community dynamics: A meta-analysis. *Sci Total Environ.* 947:174593. https://doi.org/10.1016/j.scitotenv.2024.174593.

Wang X, Lin C, Wang W, Zhang L. 2024. The non-negligible contribution of foundation species to artificial reef construction revealed by Ecopath models. *J Environ Manage.* 366:121887. https://doi.org/10.1016/j.jenvman.2024.121887.

Watson MS, Dickson J, Franken O, Govers LL, van der Heide T, Witte S, Eriksson BK. 2025. Introduced substrates trigger colonization by reef-associated fish in a degraded coastal system. *PLOS ONE.* 20(1): e0317431. https://doi.org/10.1371/journal.pone.0317431.

Werner KM, Haslob H, Reichel AF, Gimpel A, Stelzenmuller V. 2024. Offshore wind farm foundations as artificial reefs: The devil is in the detail. *Fish Res.* 272:106937. https://doi.org/10.1016/j.fishres.2024.106937.

Wu Z, Tweedley JR, Loneragan NR, Zhang X. 2019. Artificial reefs can mimic natural habitats for fish and macroinvertebrates in temperate coastal waters of the Yellow Sea. *Ecol Eng.* 139:105579. https://doi.org/10.1016/j.ecoleng.2019.08.009.

Xu Q, Zhang L, Zhang T, Zhang X, Yang H. 2017. Functional groupings and food web of an artificial reef used for sea cucumber aquaculture in northern China. *J Sea Res.* 119:1–7. https://doi.org/10.1016/j.seares.2016.10.005.

Yun DH, Kim YT. 2018. Experimental study on settlement and scour characteristics of artificial reef with different reinforcement type and soil type. *Geotext Geomembranes.* 46(4):448–454. https://doi.org/10.1016/j.geotexmem.2018.04.005.

Yuval M, Treibitz T. 2024. Releasing a dataset of 3D models of artificial reefs from the northern Red Sea for 3D printing and virtual reality applications. *Remote Sens Appl Soc Environ.* 36:101305. https://doi.org/10.1016/j.rsase.2024.101305.

Zaidi FHA, Ahmad R, Abdullah MMAB, Tahir MFM, Yahya Z, Ibrahim WMW, Sauffi AS. 2019. Performance of geopolymer concrete when exposed to marine environment. *IOP Conf Ser Mater Sci Eng.* 551:012092. https://doi.org/10.1088/1757-899X/551/1/012092.

Zhao XL, Ning DZ, Zou QP, Qiao DS, Cai SQ. 2019. Hybrid floating breakwater–WEC system: A review. *Ocean Eng.* 186:106126. https://doi.org/10.1016/j.oceaneng.2019.106126.

Zhou X, Zhao X, Zhang S, Lin J. 2019. Marine ranching construction and management in East China Sea: Programs for sustainable fishery and aquaculture. *Water.* 11(6):1237. https://doi.org/10.3390/w11061237.

# Index

Note: Page numbers in *italics* indicate a figure and page numbers in **bold** indicate a table on the corresponding page.

## D



## E



## F



## G

## S



## T



## U



## V



## W



## Y